Sampled-data Control of Logical Networks

Yang Liu • Jianquan Lu • Liangjie Sun

Sampled-data Control of Logical Networks

Springer

高等教育出版社
HIGHER EDUCATION PRESS

Yang Liu
College of Mathematics and Computer Science
Zhejiang Normal University
Jinhua, Zhejiang, China

Jianquan Lu
School of Mathematics
Southeast University
Nanjing, Jiangsu, China

Liangjie Sun
Department of Mathematics
University of Hong Kong
Hong Kong, China

ISBN 978-981-19-8263-7 ISBN 978-981-19-8261-3 (eBook)
https://doi.org/10.1007/978-981-19-8261-3

Jointly published with Higher Education Press The print edition is not for sale in China (Mainland). Customers from China (Mainland) please order the print book from: Higher Education Press

This Springer imprint is published by the registered company Springer Nature Singapore Pte Ltd.
The registered company address is: 152 Beach Road, #21-01/04 Gateway East, Singapore 189721, Singapore

Foreword

The last two decades have witnessed unprecedented progress in the research of Boolean networks as well as general logical (control) networks. The motivation for this development is mainly from the demand of various applications, including biology, game theory, coding, finite automata, graph theory, etc. Meanwhile, the semi-tensor product (STP) of matrices, proposed recently, provides a useful tool in formulating, analyzing, and designing controls for logical networks. STP enhances significantly the investigations on logical (control) networks.

Though there are already several books on the STP approach to Boolean control networks, this book is the first to introduce sampled-data control into the study of Boolean control networks. The book consists of four parts with a total of six chapters. It mainly discusses three research problems: Boolean control network under periodic sampled-data control, Boolean control network under aperiodic sampled-data control, and logical control network with event-triggered control.

The contents of this book cover the most updated results on sampled-data control of Boolean networks, including the authors' contributions as well as other researchers'. The materials are well selected and organized. The book is carefully written. All the theoretical results are clearly presented and properly proved. Many numerical examples are prepared to help readers to digest the main results.

The authors of this book are leading researchers in this field. I am confident that this book will be useful for graduate students and young scientists, who are interested in finite value dynamic (control) systems, such as Boolean networks. Readers will be benefited a lot from this book.

May 17, 2022 — Daizhan Cheng

The original version of the book has been revised: The copyright holder has been updated now in the book. A correction to this book can be found at https://doi.org/10.1007/978-981-19-8261-3_7

Preface

In recent years, logical networks, such as Boolean networks, have been extensively studied due to their wide range of applications, including systems biology, cryptography, etc. Designing controllers to achieve the desired goal is always an essential work in logical networks.

Most literature on logical control networks (Boolean control networks) consider state feedback control, whose state is updated at each discrete time. However, if the working principle of the controller is intermittently sampled to the control system, the loss of the controller can be greatly reduced and its advantages can be reflected. Moreover, in practical applications, the controller is often difficult to update at each time. Therefore, in this book, we focus on the study of logical control networks under sampled-data control. The following issues will be presented in detail: (1) Boolean control network under periodic sampled-data control, whose sampling period is constant. (2) Boolean control network under aperiodic sampled-data control, whose sampling interval is uncertain and/or time varying. (3) Logical control network under event-triggered control.

This book aims to introduce some recent research work on the logical control networks under sampled-data control. The book is organized as follows:

Chapter 1: This chapter begins with the background of logical networks. Subsequently, the research problems of this book, some important definitions, useful lemmas, and some basic knowledge about semi-tensor product of matrices are introduced.

Chapter 2: Stabilization of sampled-data Boolean control networks is investigated. Here, we first derive necessary and sufficient conditions for global stabilization by sampled-data state feedback control and present two algorithms to construct sampled-data state feedback control. Some differences between sampled-data feedback control and state feedback control for Boolean control networks are noted. Then, we study the set stabilization of Boolean control networks under sampled-data state feedback control. We find that sampled cycles allow elements to be repeated and not every element can be regarded as an initial state, and this is quite different from conventional cycles of Boolean control networks. Based on the sampled point set and the sampled point control invariant set, a necessary and sufficient condition is derived for the global set stabilization and sampled-data state feedback controllers are also designed. We also find an interesting thing that if a state enters the sampled point control invariant set as an unsampled point, then it may run out of the given set again, which is in sharp contrast to conventional Boolean control networks.

Chapter 3: Controllability, observability, and synchronization of sampled-data Boolean control networks are investigated. First, we observe some surprising phenomena that sampled-data Boolean control network can lose the controllability or observability compared to a system with the sampling period equaling one (which is a conventional Boolean control network). Necessary and sufficient conditions for the controllability and observability of sampled-data Boolean control networks are presented. We propose several algorithms for determining the observability of sampled-data Boolean control networks. Especially, an algorithm based on graph theory is designed, and its computational complexity is independent of the length of the constant sampling periods. Then, we investigate the sampled-data state feedback control for the synchronization of Boolean control networks under the configuration of drive-response coupling. Necessary and sufficient conditions for the complete synchronization are obtained. Unlike studying the synchronization problem of Boolean control networks under state feedback control, the sampling period affects the cycles of the system, so choosing an appropriate sampling period is critical to achieving synchronization.

Chapter 4: Stabilization of probabilistic Boolean control networks under sampled-data control is investigated. We first address the stabilization of probabilistic Boolean control networks under sampled-data state feedback control based on the algebraic representation of logic functions. A necessary and sufficient condition is derived for the existence of sampled-data state feedback controls for the global stabilization of probabilistic Boolean control networks. Since the considered system is a probabilistic Boolean control network, which is more general and complicated than a Boolean control network, the analysis is more difficult and challenging than that in Chap. 2. Moreover, compared with the method of designing sampled-data controller mentioned in Chap. 2, we propose a different and simpler method of designing sampled-data control. Then, we investigate the partial stabilization problem of probabilistic Boolean control networks under sample-data state-feedback control with a control Lyapunov function approach. Here, the classical control Lyapunov function approach of the dynamical systems is extended to the sampled-data state feedback stabilization problem of probabilistic Boolean control networks. It is indicated that the existence of the sampled-data state feedback control is equivalent to that of a control Lyapunov function, then a necessary and sufficient condition is obtained for the existence of control Lyapunov function. We also propose two different methods to solve all possible sampled-data state feedback controllers and corresponding structural matrices of a control Lyapunov function. Two examples are given to make a comparison with the methods. Compared with the existing results, the main advantages of our results are we consider the stabilization problem under sampled-data state feedback control, and we can design all possible sampled-data state feedback controllers as well as the corresponding control Lyapunov function using the Lyapunov theory. Finally, we investigate the finite-time and infinite-time set stabilization of probabilistic Boolean control networks under sampled-data state-feedback control. The finite-time and infinite-time set stabilization of probabilistic Boolean control networks are first studied via semi-tensor product approach. In addition, some criteria for checking each kind of set stabilization are proposed. Two algorithms are proposed to find the sampled point set and the largest sampled point control invariant set of probabilistic Boolean control networks by sampled-data state-feedback control. Then, a criterion is given for the finite-time set stabilization, and the time-optimal sampled-data state-feedback controller is designed. A necessary and sufficient criterion for infinite-time set stabilization of probabilistic Boolean control networks by sampled-data state-feedback control is obtained, and all possible sampled-data state-feedback controllers are designed. The association and distinction between finite-time and infinite-time set stabilization are illustrated by a numerical example.

Chapter 5: Stabilization of aperiodic sampled-data Boolean control networks is investigated. We first study the global stability of Boolean control networks under aperiodic sampled-data control. By converting the Boolean control network under aperiodic sampled-data control into a switched Boolean network, the global stability of Boolean control networks under aperiodic sampled-data control is first studied. Here, the switched Boolean network can only switch at sampling instants, which does not mean that the switches occur at each sampling instant. For the switched Boolean network containing both stable subsystems and unstable subsystems, we denote the activation frequencies of the stable subsystems and unstable subsystems, respectively. A sufficient condition for global stability Boolean control networks under aperiodic sampled-data control is derived and an upper bound of the cost function is determined. An algorithm is presented to construct aperiodic sampled-data controls for the global stabilization of Boolean control networks. Then, we consider the case that all subsystems of switched Boolean networks are unstable. By means of the discretized Lyapunov function and dwell time, a sufficient condition for global stability is obtained. Finally, we investigate the global stochastic stability of Boolean control networks under aperiodic sampled-data controls by a delay approach . A sufficient condition for global stochastic stability of Boolean control networks under aperiodic sampled-data control is obtained by using the Lyapunov function and augmented method. Here, when the sampling instants are uncertain and only the activation frequencies of the sampling interval are known, by transforming the aperiodic sampled-data control into delayed control, the global stochastic stability of the Boolean control network under this aperiodic sampled-data control is first considered; for sampled-data state feedback control (constant sampling interval), we can also convert it into delayed control, and then, the global stability of the Boolean control network under sampled-data state feedback control can be studied by a delay approach.

Chapter 6: Event-triggered control for logical control networks is investigated. We first investigate the global stabilization problem of k-valued logical control networks via event-triggered control, where the control inputs only work at several certain individual states. A necessary and sufficient criterion is derived for the event-triggered stabilization. And a constructive procedure is developed to design all time-optimal event-triggered stabilizers. Moreover, the switching-cost-optimal stabilizer, which is event-triggered and has a minimal number of controller executions, is designed. The labeled digraph is constructed to describe the dynamic behavior of the event-triggered controlled k-valued logical control network. Based on knowledge of graph theory, the number of controller executions is minimized through a universal procedure called minimal spanning in-tree algorithm. It deserves formulating that this algorithm can handle all circumstances and overcome the constraint of the traditional method. Then, we investigate the disturbance decoupling problem of Boolean control networks by event-triggered control. On the one hand, when the triggering mechanism is satisfied, controllers are updated such that the disturbance decoupling problem of Boolean control networks is solvable. On the other hand, in complex systems, the whole system may be controlled well by controlling only a small subset of variables directly. This strategy is referred to as the partial control. Motivated by this, event-triggered control is used to study the disturbance decoupling problem of Boolean partial control networks as well. The presented event-trigger control of Boolean partial control networks not only reduces the control time and cost but also cut down the number of controllers. Finally, we investigate the output regulation problem of probabilistic k-valued logical systems with delays by an intermittent control scheme. A sufficient and necessary condition for the output regulation problem of probabilistic k-valued logical systems with delays is obtained. Two types of approaches are given to design the event-triggered control laws.

Jinhua, China — Yang Liu
Nanjing, China — Jianquan Lu
Hong Kong, China — Liangjie Sun
April 2022

Acknowledgments

In some sense, we have been working on this book for 6 years, and we have lots of people to thank. First, we would like to thank Dr. Daizhan Cheng and his team. The semi-tensor product they proposed is the main tool for studying logical networks in this book. It is worth mentioning that the semi-tensor product is very useful in the expression and analysis of logical networks.

Then we would like to express our appreciation to Dr. Jinde Cao at Southeast University, China; Dr. Jungang Lou at Huzhou University, China; Dr. Daniel W. C. Ho at City University of Hong Kong, China; Dr. Wai-Ki Ching at the University of Hong Kong, China; Dr. Yuqian Guo and Dr. Weihua Gui both at Central South University, China; Dr. Zheng-Guang Wu at Zhejiang University, China; Dr. Jie Zhong at Zhejiang Normal University, China; Dr. Leszek Rutkowski at Czestochowa University of Technology, Poland; Dr. Bowen Li at Nanjing University of Posts and Telecommunications, China; Dr. Kit Ian Kou at the University of Macau, China; and Dr. Li Yu at Zhejiang University of Technology, China; Mr. Shiyong Zhu; Mr. Qunxi Zhu; Miss. Liyun Tong; Miss. Liqing Wang; Miss. Jingyi He; Miss. Mengxia Xu; and Mr. Jiayang Liu who have all coauthored with us a few papers, which have been included in this book. We would like to thank Mr. Ziyu Xuan, who helped a lot in proofreading the manuscripts.

Moreover, this work was supported by the National Natural Science Foundation of China under Grants 62173308, 61573102 and 11671361; the Natural Science Foundation of Zhejiang Province of China under Grants LD19A010001, LR20F030001, and D19A010003; the Natural Science Foundation of Jiangsu Province of China under Grant BK20170019; the China Postdoctoral Science Foundation under Grants 2015M580378 and 2016T90406; the National Training Programs of Innovation and Entrepreneurship under Grants 201610345020, 201710345009, 201810345005, and 201910345013; and the Jinhua Science and Technology Project under grant 2022-1-042.

Contents

Acronyms

ASDC	Aperiodic sampled-data control
BCN	Boolean control network
BN	Boolean network
KVLCN	k-valued logical control network
PBCN	Probabilistic Boolean control network
PBN	Probabilistic Boolean network
PCC	Piecewise constant control
SDSFC	Sampled-data state feedback control
SPCIS	sampled point control invariant set

Part I
Introduction

Chapter 1
Introduction

Abstract In this chapter, the research background of this book is presented. Three main research issues are discussed. Moreover, some preliminaries are given, mainly focus on the definition and properties of semi-tensor product of matrices, thc algebraic representation of Boolean functions, and so on.

1.1 Background

Recently, the rapid development of DNA microarrays has set the stage for mathematical modeling of genetic regulatory networks [1]. Generally, numerical formal types of mathematical models have been proposed to depict, simulate, and even predict the dynamic behavior of biological networks, such as Markov-type genetic networks [2] and Boolean networks (BNs). Based on experimental results, BNs, which were originally proposed by Kauffman in 1969 [3], have been capable of forecasting the dynamic sequence of protein-activation patterns within genetic regulation networks [4]. A typical biological application is a cell cycle control network in yeast [5]. In addition, the modality of BNs has constructed a natural framework to ensure a detailed understanding and insight into the dynamic behaviors represented by large-scale genetic regulation networks.

In a Boolean model, the expression of each node on a network is approximated by two levels, namely, 1 (ON) and 0 (OFF). The state update of each gene is determined by a pre-assigned logical function associated with the states of in-neighbor genes. As mentioned in [6], a recent significant discovery in systems biology is that exogenous perturbations, which can be described as "control", are almost ubiquitous in many biological systems. Thus, the concept of Boolean control networks (BCNs) has been formally generated by adding binary inputs to the BNs [7]. In capillary endothelial cells [8], for instance, a simple BCN has been established to simulate the dynamic behavior of the signaling system, where growth factors and cell shape (spreading) are both presented by two external inputs.

Y. Liu et al., *Sampled-data Control of Logical Networks*,
https://doi.org/10.1007/978-981-19-8261-3_1

In [9], the authors introduced a new class of models called probabilistic Boolean networks (PBNs), which are probabilistic generalizations of the standard BNs that offer a flexible and powerful modeling framework. Based on the probabilistic nature, PBNs are able to cope with uncertainty, which is intrinsic to biology systems. Similarly, a probabilistic Boolean control network (PBCN) is a stochastic extension of the BCN model. It can be considered as a collection of BCNs endowed with a probability structure for switching among different constituent BCNs.

1.2 Research Problems

1.2.1 Periodic Sampled-Data Control

The sampled-data control theory is a mature research field, and furthermore, it has been studied extensively over the past decades [10–13]. With the advancement of the sampled-data control theory, most of the existing results are obtained by assuming that the sampling interval is constant. In this book, we also first consider the BCN under periodic sampled-data control.

To the authors' knowledge, the problem of BCNs under state feedback control has been well investigated [14–18]. For state feedback control, the controller is updated at each moment, whereas for sampled-data state feedback control (SDSFC), the controller is updated during each sampling period. In particular, if the sampling period is one, then the SDSFC can be regarded as a normal state feedback control. In other wards, SDSFC is also a generalization of the state feedback control to some extent. In practical operation, state feedback control needs to consume large amounts of energy to achieve the desired effect, such as stabilization or synchronization. Unlike state feedback control, each sampling period of the sampled-data state feedback controller is constant. Therefore, using SDSFC can reduce the number of updates to the controller while the same desired effect can be achieved. However, the SDSFC problem for BCNs is still open and challenging, and there is no result on the construction of sampled-data controllers for BCNs as well. Therefore, in this book, we study the periodic sampled-data of BCNs (PBCNs), and mainly discuss the stabilization, controllability, observability and synchronization of BCNs (PBCNs) under periodic sampled-data control.

First, the problem of stabilization for many dynamical systems sometimes cannot be well solved by continuous state feedback controllers. Therefore, it is necessary to design time-dependent or discontinuous feedback controllers to stabilize systems. The idea of using discontinuous stabilizers instead of continuous stabilizers has been broadly discussed. For example, Coron [19] considered the use of time-dependent continuous feedback laws. And Emel'yanov et al. [20] discussed discontinuous feedback controls. Furthermore, investigations on set stability and set stabilization are meaningful in the real world. The fundamental idea of set stability and set stabilization is to study whether a system can be stabilized to a subset of the state space, which is more complex and more interesting than stabilizing to a single point or a cycle. Therefore, the study of set stability and set stabilization has been investigated in recent years. Guo et al. [21] considered the set stability and set stabilization of BCNs based on invariant subsets, and obtained a necessary and sufficient condition for global set stabilization. In [22], the set stabilization for switched BCNs has been studied. Necessary and sufficient conditions for the set stabilization of switched BCNs were presented. Chen and Francis [23] studied the set stabilization of multivalued logical systems. However, these pervious investigations on stabilization and set stabilization mainly focus on conventional state feedback controllers. Motivated by the above analysis, it is necessary to study the stabilization and set stabilization of BCNs via SDSFC.

In addition, controllability is one of the key concepts in systems biology and has drawn extensive attention in studying BCNs, whose objective is to design therapeutic interventions that steer one cell state to another cell state [24, 25]. For example, from a location corresponding to a diseased state of the biological systems to a location corresponding to a healthy state. Hence, it is necessary to avoid certain forbidden states like unhealthy state or diseased state when designing an external control sequence that steers BCNs between two states [25]. During the past few years, several kinds of controllability, such as output controllability, trajectory controllability, and so on, have been addressed, and many fundamental theoretical results have been obtained [26, 27]. In the context of BNs, the observability was first defined in [24]: for every initial state, there exists an input sequence such that this initial state can be distinguished from other states by output sequence. Then, it degenerated into the distinguishability of two initial states via the output sequence under one input sequence in [28]. Furthermore, the observability was respectively strengthened to the single-input-sequence observability [29] and the arbitrary-inputsequence observability [30]. The language completeness of finite automaton was utilized to uniformly investigate above four different types of observability [31]. Another universal approach, termed as parallel extension technique [32], was presented to interconnect a BCN in parallel with its duplicate, then the observability problem can be converted into the reachability of the corresponding augmented system from a specific state subset to another one. In [33], Zhou et al. studied the asymptotic observability in distribution and finite-time observability of PBNs, but control inputs were not concerned about. Inspired by the preceding discussions, we will study the controllability and observability of sampled-data BCNs with constant sampling periods later.

Synchronization is also an important and basic topic in control theory, which can explain and solve some natural phenomena, such as the lac operon in Escherichia coli [34], the coupled oscillations in cell cycles [35] and the apoptosis network [36]. Recently, there have been many results on the synchronization of network coupled systems. For example, synchronization problems of logical control networks have been analytically discussed for two types of controllers in [37], i.e., free sequence control and state/output feedback control; Tang et al. [38] has considered the robustness for the synchronization of complex dynamical networks and its applications; in [39], the synchronization of neural networks with distributed delays has been investigated and an improved method has been proposed to design the desired sampled-data controller by solving a set of linear matrix inequalities. Moreover, synchronization can also be studied in BCNs. In [40], there has been an analytical study of synchronization in an array of coupled deterministic BNs with time delays. An algorithm for the partial synchronization of BCNs has been studied and designed in [41]. In [42], a necessary and sufficient criterion has been presented to ensure the partial synchronization of the interconnected BNs. Pinning control has been studied for the synchronization of two coupled BNs by changing the transition matrix of the response BN in [43]. In this book, we will investigate the SDSFC for the synchronization of BCNs under the configuration of drive-response coupling.

1.2.2 Aperiodic Sampled-Data Control

Nonetheless, in some real-world applications, periodic sampling is in general hard to perform and implement. In fact, for real-world engineering problems, the sampling interval is usually not constant. Take, for example, in networked and embedded control systems, the sampling intervals are always uncertain and/or time varying. Therefore, aperiodic sampled-data control (ASDC) is proposed, and its sampling interval is uncertain and aperiodic. Moreover, as mentioned in Wu et al. [44], the utilization of ASDC can further reduce the costs of energy, computation, and communication.

Stability and stabilization are important research fields of BNs (BCNs), and they have been deeply studied [15, 16]. Most of the above results are obtained using semi-tensor product and the matrix expression of logic. Wang and Li [45, 46] constructed Lyapunov functions for BNs and presented the Lyapunov-based stability analysis for BNs. After that, the Lyapunov function has been effectively adopted to research the stability and stabilization problems of BNs (BCNs). For example, by designing a co-positive Lyapunov function, the weighted l_1-gain analysis was considered, and then the l_1 model reduction problem for BCNs was studied in [47]. Meng et al. [48] analyzed stability and l_1 gain of BNs with Markovian jump parameters. For time-dependent switched Boolean networks, Meng et al. [49] considered their stability and guaranteed the cost. In this book, we also use the techniques of switching-based Lyapunov function to investigate BCNs under ASDC. It should be noted that the global stability and stabilization of BCNs under ASDC are more complex than that under periodic sampled-data control.

1.2.3 Event-Triggered Control

The controller design strategy is always an interesting topic in complex networks [50], naturally in genetic regulatory networks. Numerical control schemes have been developed in the study of logical systems, including but not limited to output feedback control [15], state feedback control [16], and pinning control [43]. Unfortunately, in the aforementioned control paradigms, the control inputs need to be executed at each time instant; it is a waste of resources if the dynamic evolution of the original network is desirable. Motivated by this problem, another alternative control paradigm called event-triggered control has been proposed in [51]. With the advent of this triggering mechanism, the control cost can be reduced substantially; thus, event-triggered control has been applied extensively in the study of multi-agent systems [52], and smart grids [53]. In general, the event-triggered control consists of two elements, namely, a feedback controller that computes the control input, and a triggering mechanism that decides when the controller has to be updated again. In this book, we also consider the logical control networks under event-triggered control, and mainly study the stabilization, disturbance decoupling problem, and output regulation problem of logical control networks under event-triggered control.

As usual, biological signaling systems produce some outputs, such as the level of a phosphorylated protein. However, outputs are always sensitive to external environment or changes in the concentrations of the system's components [54]. Therefore, it is interesting to study the output robustness of dynamic systems. However, most of systems are not output robust, then they pose serious challenges to design controllers such that the output robustness of systems can be achieved, which is called disturbance decoupling problem. Many results have been studied about the disturbance decoupling problem of BCNs [55, 56]. The research [55] points out that the first step of the disturbance decoupling problem is to find a regular coordinate subspace which contains outputs. The second step is to design a controller such that the complement coordinate sub-basis and the disturbances are independent of the regular coordinate subspace. Based on [55], the redundant variable separation technique is used in [56, 57] to analyze the disturbance decoupling problem solvability. In this book, two kinds of event-triggered controllers are proposed to solve the disturbance decoupling problem.

Owing to plenty of control problems that can be formulated as a particular case of the output regulation problem, such as constant disturbances and set-point control, the output regulation problem has attracted much attention in control theory. Output regulation means that some fixed states are controlled for making system outputs consistent with reference signals generated by the external system. Significantly, the output regulation problem is subtly different from the tracking problem and disturbance rejection problem. What distinguishes the output regulation problem is that it adopts external signals to conduct the dynamics of the logical reference system. Many fundamental results have been obtained concerning the output regulation problem [58–60]. For example, it was also found that the output regulation problem of BCNs could be solvable based on a state feedback control method in [61]. In this book, we also introduce two types of event-triggered control to resolve the output regulation problem.

1.3 Mathematical Preliminaries

1.3.1 Matrix Products

The Kronecker product of matrices is also called the tensor product of matrices. This product is applicable to any two matrices.

Definition 1.1 Let $A = (a_{ij}) \in \mathcal{M}_{m\times n}$ and $B = (b_{ij}) \in \mathcal{M}_{p\times q}$. The Kronecker product of A and B is defined as

$$A \otimes B = \begin{bmatrix} a_{11}B & a_{12}B & \cdots & a_{1n}B \\ a_{21}B & a_{22}B & \cdots & a_{2n}B \\ \vdots & \vdots & \ddots & \vdots \\ a_{m1}B & a_{m2}B & \cdots & a_{mn}B \end{bmatrix} \in \mathcal{M}_{mp\times nq}. \tag{1.1}$$

Some basic properties of the Kronecker product are introduced as follows.

Proposition 1.1

(i) (Associative Law) $A \otimes (B \otimes C) = (A \otimes B) \otimes C.$
(ii) (Distributive Law)

$$(\alpha A + \beta B) \otimes C = \alpha(A \otimes C) + \beta(B \otimes C), \tag{1.2}$$

$$A \otimes (\alpha B + \beta C) = \alpha(A \otimes B) + \beta(A \otimes C), \quad \alpha, \beta \in \mathbb{R}. \tag{1.3}$$

Khatri-Rao product of matrices is another useful product in this paper.

Definition 1.2 Let $A \in \mathcal{M}_{m\times r}$ and $B \in \mathcal{M}_{n\times r}$. The Khatri-Rao product of A and B is defined as

$$A * B = [Col_1(A) \otimes Col_1(B)\ Col_2(A) \otimes Col_2(B)\ \cdots\ Col_r(A) \otimes Col_r(B)]. \tag{1.4}$$

Proposition 1.2

(i) (Associative Law) Let $A \in \mathcal{M}_{m\times r}$, $B \in \mathcal{M}_{n\times r}$, *and* $C \in \mathcal{M}_{p\times r}$. *Then* $(A * B) * C = A * (B * C)$.
(ii) (Distributive Law) Let $A, B \in \mathcal{M}_{m\times r}$ *and* $C \in \mathcal{M}_{n\times r}$. *Then*

$$(\alpha A + \beta B) * C = \alpha(A * C) + \beta(B * C), \tag{1.5}$$

$$A * (\alpha B + \beta C) = \alpha(A * B) + \beta(A * C), \quad \alpha, \beta \in \mathbb{R}. \tag{1.6}$$

Moreover, the set of n dimensional probabilistic vectors is denoted by γ_n.

Definition 1.3 For two probabilistic vectors $X, Y \in \gamma_n$, we assume that $X = (p_1\ p_2\ \cdots\ p_n)^{\mathrm{T}}$ and $Y = (q_1\ q_2\ \cdots\ q_n)^{\mathrm{T}}$, then $X \circ Y$ is defined by

$$X \circ Y = \begin{bmatrix} p_1 \vee q_1 \\ p_2 \vee q_2 \\ \vdots \\ p_n \vee q_n \end{bmatrix}$$

where $p_i \wedge q_i = 1$ if and only if $p_i q_i > 0$, else $p_i \wedge q_i = 0$.

1.3.2 Swap Matrix

Definition 1.4 A swap matrix $W_{[m,n]}$ is an $mn \times mn$ matrix, defined as follows. Its rows and columns are labeled by double index (i, j), the columns are arranged by the ordered multi-index $Id(i, j; m, n)$, and the rows are arranged by the ordered multi-index $Id(j, i; n, m)$. The element at position $[(I, J), (i, j)]$ is then

$$w_{(I,J),(i,j)} = \delta_{i,j}^{I,J} = \begin{cases} 1, & I = i \text{ and } J = j, \\ 0, & \text{otherwise.} \end{cases} \tag{1.7}$$

1. Letting $m = 2$ and $n = 3$, the swap matrix $W_{[m,n]}$ can be constructed as follows. Using double index (i, j) to label its columns and rows, the columns of W are labeled by $Id(i, j; 2, 3)$, that is, $(11, 12, 13, 21, 22, 23)$, and the rows of W are labeled by $Id(j, i; 3, 2)$, that is, $(11, 21, 12, 22, 13, 23)$. According to (1.7), we have

$$W_{[2,3]} = \begin{matrix} \\ (11) \\ (21) \\ (12) \\ (22) \\ (13) \\ (23) \end{matrix} \begin{matrix} (11)(12)(13)(21)(22)(23) \\ \begin{bmatrix} 1 & 0 & 0 & 0 & 0 & 0 \\ 0 & 0 & 0 & 1 & 0 & 0 \\ 0 & 1 & 0 & 0 & 0 & 0 \\ 0 & 0 & 0 & 0 & 1 & 0 \\ 0 & 0 & 1 & 0 & 0 & 0 \\ 0 & 0 & 0 & 0 & 0 & 1 \end{bmatrix} \end{matrix}$$

2. Consider $W_{[3,2]}$. Its columns are labeled by $Id(i, j; 3, 2)$, and its rows are labeled by $Id(j, i; 2, 3)$. We then have

$$W_{[3,2]} = \begin{array}{c} \\ (11) \\ (21) \\ (31) \\ (12) \\ (22) \\ (32) \end{array} \begin{array}{c} (11)(12)(21)(22)(31)(32) \\ \begin{bmatrix} 1 & 0 & 0 & 0 & 0 & 0 \\ 0 & 0 & 1 & 0 & 0 & 0 \\ 0 & 0 & 0 & 0 & 1 & 0 \\ 0 & 1 & 0 & 0 & 0 & 0 \\ 0 & 0 & 0 & 1 & 0 & 0 \\ 0 & 0 & 0 & 0 & 0 & 1 \end{bmatrix} \end{array}$$

According to the construction of swap matrix, the following two propositions are immediate consequences.

Proposition 1.3 *Let* $A \in \mathcal{M}_{m\times n}$. *Then*

$$W_{[m,n]}V_r(A) = V_c(A) \quad \text{and} \quad W_{[n,m]}V_c(A) = V_r(A). \tag{1.8}$$

Proposition 1.4 *Let* $X \in \mathbb{R}^m$ *and* $Y \in \mathbb{R}^n$ *be two column vectors. Then*

$$W_{[m,n]}(X \otimes Y) = Y \otimes X. \tag{1.9}$$

It is easy to check that $XY = X \otimes Y$, so we have

$$W_{[m,n]}XY = YX. \tag{1.10}$$

Swap matrix has some special properties, which follow from its definition immediately.

Proposition 1.5

(i) A swap matrix is an orthogonal matrix. It satisfies

$$W_{[m,n]}^{\mathrm{T}} = W_{[m,n]}^{-1} = W_{[n,m]}. \tag{1.11}$$

(ii) When $m = n$, *(1.11) becomes*

$$W_{[n,n]}^{\mathrm{T}} = W_{[n,n]}^{-1} = W_{[n,n]}. \tag{1.12}$$

(iii)

$$W_{[1,n]} = W_{[n,1]} = I_n. \tag{1.13}$$

When $m = n$, we simply denote by $W_{[n]}$ for $W_{[n,n]}$.

$W_{[m,n]}$ can be constructed in an alternative way which is convenient in some applications. Denoting by δ_n^i the ith column of the identity matrix I_n, we have the following.

Proposition 1.6

$$W_{[m,n]} = \left[\delta_n^1 \ltimes \delta_m^1 \ \cdots \ \delta_n^n \ltimes \delta_m^1 \ \cdots \ \delta_n^1 \ltimes \delta_m^m \ \cdots \ \delta_n^n \ltimes \delta_m^m\right]. \tag{1.14}$$

For convenience, we provide two more forms of swap matrix:

$$W_{[m,n]} = \begin{bmatrix} I_m \otimes {\delta_n^1}^{\mathrm{T}} \\ \vdots \\ I_m \otimes {\delta_n^n}^{\mathrm{T}} \end{bmatrix} \tag{1.15}$$

and, similarly,

$$W_{[m,n]} = \left[I_n \otimes \delta_m^1 \ \cdots \ I_n \otimes \delta_m^m\right]. \tag{1.16}$$

1.3.3 Semi-Tensor Product of Matrices

A general definition of semi-tensor product is given as follows.

Definition 1.5

(i) Let $X = (x_1 \ \cdots \ x_s)$ be a row vector, $Y = (y_1 \ \cdots \ y_t)^{\mathrm{T}}$ a column vector.

Case 1 If t is a factor of s, say, $s = t \times n$, then the n-dimensional row vector defined as

$$X \ltimes Y := \sum_{k=1}^{t} X^t y_k \in \mathbb{R}^n \tag{1.17}$$

is called the left semi-tensor inner product of X and Y, where

$$X = (X^1 \ \cdots \ X^t), \quad X^i \in \mathbb{R}^n, i = 1, 2, \ldots, t.$$

Case 2 If s is a factor of t, say, $t = s \times n$, then the n-dimensional column vector defined as

$$X \ltimes Y := \sum_{k=1}^{t} x_k Y^k \in \mathbb{R}^n \tag{1.18}$$

is called the left semi-tensor inner product of X and Y, where

$$Y = ((Y^1)^{\mathrm{T}} \ \cdots \ (Y^t)^{\mathrm{T}})^{\mathrm{T}}, \quad Y^i \in \mathbb{R}^n, i = 1, 2, \ldots, t.$$

(ii) Let $M \in \mathcal{M}_{m\times n}$ and $N \in \mathcal{M}_{p\times q}$. If n is a factor of p or p is a factor of n, then $C = M \ltimes N$ is called the left semi-tensor product of M and N, where C consists of $m \times q$ blocks as $C = (C^{ij})$, and

$$C^{ij} = M^i \ltimes N_j, \quad i = 1, 2, \ldots, m;\ j = 1, 2, \ldots, q, \tag{1.19}$$

where $M^i = Row_i(M)$ and $N_j = Col_j(N)$.

1.3.4 Boolean Matrix

Let $\mathcal{D} = \{1, 0\}$. We recall the definition of a Boolean matrix and define its operators first.

Definition 1.6 A Boolean matrix $X = (x_{ij})$ is an $m \times n$ matrix with entries $x_{ij} \in \mathcal{D}$. When $n = 1$, it is called a Boolean vector. The set of $m \times n$ Boolean matrices is denoted by $\mathscr{B}_{m\times n}$.

Next, we consider the scalar product and matrix product.

Definition 1.7

(i) Let $\alpha \in \mathcal{D}$. The scalar product of α with $X \in \mathscr{B}_{m\times n}$ is

$$\alpha X = X\alpha := \alpha \wedge X. \tag{1.20}$$

In particular, let $\alpha, \beta \in \mathcal{D}$. Then, $\alpha\beta = \alpha \wedge \beta$, which is the same as the conventional real number product.

(ii) Let $X = (x_{ij}) \in \mathscr{B}_{p\times q}$ and $Y \in \mathscr{B}_{m\times n}$ be two Boolean matrices. Then,

$$X \otimes Y = (x_{ij}Y) \in \mathscr{B}_{pm\times qn}. \tag{1.21}$$

(iii) Let $\alpha, \beta, \alpha_i \in \mathcal{D},\ i = 1, 2, \ldots, n$. Boolean addition is defined as follows:

$$\begin{cases} \alpha +_{\mathscr{B}} \beta := \alpha \vee \beta, \\ \displaystyle\sum_{i=1}^{n} {}_{\mathscr{B}}\alpha_i := \alpha_1 \vee \alpha_2 \vee \cdots \vee \alpha_n. \end{cases} \tag{1.22}$$

(iv) Let $X = (x_{ij}) \in \mathscr{B}_{m\times n}$ and $Y = (y_{ij}) \in \mathscr{B}_{n\times p}$. The Boolean product of Boolean matrices is then defined as

$$X \ltimes_{\mathscr{B}} Y := Z \in \mathscr{B}_{m\times p}, \tag{1.23}$$

where

$$z_{ij} = \sum_{k=1}^{n} {}_{\mathscr{B}} x_{ik} y_{kj}, \quad i = 1, 2, \ldots, m;\ j = 1, 2, \ldots, p.$$

(v) Assume that $A \ltimes_{\mathscr{B}} A$ is well defined. Boolean powers are then defined as follows:

$$A^{(k)} := \underbrace{A \ltimes_{\mathscr{B}} A \ltimes_{\mathscr{B}} \cdots \ltimes_{\mathscr{B}} A}_{k}. \tag{1.24}$$

1.3.5 Structure Matrix of a Logical Operator

Recall that a logical variable takes value from $\mathcal{D} = \{T, F\}$ or, equivalently, $\mathcal{D} = \{T, F\}$. To obtain a matrix expression we identify "T" and "F", respectively, with the vectors

$$T := 1 \sim \begin{bmatrix} 1 \\ 0 \end{bmatrix}, \quad T := 0 \sim \begin{bmatrix} 0 \\ 1 \end{bmatrix}. \tag{1.25}$$

To describe the vector form of logic we first recall some notation:

- δ_k^i is the ith column of the identity matrix I_k,
- $\Delta_k := \{\delta_k^i | i = 1, 2, \ldots, k\}$.

For notational ease, let $\Delta := \Delta_2$. Then,

$$\Delta = \{\delta_2^1, \delta_2^2\} = \left\{ \begin{bmatrix} 1 \\ 0 \end{bmatrix}, \begin{bmatrix} 0 \\ 1 \end{bmatrix} \right\},$$

and an r-ary logical operator is a mapping $\sigma : \Delta^r \to \Delta$.

Definition 1.8 A matrix $L \in \mathcal{M}_{n\times m}$ is called a logical matrix if $Col(L) \subset \Delta_n$. The set of $n \times m$ logical matrices is denoted by $\mathcal{L}_{n\times m}$.

If $L \in \mathcal{L}_{n\times m}$, then it has the form

$$L = \left[\delta_n^{i_1}\ \delta_n^{i_2}\ \cdots\ \delta_n^{i_m} \right].$$

For notational compactness we write this as

$$L = \delta_n[i_1\, i_2\, \cdots\, i_m].$$

Definition 1.9 A 2×2^r matrix M_σ is said to be the structure matrix of the r-ary logical operator σ if

$$\sigma(p_1, \ldots, p_r) = M_\sigma \ltimes p_1 \ltimes \cdots \ltimes p_r := M_\sigma \ltimes_{i=1}^r p_i. \tag{1.26}$$

Note that throughout this book we assume the matrix product is the (left) semi-tensor product, and hereafter the symbol "$\ltimes$" will be omitted in most cases. However, we use

$$\ltimes_{i=1}^r p_i := p_1 \ltimes p_2 \ltimes \cdots \ltimes p_r.$$

We start by constructing the structure matrices for some fundamental logical operators. We define the structure matrix for negation, $\neg$, denoted by M_n, as

$$M_n = \begin{bmatrix} 0 & 1 \\ 1 & 0 \end{bmatrix} = \delta_2[2\ 1]. \tag{1.27}$$

It is then easy to check that when a logical variable p is expressed in vector form, we have

$$\neg p = M_n p. \tag{1.28}$$

To see this, when $p = T$,

$$p = T \sim \delta_2^1 \Rightarrow M_n p = \delta_2^2 \sim F,$$

and when $p = F$,

$$p = F \sim \delta_2^2 \Rightarrow M_n p = \delta_2^1 \sim T.$$

Similarly, for conjunction, $\wedge$, disjunction, $\vee$, conditional, $\rightarrow$, and biconditional, $\leftrightarrow$, we define their corresponding structure matrices, denoted by $M_c,\ M_d,\ M_i$, and M_e, respectively, as follows:

$$M_c = \delta_2[1\ 2\ 2\ 2], \tag{1.29}$$

$$M_d = \delta_2[1\ 1\ 1\ 2], \tag{1.30}$$

$$M_i = \delta_2[1\ 2\ 1\ 1], \tag{1.31}$$

$$M_e = \delta_2[1\ 2\ 2\ 1]. \tag{1.32}$$

A straightforward computation then shows that for any two logical variables p and q, we have

$$p \wedge q = M_c pq, \tag{1.33}$$

$$p \vee q = M_d pq, \tag{1.34}$$

$$p \rightarrow q = M_i pq, \tag{1.35}$$

$$p \leftrightarrow q = M_e pq. \tag{1.36}$$

In the following we will show that for any logical function, f , there exists a unique structure matrix M_f of f such that (1.26) holds.

Moreover, define a matrix Φ_n, called the power-reducing matrix, as

$$\Phi_n = \delta_{2^{2n}}[1\ 2^n + 2\ 2 \cdot 2^n + 3\ \cdots\ (2^n - 2) \cdot 2^n + 2^n - 1\ 2^{2n}].$$

Lemma 1.1 *Let $p = \ltimes_{i=1}^n p_i$ with $p_i \in \Delta$, then $p^2 = \Phi_n p$.*

Theorem 1.1 *Given a logical function $f(p_1, p_2, \ldots, p_r$ with logical variables $p_1, p_2, \ldots, p_r$, there exists a unique 2×2^r matrix M_f, called the structure matrix of f, such that $f(p_1, p_2, \ldots, p_r) = M_f p_1 p_2 \cdots p_r$. Moreover, $M_f \in \mathcal{L}_{2\times 2^r}$.*

References

1. Davidson, E.H., Rast, J.P., Oliveri, P., et al.: A genomic regulatory network for development. Science **295**(5560), 1669–1678 (2002)
2. Liang, J., Lam, J., Wang, Z.: State estimation for Markov-type genetic regulatory networks with delays and uncertain mode transition rates. Phys. Lett. A. **373**(47), 4328–4337 (2009)
3. Kauffman, S.A.: Metabolic stability and epigenesis in randomly constructed genetic nets. J. Theor. Biol. **22**(3), 437–467 (1969)
4. Goodwin, B.C.: Temporal Organization in Cells: A Dynamic Theory of Cellular Control Processes. Academic Press, London (1963)
5. Davidich, M.I., Bornholdt, S.: Boolean network model predicts cell cycle sequence of fission yeast. PLoS One **3**(2), e1672 (2008)
6. Ideker, T., Galitski, T., Hood, L.: A new approach to decoding life: systems biology. Ann. Rev. Genomics Hum. Genet. **2**(1), 343–372 (2001)
7. Akutsu, T., Hayashida, M., Ching, W.K., et al.: Control of Boolean networks: hardness results and algorithms for tree structured networks. J. Theor. Biol. **244**(4), 670–679 (2007)
8. Huang, S., Ingber, D.E.: Shape-dependent control of cell growth, differentiation, and apoptosis: switching between attractors in cell regulatory networks. Exp. Cell Res. **261**(1), 91–103 (2000)
9. Shmulevich, I., Dougherty, E.R., Kim, S., et al.: Probabilistic Boolean networks: a rule-based uncertainty model for gene regulatory networks. Bioinformatics **18**(2), 261–274 (2002)
10. Fujioka, H.: Stability analysis of systems with aperiodic sample-and-hold devices. Automatica **45**(3), 771–775 (2009)
11. Oishi, Y., Fujioka, H.: Stability and stabilization of aperiodic sampled-data control systems using robust linear matrix inequalities. Automatica **46**(8), 1327–1333 (2010)

12. Fridman, E.: A refined input delay approach to sampled-data control. Automatica **46**(2), 421–427 (2010)
13. Fujioka, H., Nakai, T.: Stabilising systems with aperiodic sample-and-hold devices: state feedback case. IET Control Theory Appl. **4**(2), 265–272 (2010)
14. Fornasini, E., Valcher, M.E.: On the periodic trajectories of Boolean control networks. Automatica **49**(5), 1506–1509 (2013)
15. Li, H., Wang, Y.: Output feedback stabilization control design for Boolean control networks. Automatica **49**(12), 3641–3645 (2013)
16. Li, R., Yang, M., Chu, T.: State feedback stabilization for Boolean control networks. IEEE Trans. Autom. Control **58**(7), 1853–1857 (2013)
17. Li, R., Yang, M., Chu, T.: State feedback stabilization for probabilistic Boolean networks. Automatica **50**(4), 1272–1278 (2014)
18. Zhao, Y., Cheng, D.: On controllability and stabilizability of probabilistic Boolean control networks. Sci. China Inf. Sci. **57**(1), 1–14 (2014)
19. Coron, J.M.: Global asymptotic stabilization for controllable systems without drift. Math. Control Signals Syst. **5**(3), 295–312 (1992)
20. Emel'yanov, S.V.E., Korovin, S.K., Nikitin, S.V.: Global controllability and stabilization of nonlinear systems. Mat. Model. **1**(1), 51–90 (1989)
21. Guo, Y., Wang, P., Gui, W., et al: Set stability and set stabilization of Boolean control networks based on invariant subsets. Automatica **61**, 106–112 (2015)
22. Li, F., Tang, Y.: Set stabilization for switched Boolean control networks. Automatica **78**, 223–230 (2017)
23. Chen, T., Francis, B.A.: Optimal Sampled-Data Control Systems. Springer Science & Business Media, New York (2012)
24. Cheng, D., Qi, H.: Controllability and observability of Boolean control networks. Automatica **45**(7), 1659–1667 (2009)
25. Laschov, D., Margaliot, M.: Controllability of Boolean control networks via the Perron-Frobenius theory. Automatica **48**(6), 1218–1223 (2012)
26. Chen, H., Sun, J.: Output controllability and optimal output control of state-dependent switched Boolean control networks. Automatica **50**(7), 1929–1934 (2014)
27. Lu, J., Zhong, J., Ho, D.W.C., et al.: On controllability of delayed Boolean control networks. SIAM J. Control Optim. **54**(2), 475–494 (2016)
28. Zhao, Y., Qi ,H., Cheng, D.: Input-state incidence matrix of Boolean control networks and its applications. Syst. Control Lett. **59**(12), 767–774 (2010)
29. Cheng, D., Zhao, Y.: Identification of Boolean control networks. Automatica **47**(4), 702–710 (2011)
30. Fornasini, E., Valcher, M.E.: Observability, reconstructibility and state observers of Boolean control networks. IEEE Trans. Autom. Control **58**(6), 1390–1401 (2012)
31. Zhang, K., Zhang, L.: Observability of Boolean control networks: A unified approach based on finite automata. IEEE Trans. Autom. Control **61**(9), 2733–2738 (2015)
32. Guo, Y.: Observability of Boolean control networks using parallel extension and set reachability. IEEE Trans. Neural Netw. Learning Syst. **29**(12), 6402–6408 (2018)
33. Zhou, R., Guo, Y., Gui, W.: Set reachability and observability of probabilistic Boolean networks. Automatica **106**, 230–241 (2019)
34. Veliz-Cuba, A., Stigler, B.: Boolean models can explain bistability in the lac operon. J. Comput. Biol. **18**(6), 783–794 (2011)
35. Heidel, J., Maloney, J., Farrow, C., et al.: Finding cycles in synchronous Boolean networks with applications to biochemical systems. Int. J. Bifurcation Chaos **13**(03), 535–552 (2003)
36. Chaves, M.: Methods for qualitative analysis of genetic networks. In: 2009 European Control Conference (ECC), pp. 671–676 (2009)
37. Zhong, J., Lu, J., Huang, T., et al.: Controllability and synchronization analysis of identical-hierarchy mixed-valued logical control networks. IEEE Trans. Cybern. **47**(11), 3482–3493 (2016)

38. Tang, Y., Qian, F., Gao, H., et al.: Synchronization in complex networks and its application-a survey of recent advances and challenges. Ann. Rev. Control **38**(2), 184–198 (2014)
39. Wu, Z., Shi, P., Su, H., et al.: Exponential synchronization of neural networks with discrete and distributed delays under time-varying sampling. IEEE Trans. Neural Netw. Learn. Syst. **23**(9), 1368–1376 (2012)
40. Zhong, J., Lu, J., Liu, Y., et al.: Synchronization in an array of output-coupled Boolean networks with time delay. IEEE Trans. Neural Netw. Learn. Syst. **25**(12), 2288–2294 (2014)
41. Liu, Y., Sun, L., Lu, J., et al.: Feedback controller design for the synchronization of Boolean control networks. IEEE Trans. Neural Netw. Learn. Syst. **27**(9), 1991–1996 (2015)
42. Chen, H., Liang, J., Lu, J.: Partial synchronization of interconnected Boolean networks. IEEE Trans. Cybern. **47**(1), 258–266 (2016)
43. Li, F.: Pinning control design for the synchronization of two coupled Boolean networks. IEEE Trans. Circuits Syst. II Express Briefs **63**(3), 309–313 (2015)
44. Wu, Y., Su, H., Shi, P., et al.: Consensus of multiagent systems using aperiodic sampled-data control. IEEE Trans. Cybern. **46**(9), 2132–2143 (2015)
45. Wang, Y., Li, H.: On definition and construction of Lyapunov functions for Boolean networks. In: Proceedings of the 10th World Congress on Intelligent Control and Automation, pp. 1247–1252 (2012)
46. Li, H., Wang, Y.: Lyapunov-based stability and construction of Lyapunov functions for Boolean networks. SIAM J. Control Optim. **55**(6), 3437–3457 (2017)
47. Meng, M., Lam, J., Feng, J., et al.: l_1-gain analysis and model reduction problem for Boolean control networks. Inf. Sci. **348**, 68–83 (2016)
48. Meng, M., Liu, L., Feng, G.: Stability and l_1 gain analysis of Boolean networks with Markovian jump parameters. IEEE Trans. Autom. Control **62**(8), 4222–4228 (2017)
49. Meng, M., Lam, J., Feng, J., et al.: Stability and guaranteed cost analysis of time-triggered Boolean networks. IEEE Trans. Neural Netw. Learn. Syst. **29**(8), 3893–3899 (2017)
50. Cao, Y., Zhang, L., Li, C., et al.: Observer-based consensus tracking of nonlinear agents in hybrid varying directed topology. IEEE Trans. Cybern. **47**(8), 2212–2222 (2016)
51. Heemels, W.P.M.H, Johansson, K.H., Tabuada, P.: An introduction to event-triggered and self-triggered control. In: Proceeding of 51st IEEE Conference on Decision and Control, pp. 3270–3285 (2012)
52. Tan, X., Gao, J., Li, X.: Consensus of leader-following multiagent systems: a distributed event-triggered impulsive control strategy. IEEE Trans. Cybern. **49**(3), 792–801 (2018)
53. Li, C., Yu, X., Yu, W., et al.: Distributed event-triggered scheme for economic dispatch in smart grids. IEEE Trans. Ind. Inf. **12**(5), 1775–1785 (2015)
54. Shinar, G., Milo, R., Martínez, M.R., et al.: InputÍCoutput robustness in simple bacterial signaling systems. Proc. Natl. Acad. Sci. **104**(50), 19931–19935 (2007)
55. Cheng, D.: Disturbance decoupling of Boolean control networks. IEEE Trans. Autom. Control **56**(1), 2–10 (2010)
56. Yang, M., Li, R., Chu, T.: Controller design for disturbance decoupling of Boolean control networks. Automatica **49**(1), 273–277 (2013)
57. Li, H., Wang, Y., Xie, L., et al.: Disturbance decoupling control design for switched Boolean control networks. Syst. Control Lett. **72**, 1–6 (2014)
58. Li, H., Wang, Y., Guo, P.: State feedback based output tracking control of probabilistic Boolean networks. Inf. Sci. **349**, 1–11 (2016)
59. Liu, T., Huang, J.: Robust output regulation of discrete-time linear systems by quantized output feedback control. Automatica **107**, 587–590 (2019)
60. Byrnes, C.I., Isidori, A.: Output regulation for nonlinear systems: an overview. Int. J. Robust Nonlinear Control: IFAC-Affiliated J. **10**(5), 323–337 (2000)
61. Li, H., Xie, L., Wang, Y.: Output regulation of Boolean control networks. IEEE Trans. Autom. Control **62**(6), 2993–2998 (2016)

Part II
Periodic Sampled-Data Control

Chapter 2
Stabilization of Sampled-Data Boolean Control Networks

Abstract In this chapter, we study stabilization and set stabilization of Boolean control networks (BCNs) under sampled-data state feedback control (SDSFC).

2.1 Sampled-Data State Feedback Stabilization of Boolean Control Networks

In this section, we investigate the sampled-data state feedback control (SDSFC) problem of Boolean control networks (BCNs). Some necessary and sufficient conditions are obtained for the global stabilization of BCNs by SDSFC. Different from conventional state feedback controls, new phenomena are observed in the study of SDSFC. Based on the controllability matrix, we derive some necessary and sufficient conditions under which the trajectories of BCNs can be stabilized to a fixed point by piecewise constant control (PCC). It is proved that the global stabilization of BCNs under SDSFC is equivalent to that by PCC. Moreover, algorithms are given to construct the sampled-data state feedback controllers.

2.1.1 Problem Formulation

In this section, we consider the BCN as follows:

$$\begin{cases} x_1(t+1) = f_1(x_1(t), \ldots, x_n(t), u_1(t), \ldots, u_m(t)), \\ \qquad\qquad\vdots \\ x_n(t+1) = f_n(x_1(t), \ldots, x_n(t), u_1(t), \ldots, u_m(t)), \end{cases} \tag{2.1}$$

for $t \geq 0$, where $f_i : \mathcal{D}^n \to \mathcal{D},\ i = 1, 2, \ldots, n$ are logical functions. $x_i \in \mathcal{D},\ i = 1, 2, \ldots, n$ and $u_j \in \mathcal{D},\ j = 1, 2, \ldots, m$ are states and control inputs, respectively.

Y. Liu et al., *Sampled-data Control of Logical Networks*,
https://doi.org/10.1007/978-981-19-8261-3_2

For each logical function f_i, $i = 1, 2, \ldots, n$, we can find its unique structure matrix M_i. Let $x(t) = \ltimes_{i=1}^{n} x_i(t)$ and $u(t) = \ltimes_{l=1}^{m} u_l(t)$.

Then system (2.1) can be converted into an algebraic form as

$$x_i(t+1) = M_i u(t) x(t), \quad i = 1, \ldots, n. \tag{2.2}$$

Multiplying the equations in (2.2) together yields

$$x(t+1) = L u(t) x(t), \quad t \geq 0, \tag{2.3}$$

where

$$L = M_1 \ltimes_{i=2}^{n} [(I_{2^{n+m}} \otimes M_i)\Phi_{m+n}] \in \mathcal{L}_{2^n \times 2^{m+n}}.$$

The feedback law to be determined for system (2.1) is in the following form,

$$\begin{cases} u_1(t) = e_1(x_1(t_l), \ldots, x_n(t_l)), \\ u_2(t) = e_2(x_1(t_l), \ldots, x_n(t_l)), \\ \quad\vdots \\ u_m(t) = e_m(x_1(t_l), \ldots, x_n(t_l)), \end{cases} \tag{2.4}$$

for $t_l \leq t < t_{l+1}$, where $e_j : \mathcal{D}^n \to \mathcal{D}$, $j = 1, 2, \ldots, m$ are Boolean functions, $t_l = l\tau \geq 0$ for $l = 0, 1, \ldots$ are sampling instants, and $t_{l+1} - t_l = \tau$ denotes the constant sampling period.

Similar to the analysis above, Eq. (2.4) is equivalent to the following algebraic form,

$$u(t) = E x(t_l), \quad t_l \leq t < t_{l+1}, \tag{2.5}$$

where $E = \tilde{M}_1 \ltimes_{j=2}^{m} [(I_{2^n} \otimes \tilde{M}_j)\Phi_n] \in \mathcal{L}_{2^m \times 2^n}$ and $\tilde{M}_j$ is the structure matrix of e_j.

If $\tau = 1$, then Eq. (2.5) can be regarded as a normal state feedback control as in [1–5] and others.

Definition 2.1 For a given state $X_e = (x_1^e, \ldots, x_n^e) \in \mathcal{D}^n$, system (2.3) is said to be globally stabilizable to X_e, if there exist a logical control sequence $U = \{u(t), t = 0, 1, 2, \ldots\}$ and an integer $T > 0$, such that $X(t; X_0; U) = X_e$ for $\forall X_0 \in \mathcal{D}^n$ and $\forall t \geq T$.

2.1.2 Sampled-Data State Feedback Control for Boolean Control Networks

Considering system (2.3) and the SDSFC (2.5), for $0 \leq t \leq \tau$, we have

$$\begin{aligned} x(1) =& LEx(0)x(0) = LEW_{[2^n]}x(0)x(0) = LEW_{[2^n]}\Phi_n x(0), \\ x(2) =& LEx(0)x(1) = LEW_{[2^n]}x(1)x(0) \\ =& LEW_{[2^n]}\left(LEW_{[2^n]}\Phi_n x(0)\right)x(0) = \left(LEW_{[2^n]}\right)^2 \Phi_n^2 x(0), \\ &\vdots \\ x(t) =& \left(LEW_{[2^n]}\right)^t \Phi_n^t x(0). \end{aligned} \tag{2.6}$$

Similarly, regarding $x(\tau)$ as an initial state for $\tau < t \leq 2\tau$, one gets from (2.6) that

$$\begin{aligned} x(t) &= \left(LEW_{[2^n]}\right)^{t-\tau} \Phi_n^{t-\tau} x(\tau) \\ &= \left(LEW_{[2^n]}\right)^{t-\tau} \Phi_n^{t-\tau} \left(\left(LEW_{[2^n]}\right)^{\tau} \Phi_n^{\tau}\right) x(0). \end{aligned} \tag{2.7}$$

Consequently, for $l\tau < t \leq (l+1)\tau$, we have by the induction that

$$x(t) = \left(LEW_{[2^n]}\right)^{t-l\tau} \Phi_n^{t-l\tau} \left(\left(LEW_{[2^n]}\right)^{\tau} \Phi_n^{\tau}\right)^l x(0). \tag{2.8}$$

Theorem 2.1 *System (2.3) can be globally stabilized to $\delta_{2^n}^r$ by SDSFC in the form of (2.5), if and only if there exists $k > 0$, such that*

$$\begin{cases} \left(\left(LEW_{[2^n]}\right)^{\tau} \Phi_n^{\tau}\right)^k = \delta_{2^n}[\underbrace{r\ r\ \cdots\ r}_{2^n}], \\ \left(LEW_{[2^n]}\Phi_n\right)_{rr} = 1, \end{cases} \tag{2.9}$$

where $(LEW_{[2^n]}\Phi_n)_{rr}$ is the (r,r)-th element of the matrix $LEW_{[2^n]}\Phi_n$.

Proof (Sufficiency) Suppose that (2.9) holds. It then follows from (2.8) that

$$x(k\tau) = \left(\left(LEW_{[2^n]}\right)^{\tau} \Phi_n^{\tau}\right)^k x(0) = \delta_{2^n}^r,$$

for all $x(0) \in \Delta_{2^n}$. Then $x(k\tau+1) = LEW_{[2^n]}\Phi_n x(k\tau) = \delta_{2^n}^r$ from (2.9), and so does $x(k\tau+2)$. Similarly, we have $x(t) = \delta_{2^n}^r$ for $k\tau \leq t \leq (k+1)\tau$. By induction, one can see that when $t \geq T \triangleq k\tau,\ x(t) \equiv \delta_{2^n}^r$ for all $x(0) \in \Delta_{2^n}$. Sufficiency is proved.

(Necessity) Assume that system (2.3) is globally stabilizable to $\delta_{2^n}^r$ by SDSFC in the form of (2.5). Then there exits $T > 0$ such that $\forall t \geq T,\ x(t) = \delta_{2^n}^r$. Without loss of generality, supposing that $T = k\tau$ for some $k > 0$, we have $x(k\tau) = \left(\left(LEW_{[2^n]}\right)^{\tau} \Phi_n^{\tau}\right)^k x(0) \equiv \delta_{2^n}^r$ for $\forall x(0) = \delta_{2^n}^i$, $i = 1, 2, \ldots, 2^n$. Therefore, the first equation in (2.9) holds. Furthermore, $x(k\tau + 1) = LEW_{[2^n]}\Phi_n x(k\tau) = LEW_{[2^n]}\Phi_n \delta_{2^n}^r \equiv \delta_{2^n}^r$. Hence the second equation in (2.9) holds as well. Necessity is now proved. □

Remark 2.1 When $\tau = 1$, the SDSFC reduces to state feedback control. Then $\left(\left(LEW_{[2^n]}\right)^{\tau} \Phi_n^{\tau}\right)^k = \delta_{2^n}[r\ r\ \cdots\ r]$ implies $\left(LEW_{[2^n]}\Phi_n\right)_{rr} = 1$. In fact,

$$\begin{aligned}\left(LEW_{[2^n]}\Phi_n\right)^{k+1} x(0) &= \left(LEW_{[2^n]}\Phi_n\right)\left(LEW_{[2^n]}\Phi_n\right)^k x(0)\\ &= \left(LEW_{[2^n]}\Phi_n\right)\delta_{2^n}[r\ r\ \cdots\ r]x(0)\\ &= \left(LEW_{[2^n]}\Phi_n\right)\delta_{2^n}^r.\end{aligned} \tag{2.10}$$

On the other hand,

$$\begin{aligned}\left(LEW_{[2^n]}\Phi_n\right)^{k+1} x(0) &= \left(LEW_{[2^n]}\Phi_n\right)^k \left(LEW_{[2^n]}\Phi_n\right) x(0)\\ &=\delta_{2^n}[r\ r\ \cdots\ r]\left(LEW_{[2^n]}\Phi_n\right) x(0)\\ &=\delta_{2^n}^r.\end{aligned} \tag{2.11}$$

Combining (2.10) and (2.11) gives that $\left(LEW_{[2^n]}\Phi_n\right)\delta_{2^n}^r = \delta_{2^n}^r$, that is,

$$\left(LEW_{[2^n]}\Phi_n\right)_{rr} = 1.$$

When $\tau > 1$, this argument does not work, which means that $\left(LEW_{[2^n]}\Phi_n\right)_{rr} = 1$ cannot be guaranteed by $\left(\left(LEW_{[2^n]}\right)^{\tau} \Phi_n^{\tau}\right)^k = \delta_{2^n}[r\ r\ \cdots\ r]$. This is different from the case $\tau = 1$. Therefore, both conditions in (2.9) are necessary for $\tau > 1$ in Theorem 2.1. The following example illustrates this point.

Let $\tau = 2,\ n = 1,\ r = 1$ and

$$LEW_{[2]} = \begin{bmatrix} 0\,1\,1\,1 \\ 1\,0\,0\,0 \end{bmatrix}, \quad \Phi_1 = \begin{bmatrix} 1\,0\,0\,0 \\ 0\,0\,0\,1 \end{bmatrix}^{\mathrm{T}}.$$

Then

$$(LEW_{[2]})^{\tau} = (LEW_{[2]})^2 = \begin{bmatrix} 1\,1\,0\,1\,0\,1\,0\,1 \\ 0\,0\,1\,0\,1\,0\,1\,0 \end{bmatrix},$$

and

$$\Phi_1^{\tau} = \Phi_1^2 = \begin{bmatrix} 1\,0\,0\,0\,0\,0\,0\,0 \\ 0\,0\,0\,0\,0\,0\,0\,1 \end{bmatrix}^{\mathrm{T}}.$$

As a result,

$$\left((LEW_{[2]})^2\Phi_1^2\right)^1 = \begin{bmatrix} 1 & 1 \\ 0 & 0 \end{bmatrix}.$$

However, it is noticed that

$$LEW_{[2]}\Phi_1 = \begin{bmatrix} 0 & 1 \\ 1 & 0 \end{bmatrix},$$

and $\left(LEW_{[2]}\Phi_1\right)_{11} \neq 1$. Therefore, $\left(LEW_{[2^n]}\Phi_n\right)_{rr} = 1$ is not necessarily guaranteed by $\left((LEW_{[2^n]})^\tau \Phi_n^\tau\right)^k = \delta_{2^n}[r\ r\ \cdots\ r]$ for $\tau > 1$.

It should be noticed that only when t is sufficiently large ($t \geq k\tau$, where k is given by (2.9)), then the state $\delta_{2^n}^r$ is fixed. Otherwise, when $t < k\tau$, although the state of the system reaches $\delta_{2^n}^r$ by the SDSFC (i.e., $x(t) = \delta_{2^n}^r$), we cannot guarantee $x(t+1) = \delta_{2^n}^r$. It is a very different fact from the normal state feedback controlled system with $\tau = 1$. We use the following example to illustrate the statement.

Let $\tau = 3,\ n = 1,\ r = 1$ and

$$LEW_{[2]} = \begin{bmatrix} 1 & 0 & 1 & 1 \\ 0 & 1 & 0 & 0 \end{bmatrix}, \quad \Phi_1 = \begin{bmatrix} 1 & 0 & 0 & 0 \\ 0 & 0 & 0 & 1 \end{bmatrix}^{\mathrm{T}}.$$

Then

$$((LEW_{[2]})^3{\Phi_1}^3)^1 = \begin{bmatrix} 1 & 1 \\ 0 & 0 \end{bmatrix},$$

and $(LEW_{[2]}\Phi_1)_{11} = 1$. So the system can be globally stabilized to δ_2^1 for $t \geq 3$ from Theorem 2.1. When $x(0) = \delta_2^2$, $x(1) = LEW_{[2]}\Phi_1 x(0) = \delta_2^1$, while $x(2) = (LEW_{[2]})^2{\Phi_1}^2 x(0) = \delta_2^2 \neq \delta_2^1$. It means that δ_2^1 is not fixed for $t = 2 < 3$.

Lemma 2.1 *Let* $L = \delta_{2^n}[\alpha_1\ \alpha_2\ \cdots\ \alpha_{2^{n+m}}]$ *and assume* $E = \delta_{2^m}[p_1\ p_2\ \cdots\ p_{2^n}]$. *Define sets by*

$$S_l(i) = \{x_0 \in \Delta_{2^n} :\ \left((LEW_{[2^n]})^\tau \Phi_n^\tau\right)^l x_0 = \delta_{2^n}^i\}, \quad l \geq 1,\ 1 \leq i \leq 2^n. \tag{2.12}$$

Then we have the following results.

(I) $S_1(r)$ *can be rewritten by a set*

$$\{\delta_{2^n}^i :\ 1 \leq i \leq 2^n,\ \beta_i^\tau = r,\ \textit{for some}\ 1 \leq p_i \leq 2^m\},$$

where

$$\begin{cases} \beta_i^1 = \alpha_{(p_i-1)2^n+i}, \\ \beta_i^{l+1} = \alpha_{(p_i-1)2^n+\beta_i^l}, \quad l \geq 1. \end{cases} \tag{2.13}$$

(II) $S_{l+1}(r) = \cup\{S_1(i) : 1 \leq i \leq 2^n,\ \delta_{2^n}^i \in S_l(r)\}$ *for all* $l \geq 1$.

Proof

(I) Assuming $\delta_{2^n}^i \in S_1(r)$, we have from (2.12) that

$$\begin{aligned}
\delta_{2^n}^r &= \left(LEW_{[2^n]}\right)^\tau \Phi_n^\tau \delta_{2^n}^i \\
&= \left(LEW_{[2^n]}\right)^\tau (\delta_{2^n}^i)^{\tau+1} \\
&= \left(LEW_{[2^n]}\right)^{\tau-1} \left[\left(LEW_{[2^n]}(\delta_{2^n}^i)^2\right)\right] (\delta_{2^n}^i)^{\tau-1} \\
&= \left(LEW_{[2^n]}\right)^{\tau-1} \left[\delta_{2^n}^{\alpha_{(p_i-1)2^n+i}}\right] (\delta_{2^n}^i)^{\tau-1} \\
&= \left(LEW_{[2^n]}\right)^{\tau-1} \left[\delta_{2^n}^{\beta_i^1}\right] (\delta_{2^n}^i)^{\tau-1}.
\end{aligned}$$

Moreover, for $1 \leq k \leq \tau - 1$,

$$\begin{aligned}
&\left(LEW_{[2^n]}\right)^{\tau-k} \left[\delta_{2^n}^{\beta_i^k}\right] (\delta_{2^n}^i)^{\tau-k} \\
=&\left(LEW_{[2^n]}\right)^{\tau-(k+1)} \left[\left(LEW_{[2^n]}\delta_{2^n}^{\beta_i^k}\delta_{2^n}^i\right)\right] (\delta_{2^n}^i)^{\tau-(k+1)} \\
=&\left(LEW_{[2^n]}\right)^{\tau-(k+1)} \left[\delta_{2^n}^{\alpha_{(p_i-1)2^n+\beta_i^k}}\right] (\delta_{2^n}^i)^{\tau-(k+1)} \\
=&\left(LEW_{[2^n]}\right)^{\tau-(k+1)} \left[\delta_{2^n}^{\beta_i^{k+1}}\right] (\delta_{2^n}^i)^{\tau-(k+1)}.
\end{aligned}$$

By induction, we have $\delta_{2^n}^r = \delta_{2^n}^{\beta_i^\tau}$ for $k = \tau - 1$. Therefore, $\beta_i^\tau = r$ and we prove the first part of the Lemma.

(II) From the definition of $S_l(r)$ in (2.12),

$$S_{l+1}(r) = \{x_0 \in \Delta_{2^n} : \ \left(\left(LEW_{[2^n]}\right)^\tau \Phi_n^\tau\right)^{l+1} x_0 = \delta_{2^n}^r\}.$$

We notice that

$$\left(\left(LEW_{[2^n]}\right)^\tau \Phi_n^\tau\right)^{l+1} x_0 = \left(\left(LEW_{[2^n]}\right)^\tau \Phi_n^\tau\right)^l \left(\left(LEW_{[2^n]}\right)^\tau \Phi_n^\tau\right) x_0.$$

Therefore, $S_{l+1}(r) = \{x_0 \in \Delta_{2^n} : \left(\left(LEW_{[2^n]}\right)^{\tau} \Phi_n^{\tau}\right) x_0 = \delta_{2^n}^i, \ \delta_{2^n}^i \in S_l(r)\} = \cup\{S_1(i) : 1 \le i \le 2^n, \ \delta_{2^n}^i \in S_l(r)\}$.

□

Theorem 2.2 *System (2.3) is globally stabilizable to $\delta_{2^n}^r$ by SDSFC (2.5), then*

(I) $\delta_{2^n}^r \in S_1(r)$.
(II) There exists $1 \le N \le 2^n$ *such that* $\Delta_{2^n} = \cup_{1 \le l \le N} S_l(r) \setminus S_{l-1}(r)$ *with* $S_0(r)$ *denoted by* $\varnothing$.

Proof

(I) If system (2.3) is globally stabilizable to $\delta_{2^n}^r$ by (2.5), then there exists $k > 0$, such that $x(t) \equiv \delta_{2^n}^r$ for all $t \ge k\tau$ and $x(0) \in \Delta_{2^n}$. Therefore, $\delta_{2^n}^r = x((k+1)\tau) = \left(LEW_{[2^n]}\right)^{\tau} \Phi_n^{\tau} x(k\tau) = \left(LEW_{[2^n]}\right)^{\tau} \Phi_n^{\tau} \delta_{2^n}^r$. From (2.12), we conclude that

$$\delta_{2^n}^r \in S_1(r).$$

(II) If $S_1(r) = \{\delta_{2^n}^r\}$, since $S_{l+1}(r) = \cup\{S_1(i) : 1 \le i \le 2^n, \ \delta_{2^n}^i \in S_l(r)\}$ for all $l \ge 1$, it follows from Lemma 2.1 that $S_l(r) = \{\delta_{2^n}^r\}$ for any $l > 1$. It is a contradiction to the condition that system (2.3) can be globally stabilizable to $\delta_{2^n}^r$. Therefore, there must be $\delta_{2^n}^i \in S_1(r)$ besides $\delta_{2^n}^r$.

Due to $S_2(r) = \cup\{S_1(i) : 1 \le i \le 2^n, \ \delta_{2^n}^i \in S_1(r)\}$ and $\delta_{2^n}^r \in S_1(r)$, then $S_1(r) \subset S_2(r)$. If $S_2(r) \setminus S_1(r) = \varnothing$, then $S_l(r) = S_1(r)$ for any $l > 1$. If $S_1(r) = \Delta_{2^n}$, then $N = 1$. Otherwise, it contradicts to the fact that system (2.3) can be globally stabilizable to $\delta_{2^n}^r$. Therefore, $S_2(r) \setminus S_1(r) \ne \varnothing$.

Similarly, we can conclude that $S_l(r) \setminus S_{l-1}(r) \ne \varnothing$ for $l = 1, 2, \ldots$. The upper boundary of l, denoted by N, will be less than or equal 2^n, since the entire set is Δ_{2^n} containing 2^n elements. Therefore, $\Delta_{2^n} = \cup_{1 \le l \le N} S_l(r) \setminus S_{l-1}(r)$.

□

Now we are ready to solve (2.9) to get the SDSFC.

Theorem 2.3 *If the SDSFC in the form of (2.5) exists such that system (2.3) can be globally stabilized to* $\delta_{2^n}^r$, *then for every* $1 \le i \le 2^n$ *there is a unique integral* $1 \le l_i \le N$ *such that* $\delta_{2^n}^i \in S_{l_i}(r) \setminus S_{l_i-1}(r)$ *with* $S_0(r) = \varnothing$. *Let* p_i *be the solution such that*

$$\begin{cases} \alpha_{(p_r-1)2^n+r} = r, \\ \beta_i^{\tau} = r \textit{ for } l_i = 1 \textit{ with } i \ne r, \\ \delta_{2^n}^{\beta_i^{\tau}} \in S_{l_i-1}(r) \setminus S_{l_i-2}(r) \textit{ for } l_i \ge 2, \end{cases} \tag{2.14}$$

then the SDSFC can be determined by $E = \delta_{2^m}[p_1 \ p_2 \ \cdots \ p_{2^n}]$.

Proof From Theorem 2.2, $\Delta_{2^n} = \cup_{1\le l\le N} S_l(r) \setminus S_{l-1}(r)$. Then for any $1 \le i \le 2^n$, it is easy to see that there exists a unique integral $1 \le l_i \le N$ such that $\delta_{2^n}^i \in S_{l_i}(r) \setminus S_{l_i-1}(r)$.

Since $\left(LEW_{[2^n]}\Phi_n\right)_{rr} = 1$ from Theorem 2.1,

$$\delta_{2^n}^r = LEW_{[2^n]}\Phi_n\delta_{2^n}^r = LE(\delta_{2^n}^r)^2 = \delta_{2^n}^{\alpha(p_r-1)2^n+r}.$$

Therefore, $\alpha_{(p_r-1)2^n+r} = r$.

For $l_i = 1$ and $i \ne r$, from Lemma 2.1, we need to find a possible p_i such that $\beta_i^\tau = r$. By the definition of β_i^l, $l \ge 1$ in (2.13), if $\beta_i^\tau = r$, then $\alpha_{(p_i-1)2^n+\beta_i^{\tau-1}} = r$. Define $D(i) = \{j : \alpha_j = i,\ 1 \le j \le 2^{n+m}\}$ for $1 \le i \le 2^n$.

Step 1 Since $1 \le \beta_i^{\tau-1} \le 2^n$, then for any $j_1 \in D(r)$ there exists a unique $1 \le p_i \le 2^n$ such that $(p_i - 1)2^n + \beta_i^{\tau-1} = j_1$.

Step 2 Consider $\beta_i^{\tau-1} = j_1 - (p_i - 1)2^n$ which is between 1 and 2^n. Then $\alpha_{(p_i-1)2^n+\beta_i^{\tau-2}} = j_1 - (p_i - 1)2^n$ from (2.13). To determine $\beta_i^{\tau-2}$, we consider an element $j_2 \in D(j_1 - (p_i - 1)2^n)$ such that $1 \le j_2 - (p_i - 1)2^n \le 2^n$. Then $j_2 = (p_i - 1)2^n + \beta_i^{\tau-2}$, and therefore, $\beta_i^{\tau-2} = j_2 - (p_i - 1)2^n$

...

Step $\tau - 1$ Since $S_1(r) \setminus \{\delta_{2^n}^r\} \ne \varnothing$, there exists at least $j_{\tau-1} \in D(j_{\tau-2} - (p_i - 1)2^n)$ such that $1 \le j_{\tau-1} - (p_i - 1)2^n \le 2^n$. Then $j_{\tau-1} = (p_i - 1)2^n + \beta_i^1$, that is, $\beta_i^1 = \alpha_{(p_i-1)2^n+i} = j_{\tau-1} - (p_i - 1)2^n$.

Step τ Solving the equation $\alpha_{(p_i-1)2^n+i} = j_{\tau-1} - (p_i - 1)2^n$ gives at least one possible i such that p_i satisfies $\beta_i^\tau = r$.

For $l_i \ge 2$, we only need to regard $\delta_{2^n}^j \in S_{l_i-1}(r) \setminus S_{l_i-2}(r)$ to be $\delta_{2^n}^r$ for $l_i = 1$. Following this way, consider all $\delta_{2^n}^j \in S_{l_i-1}(r) \setminus S_{l_i-2}(r)$, and then p_i can be obtained with a similar argument as the one above. Consequently, the matrix E is constructed. □

Based on Theorem 2.3, we give the following algorithm to get E.

Algorithm 1 Get the matrix E

Step 1. Solving $\alpha_{(p_r-1)2^n+r} = r$ to get p_r. If there is no solution, E does not exist.

Step 2. Solving $\beta_i^\tau = r$ for $l_i = 1$ with $i \ne r$ to get one solution of p_i for $\delta_{2^n}^i \in S_1(r) \setminus \{\delta_{2^n}^r\}$. If there is no solution of such p_i, E does not exist.

Step 3. Regard $\delta_{2^n}^j \in S_{l_i-1}(r) \setminus S_{l_i-2}(r)$ to be $\delta_{2^n}^r$ for $l_i = 1$, and solve $\delta_{2^n}^{\beta_i^\tau} \in S_{l_i-1}(r) \setminus S_{l_i-2}(r)$ for $l_i \ge 2$. Get one solution of each p_i. If there is no solution of such p_i, E does not exist.

Step 4. Get u with $E = \delta_{2^m}[p_1\ p_2\ \cdots\ p_{2^n}]$.

Theorem 2.3 presents an approach to construct SDSFC such that system (2.3) can be globally stabilized. If the SDSFC exists, it can be constructed by (2.14). On the other hand, if there is no solution of p_i to Eqs. (2.14), then the SDSFC does not exist. In other words, we do not know if the SDSFC exists or not unless we solve all equations in (2.14).

Let us consider the BCN model presented in [6]. It is a reduced Boolean model for the lac operon in the bacterium Escherichia coli, which shows that lac mRNA and lactose form the core of the lac operon. The model is given as follows:

$$\begin{cases} x_1(t+1) = \neg u_1(t) \wedge (x_2(t) \vee x_3(t)), \\ x_2(t+1) = \neg u_1(t) \wedge u_2(t) \wedge x_1(t), \\ x_3(t+1) = \neg u_1(t) \wedge (u_2(t) \vee (u_3(t) \wedge x_1(t))), \end{cases} \tag{2.15}$$

where $x_1,\ x_2$ and x_3 are state variables denoting the lac mRNA, the lactose in high concentrations, and the lactose in medium concentrations, respectively; $u_1,\ u_2$ and u_3 are control inputs that represent the extracellular glucose, the high extracellular lactose, and the medium extracellular lactose, respectively.

Using the vector form of logical variables and setting $x(t) = \ltimes_{i=1}^{3} x_i(t)$ and $u(t) = \ltimes_{i=1}^{3} u_i(t)$, by the semi-tensor product, we can express the system (2.15) in its algebraic from as

$$x(t+1) = Lu(t)x(t),$$

where

$$\begin{aligned} L = \delta_8[&8\ 8\ 8\ 8\ 8\ 8\ 8\ 8\ 8\ 8\ 8\ 8\ 8\ 8\ 8\ 8\ 8\ 8\ 8\ 8\ 8\ 8\ 8\ 8\ 8\ 8\ 8\ 8\ 8\ 8\ 8\ 8 \\ &1\ 1\ 1\ 5\ 3\ 3\ 3\ 7\ 1\ 1\ 1\ 5\ 3\ 3\ 3\ 7\ 3\ 3\ 3\ 7\ 4\ 4\ 4\ 8\ 4\ 4\ 4\ 8\ 4\ 4\ 4\ 8]. \end{aligned}$$

The feedback law to be determined for system (2.15) is in the form (2.5) as

$$u(t) = Ex(t_l), \quad t_l \le t < t_{l+1},$$

where $E = \tilde{M}_1 \ltimes_{j=2}^{3} [(I_{2^3} \otimes \tilde{M}_j)\Phi_3] \in \mathcal{L}_{2^3 \times 2^3}$ and $\tilde{M}_j$ is the structure matrix of e_j. Assume that the state feedback matrix E is given by

$$E = \delta_8[p_1\ p_2\ \cdots\ p_8]. \tag{2.16}$$

Let $\tau = 2$, we now construct the SDSFC such that system (2.15) is globally stabilizable to δ_8^1.

Step 1 Solving $\alpha_{(p_1-1)2^3+1} = 1$, we get $p_1 = 5$ *or* 6.

Step 2 Since $\beta_i^2 = 1$ and $\alpha_{j_1} = 1$, then $j_1 \in D(1) = \{33, 34, 35, 41, 42, 43\}$. Solve $\alpha_{(p_i-1)2^3+i} = j_1 - (p_i - 1)2^3$. If $j_1 = 33$, then $p_2 = p_3 = 5$; if $j_1 = 35$, then $p_5 = p_6 = p_7 = 5$; if $j_1 = 41$, then $p_2 = p_3 = 6$; if $j_1 = 43$, then $p_5 = p_6 = p_7 = 6$. We have $\delta_8^2, \delta_8^3, \delta_8^5, \delta_8^6, \delta_8^7 \in S_1(1) \setminus \{\delta_8^1\}$.

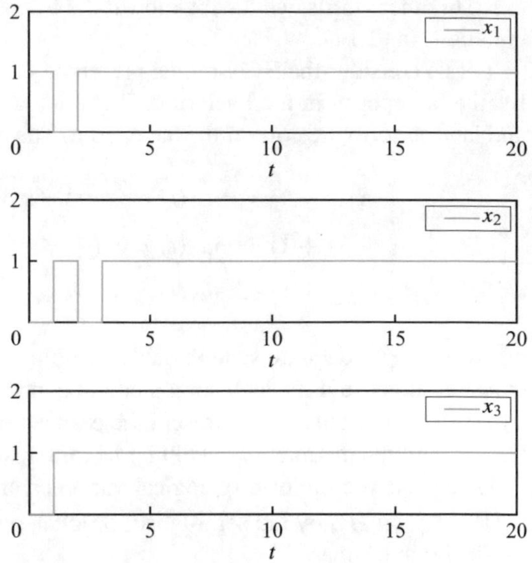

Fig. 2.1 The trajectories of system (2.15) with initial states $x_1(0) = \delta_2^1, x_2(0) = \delta_2^2$, $x_3(0) = \delta_2^2$ is globally stabilized to δ_8^1 under the SDSFC as shown in Fig. 2.2

Step 3 Regard $\delta_8^4, \delta_8^8 \in S_2(1) \setminus S_1(1)$. Assume $\beta_i^2 = 3$, $\alpha_{j_1} = 3$, then $j_1 \in D(3) = \{37, 38, 39, 45, 46, 47, 49, 50, 51\}$. Solve $\alpha_{(p_i-1)2^3+i} = j_1 - (p_i - 1)2^3$. If $j_1 = 37$, then $p_4 = 5$; if $j_1 = 39$, then $p_8 = 5$.

Consequently, one of the possible E is given by

$$E = \delta_8[5\ 6\ 6\ 5\ 5\ 6\ 6\ 5],$$

and the corresponding SDSFC is in the form of

$$\begin{cases} u_1(t) = 0, \\ u_2(t) = 1, \qquad 2l \leq t < 2(l+1), \\ u_3(t) = x_2(2l) \leftrightarrow x_3(2l), \end{cases} \tag{2.17}$$

for $l = 0, 1, 2, \ldots$.

Figure 2.1 shows the trajectories of system (2.15) with initial states $x_1(0) = \delta_2^1$, $x_2(0) = \delta_2^2$, $x_3(0) = \delta_2^2$ under the constructed SDSFC as shown in Fig. 2.2.

2.1.3 Piecewise Constant Control of Boolean Control Networks

Piecewise constant control (PCC) [7, 8] is another kind of discontinuous control, which is defined on each sampling interval to stabilize some dynamical systems or networks. When a control parameter is switched instantaneously from one value to

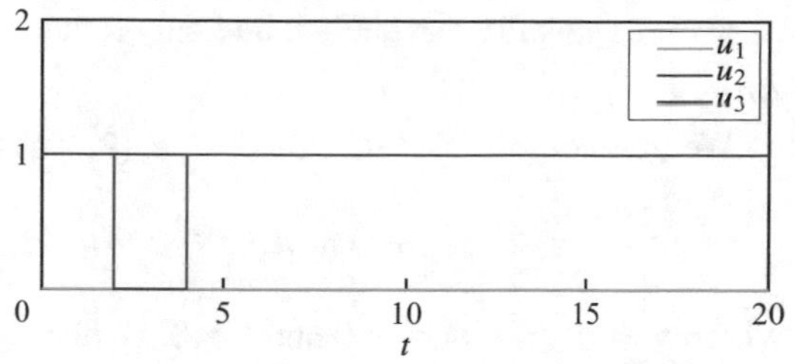

Fig. 2.2 An illustration of the designed $u(t)$ to achieve system (2.15) stabilization

another, it can take the solution as the new initial state and evolve it further until the next switching event occurs. The PCC allows one to account for higher-order terms normally neglected while using a discretizing sampled data controller or a continuous feedback controller [9]. There are many interesting results on PCC of systems. For example, constructing a PCC for a continuous Roesser system has been studied in [10]. An algorithm for PCC design in minimal time has been proposed by Lefebvre [11].

Considering system (2.3) and the fact of $Lu(t) \in \mathcal{L}_{2^n \times 2^n}$, we split L into 2^m equal blocks as

$$L = [{}^1L \; {}^2L \; \cdots \; {}^{2^m}L],$$

where ${}^iL \in \mathcal{L}_{2^n \times 2^n}$, $i \in \Omega_m$. Denote the controllability matrix

$$\hat{L}^l = \bigvee_{j=1}^{2^m} ({}^jL)^l, \quad l \geq 1. \tag{2.18}$$

Assume that $u(t) = \delta_{2^m}^j$, then $Lu(t) = {}^jL$. Therefore, system (2.3) can be regarded as a Boolean switched system [1],

$$x(t+1) = {}^{\sigma(t)}Lx(t), \quad t \geq 0, \tag{2.19}$$

where $1 \leq \sigma(t) \leq 2^m$ is a switching sequence and $x(t+1) = {}^jLx(t)$ is the subsystem of Boolean switched system (2.19).

Definition 2.2 Consider system (2.3). Let $X_0 \in \mathcal{D}^n$. $X_d \in \mathcal{D}^n$ is said to be reachable from X_0, if there exist $l > 0$ and a control sequence $U = \{u(0), \ldots, u(l-1)\}$ that steers system (2.3) from X_0 to X_d. $X_d \in \mathcal{D}^n$ is said to be globally reachable, if X_d is reachable from any $X_0 \in \mathcal{D}^n$. System (2.3) is said to be controllable if any $X_d \in \mathcal{D}^n$ is globally reachable.

It is learnt from [12] that $X_l = \delta_{2^n}^j$ is reachable from $X_0 = \delta_{2^n}^i$ at the l-th step if and only if $(\hat{L}^1)^l_{ji} > 0$, where $\hat{L}^1$ is given by (2.18). $X = \delta_{2^n}^j$ is reachable from $x(0) = \delta_{2^n}^i$ if and only if there exists a positive integer l such that $(\hat{L}^1)^l_{ji} > 0$.

$X = \delta_{2^n}^r$ is globally reachable if and only if there exists a positive integer l such that $Row_r\left(\bigvee_{j=1}^{l}(\hat{L}^1)^j\right) > 0$.

We now consider the PCC for system (2.3) as follows:

$$u(t) = u(t_l) \in \Delta_{2^m}, \quad t_l \le t < t_{l+1}, \tag{2.20}$$

where $t_l = l\tau \ge 0$ for $\tau > 0$ and $l \ge 0$. Define

$$\begin{aligned} S_l^0(i) = \{&x_0 \in \Delta_{2^n} : \exists u(0), \ldots, u((l-1)\tau) \in \Delta_{2^m} \text{ such that} \\ &(Lu((l-1)\tau))^\tau \cdots (Lu(0))^\tau x_0 = \delta_{2^n}^i\}, \quad 1 \le i \le 2^n,\ l \ge 1. \end{aligned} \tag{2.21}$$

Theorem 2.4 *There exists a sequence of PCCs in the form of (2.20) such that system (2.3) can be globally stabilized to $\delta_{2^n}^r$ if and only if there exists $1 \le N \le 2^n$, such that*

$$Row_r\left(\bigvee_{j=1}^{N}(\hat{L}^\tau)^j\right) > 0 \text{ and } \left(\hat{L}^1\right)_{rr} = 1. \tag{2.22}$$

Proof (Sufficiency) From (2.18), $\left(\hat{L}^1\right)_{rr} = 1$ implies that there exists at least one $1 \le j \le 2^m$ such that $\left({}^jL\right)_{rr} = 1$,

$$\delta_{2^n}^r = L\delta_{2^m}^j\delta_{2^n}^r. \tag{2.23}$$

It is easy to see from (2.23) that $\delta_{2^n}^r \in S_1^0(r)$.

On the other hand, $Row_r\left(\bigvee_{j=1}^{N}(\hat{L}^\tau)^j\right) > 0$ in (2.22) means that

$$1 = (\delta_{2^n}^r)^{\mathrm{T}}\left(\bigvee_{j=1}^{N}(\hat{L}^\tau)^j\right)\delta_{2^n}^i$$

for any $1 \le i \le 2^n$.

Step 1 Since $\delta_{2^n}^r \in S_1^0(r)$, $S_1^0(r) \neq \emptyset$. There exist at least one $x(0) = \delta_{2^n}^i$ and a corresponding $u(0)$, denoted by $u^*(0, \delta_{2^n}^i, \delta_{2^n}^r)$, such that $\delta_{2^n}^r = \left(Lu^*(0, \delta_{2^n}^i, \delta_{2^n}^r)\right)^\tau \delta_{2^n}^i$, i.e.,

$$1 = (\delta_{2^n}^r)^{\mathrm{T}}\left(Lu^*(0, \delta_{2^n}^i, \delta_{2^n}^r)\right)^\tau \delta_{2^n}^i.$$

It also implies that $1 = (\delta_{2^n}^r)^{\mathrm{T}}(\hat{L}^\tau)\delta_{2^n}^i > 0$. From the controllability criteria obtained by Laschov and Margaliot [12] and the definition of $S_l^0(r)$ in (2.21), we conclude that all such $\delta_{2^n}^i$ consists $S_1^0(r)$.

If $S_1^0(r) = \Delta_{2^n}$, that is, $(\delta_{2^n}^r)^{\mathrm{T}}(\hat{L}^\tau)\delta_{2^n}^i = 1$ holds for all $1 \leq i \leq 2^n$, then $Row_r(\hat{L}^\tau) > 0$ and $N = 1$ here. Considering (2.23), $\delta_{2^n}^r$ can be fixed for any $t \geq \tau$. Then system (2.3) can be globally stabilized to $\delta_{2^n}^r$. The proof is finished. Otherwise, $\Delta_{2^n} \setminus S_1^0(r) \neq \varnothing$ and $Row_r(\hat{L}^\tau) > 0$ does not hold any more.

Step 2 It is easy to see from $x(2\tau) = (Lu(\tau))^\tau (Lu(0))^\tau x(0)$ that $S_2^0(r) = \cup\{S_1(i) : 1 \leq i \leq 2^n,\ \delta_{2^n}^i \in S_1^0(r)\}$. Therefore, $S_1^0(r) \subset S_2^0(r)$ due to $\delta_{2^n}^r \in S_1^0(r)$. There exist $u(0)$, $u(\tau)$ and $x(0) = \delta_{2^n}^i$ such that $\delta_{2^n}^r$ is reachable. Denote such controls by $u^*(0, x(0), x(\tau))$, $u^*(\tau, x(\tau), x(2\tau))$. Similar to Step 1, we have

$$1 = (\delta_{2^n}^r)^{\mathrm{T}} \left(Lu^*(\tau, x(\tau), x(2\tau))\right)^\tau \left(Lu^*(0, x(0), x(\tau))\right)^\tau \delta_{2^n}^i.$$

It implies that $1 = (\delta_{2^n}^r)^{\mathrm{T}}(\hat{L}^\tau)^2\delta_{2^n}^i > 0$ and as in Step 1, all such $\delta_{2^n}^i$ consists of $S_2^0(r)$.

If $S_2^0(r) = \Delta_{2^n}$, that is, $Row_r\left(\bigvee_{j=1}^2(\hat{L}^\tau)^j\right) > 0$, then $\delta_{2^n}^r$ is globally reachable with $N = 2$. From (2.23) again, $\delta_{2^n}^r$ is fixed for any $t \geq 2\tau$ and system (2.3) is globally stabilized to $\delta_{2^n}^r$. We finish the proof. Otherwise, $\Delta_{2^n} \setminus S_2^0(r) \neq \varnothing$. Moreover, $S_2^0(r) \setminus S_1^0(r) \neq \varnothing$. Otherwise, $S_2^0(r) = S_1^0(r)$. In this way, one has $S_N^0(r) = S_{N-1}^0(r) = \cdots = S_1^0(r)$, which means that $Row_r(\hat{L}^\tau)^N = Row_r(\hat{L}^\tau)^{N-1} = \cdots = Row_r(\hat{L}^\tau)$. As a result, $Row_r\left(\bigvee_{j=1}^N(\hat{L}^\tau)^j\right) = Row_r(\hat{L}^\tau)$. Since $\Delta_{2^n} \setminus S_1^0(r) \neq \varnothing$, which means that $Row_r(\hat{L}^\tau) > 0$ is not satisfied. It is a contradiction to

$$Row_r\left(\bigvee_{j=1}^N(\hat{L}^\tau)^j\right) > 0.$$

Step 3 Similarly, one can also prove that $S_{j+1}^0(r) \setminus S_j^0(r) \neq \varnothing$ with $S_0^0(r)$ denoted by $\varnothing$. Then one can conclude that there exists $N \leq 2^n$ such that $\Delta_{2^n} = \cup_{0\leq j\leq N-1}(S_{j+1}^0(r) \setminus S_j^0(r))$. From the definition of $S_l^0(r)$, we obtain that $\delta_{2^n}^r$ is globally reachable. Moreover, for $t \geq N\tau$, $\delta_{2^n}^r$ is fixed from (2.23). Therefore, system (2.3) is globally stabilizable to $\delta_{2^n}^r$.

The necessary part is easy to get from the proof of the sufficiency, and we omit it here. □

When $\tau = 1$, the PCC can be regarded as a normal control sequence, see [1, 12–15]. Then we get the following corollary from Theorem 2.4.

Corollary 2.1 *System (2.3) can be globally stabilized to $\delta_{2^n}^r$ if and only if there exists $1 \leq N \leq 2^n$, such that*

$$Row_r\left(\bigvee_{j=1}^{N}(\hat{L}^1)^j\right) > 0 \; and \; \left(\hat{L}^1\right)_{rr} = 1. \tag{2.24}$$

Remark 2.2 It is easy to see that condition (2.24) in Corollary 2.1 is actually equivalent to conditions 1 and 2 in Proposition 3 of [1], when the cycle considered there is an equilibrium point. Motivated by [1], it is also possible to consider the stabilization of system (2.3) to a limit cycle by the PCC.

The following result presents the relationship between SDSFCs and PCCs for system (2.3).

Theorem 2.5 *System (2.3) can be globally stabilized to $\delta_{2^n}^r$ by PCC (2.20) if and only if it is globally stabilizable by means of SDSFC (2.5).*

Proof The sufficiency is obvious by letting $u(t_k) = Ex(t_k)$. For the necessary part, from condition (2.22) of Theorem 2.4, $\left(\hat{L}^1\right)_{rr} = 1$, there is $1 \leq j \leq 2^m$ such that $\left({}^jL\right)_{rr} = 1$. Then $\delta_{2^n}^r = {}^jL\delta_{2^n}^r = L\delta_{2^m}^j\delta_{2^n}^r$, which means that $u(t_k) = \delta_{2^m}^j = E\delta_{2^n}^r = Ex(t_k)$ for some k, and it implies $Col_r(E) = \delta_{2^m}^j$.

From Step 3 in the proof of Theorem 2.4, there exists $N \leq 2^n$ such that $\Delta_{2^n} = \cup_{0\leq j\leq N-1}(S_{j+1}^0(r) \setminus S_j^0(r))$. For any $x(t_j) = \delta_{2^n}^i \in S_{j+1}^0(r) \setminus S_j^0(r)$, there exists $u(t_j) = \delta_{2^m}^j$ such that $(L\delta_{2^m}^j)^\tau\delta_{2^n}^i \in S_j^0(r)$, which means that $u(t_k) = \delta_{2^m}^j = E\delta_{2^n}^i = Ex(t_j)$, and further implies $Col_i(E) = \delta_{2^m}^j$. Considering every $\delta_{2^n}^i \in \cup_{0\leq j\leq N-1}(S_{j+1}^0(r) \setminus S_j^0(r))$, all the columns of E can be determined.

Consequently, the global stabilization by PCC (2.20) is equivalent to the global stabilization by means of SDSFC in the form of (2.5). □

When $\tau = 1$, we have the following corollary.

Corollary 2.2 *System (2.3) can be globally stabilized to $\delta_{2^n}^r$ if and only if it is globally stabilizable by state feedback control.*

Based on Theorem 2.5, we give another approach to construct the SDSFC.

Remark 2.3 Compared with Algorithm 1, Algorithm 2 is much more effective. First, we can judge the existence of the SDSFC by checking the existence of N such that (2.22) is satisfied. However, in Algorithm 1, we need to compute each solution of p_i to learn the existence of E until all of them are found. Second, we notice that in Algorithm 1, one should solve the equation $\beta_i^\tau = r$ to get p_i. Although the equation looks very simple, it actually includes τ equations in the induction (see the proof of Theorem 2.2). When τ is a big number, the computation will be complex. For Algorithm 2, to get the solution of p_i, one needs only to solve the equation $\delta_{2^m}^{j_i} = E\delta_{2^n}^i$, and it is very easy to get $p_i = \delta_{2^m}^{j_i}$.

Algorithm 2 Construct the SDSFC

Step 1. Judge whether $\left(\hat{L}^1\right)_{rr} = 1$. If yes, find j such that $({}^jL)_{rr} = 1$, then $Col_r(E) = \delta_{2^m}^j$. Otherwise, E does not exist.
Step 2. Find $1 \le N \le 2^n$, such that $Row_r\left(\bigvee_{j=1}^N (\hat{L}^\tau)^j\right) > 0$. If such N exists, then E can be constructed. Otherwise, there is no E.
Step 3. Find all i, the set of which is denoted by $s_1(r)$ such that the $(\delta_{2^n}^r)^{\mathrm{T}}(\hat{L}^\tau)\delta_{2^n}^i > 0$. For each $i \in s_1(r) \setminus \{r\}$, find j_i, such that $(\delta_{2^n}^r)^{\mathrm{T}}({}^{j_i}L^\tau)\delta_{2^n}^i > 0$. Then $\delta_{2^m}^{j_i} = E\delta_{2^n}^i$, and as a result $p_i = \delta_{2^m}^{j_i}$ for all $i \in s_1(r)$.
Step 4. For each $j \in s_1(r)$, we get a set $s_1(j)$ as in Step 3. Then we define $s_2(r) = \cup_{j \in s_1(r)} s_1(j)$. Repeat Step 3 for any $i \in s_2(r) \setminus s_1(r)$. Then p_i for all $i \in s_2(r) \setminus s_1(r)$ can be determined.
...
Step $N+2$. For each $j \in s_{N-1}(r)$, we get a set $s_1(j)$. Then we define $s_N(r) = \cup_{j \in s_{N-1}(r)} s_1(j)$. Repeat Step 3 for any $i \in s_N(r) \setminus s_{N-1}(r)$. Then p_i for all $i \in s_N(r) \setminus s_{N-1}(r)$ can be determined.
Step $N+3$. Get the SDSFC u with $E = \delta_{2^m}[p_1 \; p_2 \; \cdots \; p_{2^n}]$.

Consider

$$\begin{aligned} L = \delta_8[&8\,8\,8\,8\,8\,8\,8\,8\,8\,8\,8\,8\,8\,8\,8\,8\,8\,8\,8\,8\,8\,8\,8\,8\,8\,8\,8\,8\,8\,8\,8\,8 \\ &1\,1\,1\,5\,3\,3\,3\,7\,1\,1\,1\,5\,3\,3\,3\,7\,3\,3\,3\,7\,4\,4\,4\,8\,4\,4\,4\,8\,4\,4\,4\,8]. \end{aligned}$$

Assume that the state feedback matrix E is given by

$$E = \delta_8[p_1 \; p_2 \; \cdots \; p_8]. \tag{2.25}$$

Let $\tau = 2$, we now construct the SDSFC such that the system is globally stabilizable to δ_8^1 by Algorithm 2. We can get

$$\hat{L}^1 = \begin{bmatrix} 1\,1\,1\,0\,0\,0\,0\,0 \\ 0\,0\,0\,0\,0\,0\,0\,0 \\ 1\,1\,1\,0\,1\,1\,1\,0 \\ 1\,1\,1\,0\,1\,1\,1\,0 \\ 0\,0\,0\,1\,0\,0\,0\,0 \\ 0\,0\,0\,0\,0\,0\,0\,0 \\ 0\,0\,0\,1\,0\,0\,0\,1 \\ 1\,1\,1\,1\,1\,1\,1\,1 \end{bmatrix}.$$

Step 1 It is found that $\left(\hat{L}^1\right)_{11} = 1$. For $j = 5$ *or* 6, $({}^jL)_{11} = 1$, then $Col_1(E) = p_1 = \delta_8^5$ or $p_1 = \delta_8^6$.

Step 2 We have

$$\hat{L}^2 = \begin{bmatrix} 1\,1\,1\,0\,1\,1\,1\,0 \\ 0\,0\,0\,0\,0\,0\,0\,0 \\ 1\,1\,1\,1\,0\,0\,0\,1 \\ 0\,0\,0\,1\,0\,0\,0\,0 \\ 0\,0\,0\,0\,0\,0\,0\,0 \\ 0\,0\,0\,0\,0\,0\,0\,0 \\ 0\,0\,0\,0\,1\,1\,1\,0 \\ 1\,1\,1\,1\,1\,1\,1\,1 \end{bmatrix}, \quad (\hat{L}^2)^2 = \begin{bmatrix} 2\,2\,2\,1\,2\,2\,2\,1 \\ 0\,0\,0\,0\,0\,0\,0\,0 \\ 3\,3\,3\,3\,2\,2\,2\,2 \\ 0\,0\,0\,1\,0\,0\,0\,0 \\ 0\,0\,0\,0\,0\,0\,0\,0 \\ 0\,0\,0\,0\,0\,0\,0\,0 \\ 0\,0\,0\,0\,1\,1\,1\,0 \\ 3\,3\,3\,3\,3\,3\,3\,2 \end{bmatrix}.$$

So when $N = 2$, then $Row_1\left(\bigvee_{j=1}^{2}(\hat{L}^2)^j\right) > 0$ is satisfied. Then E can be constructed.

Step 3 Since $(\delta_8^1)^{\mathrm{T}}(\hat{L}^2)\delta_8^i > 0$, we can find $i = 1, 2, 3, 5, 6, 7$. For $i = 2$, we can find $j_2 = 5$ or $j_2 = 6$, such that $(\delta_8^1)^{\mathrm{T}}({}^5L^2)\delta_8^2 > 0$ and $(\delta_8^1)^{\mathrm{T}}({}^6L^2)\delta_8^2 > 0$. Then $\delta_8^5 = E\delta_8^2$ or $\delta_8^6 = E\delta_8^2$ and hence $p_2 = 5$ *or* 6. Repeating the same method, finally, we can get $p_3 = 5$ *or* 6, $p_5 = 5$ *or* 6, $p_6 = 5$ *or* 6 and $p_7 = 5$ *or* 6.

Step 4 Since $3 \in s_1(1)$, we get a set $s_1(3)$ as in Step 3. Repeating Step 3, $(\delta_8^3)^{\mathrm{T}}(\hat{L}^2)\delta_8^i > 0$, then $i = 4, 8$. For $i = 4$, we can find $j_4 = 5$ or $j_4 = 6$, such that $(\delta_8^3)^{\mathrm{T}}({}^5L^2)\delta_8^2 > 0$ and $(\delta_8^3)^{\mathrm{T}}({}^6L^2)\delta_8^2 > 0$. Then $\delta_8^5 = E\delta_8^4$ or $\delta_8^6 = E\delta_8^4$ and hence $p_4 = 5$ *or* 6. We can also get $p_8 = 5$ *or* 6.

Consequently, we can construct the SDSFC with E given by $E = \delta_8[5\ 6\ 6\ 5\ 5\ 6\ 6\ 5]$.

In this example, Algorithm 2 is used to get the solution of p_i for system (2.15). It is simple to solve the equation $\delta_{2^m}^{j_i} = E\delta_{2^n}^{i}$, and obtain $p_i = \delta_{2^m}^{j_i}$.

2.2 Sampled-Data State Feedback Control for the Set Stabilization of Boolean Control Networks

In this section, we study the set stabilization of Boolean control networks (BCNs) under sampled-data state feedback control (SDSFC). The main research content is divided into two parts. First, the topological structure of BCNs under given SDSFC is investigated. The fixed point and sampled cycle are defined, respectively. It is found that sampled cycles allow elements to be repeated and not every element can be regarded as an initial state, and this is quite different from conventional cycles of BCNs. A theorem is presented to calculate the number of fixed points and an algorithm is given to find all fixed points and sampled cycles. Second, the set stabilization problem of BCNs by SDSFC is investigated based on the sampled point set and the sampled point control invariant set (SPCIS). A necessary and sufficient condition is derived for the global set stabilization of BCNs by SDSFC, and further sampled-data state feedback controllers are also designed. The interesting thing is that if a state enters the SPCIS as an unsampled point, then it may run out of the given set again, which is in sharp contrast to conventional BCNs. Finally, an example is given to illustrate the efficiency of the obtained results.

2.2.1 Problem Formulation

In this part, we consider the following BCN:

$$\begin{cases} x_1(t+1) = f_1(x_1(t), \ldots, x_n(t), u_1(t), \ldots, u_m(t)), \\ x_2(t+1) = f_2(x_1(t), \ldots, x_n(t), u_1(t), \ldots, u_m(t)), \\ \quad \vdots \\ x_n(t+1) = f_n(x_1(t), \ldots, x_n(t), u_1(t), \ldots, u_m(t)), \end{cases} \tag{2.26}$$

for $t \geq 0$, where $x_1(t),\ x_2(t),\ \ldots,\ x_n(t) \in \mathcal{D}$ are states, $u_1(t),\ u_2(t),\ \ldots,\ u_m(t) \in \mathcal{D}$ are control inputs, and $f_i : \mathcal{D}^n \to \mathcal{D}$, $i = 1, 2, \ldots, n$, are logical functions.

Let M_i denote the structure matrix of each f_i and $x(t) = \ltimes_{i=1}^{n} x_i(t),\ u(t) = \ltimes_{j=1}^{n} u_j(t)$, the algebraic form of system (2.26) can be expressed as

$$x_i(t+1) = M_i u(t) x(t), \quad i = 1, 2, \ldots, n. \tag{2.27}$$

Further, multiply the equations in (2.27) together this yields

$$x(t+1) = L u(t) x(t), \quad t \geq 0, \tag{2.28}$$

where $L = M_1 \ltimes_{i=2}^{n} [(I_{2^{n+m}} \otimes M_i)\Phi_{m+n}] \in \mathcal{L}_{2^n \times 2^{m+n}}$.

As for controllers for system (2.26), the sampled-data state feedback law is as follows:

$$\begin{cases} u_1(t) = e_1(x_1(t_l), \ldots, x_m(t_l)), \\ u_2(t) = e_2(x_1(t_l), \ldots, x_m(t_l)), \\ \quad \vdots \\ u_m(t) = e_m(x_1(t_l), \ldots, x_m(t_l)), \end{cases} \quad t_l \leq t < t_{l+1}, \tag{2.29}$$

where constant sampling period $\tau := t_{l+1} - t_l \in \mathbb{Z}_+,\ t_l = l\tau \geq 0,\ l = 0, 1, \ldots$ are sampling instants, and Boolean function $e_j : \mathcal{D}^n \to \mathcal{D},\ j = 1, 2, \ldots, m$ are mappings.

Similarly, it holds that

$$u_j(t) = E_j x(t_l), \quad t_l \leq t < t_{l+1},\ j = 1, 2, \ldots, m.$$

where E_j is the structure matrix of logical function e_j. Letting $u(t) = \ltimes_{i=1}^{m} u_i(t)$ and multiplying the above equation together, controller (2.29) can be converted into the algebraic form as follows:

$$u(t) = Ex(t_l), \quad t_l \leq t < t_{l+1}, \tag{2.30}$$

where $E = E_1 \ltimes_{j=2}^{m} [(I_{2^n} \otimes E_j)\Phi_n] \in \mathcal{L}_{2^m \times 2^n}$.

It should be noted that when $\tau = 1$, sampled-data state feedback controller (2.30) shows as a conventional one studied in [5, 13, 16, 17].

Finally, some necessary definitions are presented.

Definition 2.3 Under given SDSFC (2.30), a state $x_0 \in \Delta_{2^n}$ is called a fixed point of system (2.28) if $x(0) = x_0$, $x(t) = x_0$ holds for all $t \in \mathbb{Z}_+$.

Definition 2.4 Under given SDSFC (2.30), $C = \{\delta_{2^n}^{i_1}, \delta_{2^n}^{i_2}, \ldots, \delta_{2^n}^{i_k}\}$ is a sampled cycle with length k of system (2.28), if there exists an integer $l \in [1, k]$, such that $x(0) = \delta_{2^n}^{i_l}$, then $x(t) = \delta_{2^n}^{i_j}$, where $j \in [1, k]$ and $j = (t + l) \bmod k$. Here k is the least positive period, that is, $k \backslash l$ holds for any l satisfying the above requirement.

Definition 2.5 Given a set $S \subseteq \Delta_{2^n}$, the set $S^* \subseteq S$ is called the sampled point set of system (2.28) with sampling period τ, if for any $x(0) \in S^*$, there exists a control sequence $\mathbf{u}$ such that $x(1) \in S, x(2) \in S, \ldots, x(\tau) \in S$.

Definition 2.6 ([18]) Given a set $S \subseteq \Delta_{2^n}$, BCN (2.28) is said to be S-stabilization if, for any initial state $x(0) \in \Delta_{2^n}$, there exists a control sequence $\mathbf{u}$ and an integer $T \geq 0$ such that $x(t; x(0), \mathbf{u}) \in S$, for all $t \geq T$.

2.2.2 Topological Structure of Boolean Control Networks Under Sampled-Data State Feedback Control

Considering system (2.28) with SDSFC (2.30), we have the following system:

$$x(t+1) = Lu(t)x(t) = LEx(t_l)x(t) = \widetilde{L}x(t_l)x(t), \quad t_l \leq t < t_{l+1}, \tag{2.31}$$

where $t_{l+1} - t_l = \tau$ and $\widetilde{L} = L \ltimes E \in \mathcal{L}_{2^n \times 2^{2n}}$.

Remark 2.4 The difference between the sampled cycle and the conventional cycle is that the sampled cycle allows repeated elements but the conventional cycle does not allow. Moreover, not every $\delta_{2^n}^{i_l} \in C$ can be regarded as an initial sampled state $x(0)$, or the trajectory may possibly jump out of the sampled cycle, which is shown by the following two examples.

Let $\tau = 3$, and

$$L = \delta_8[2\ 3\ 4\ 1\ 2\ 2\ 2\ 2\ 2\ 3\ 1\ 2\ 2\ 4\ 1\ 1],$$
$$E = \delta_2[1\ 2\ 1\ 2\ 1\ 2\ 1\ 2].$$

Using the semi-tensor product, matrix $\widetilde{L}$ can be obtained as follows:

$$\widetilde{L} = L \ltimes E = \delta_8[2\ 3\ 4\ 1\ 2\ 2\ 2\ 2\ 2\ 3\ 1\ 2\ 2\ 4\ 1\ 1\ 2\ 3\ 4\ 1\ 2\ 2\ 2\ 2\ 2\ 3\ 1\ 2\ 2\ 4\ 1\ 1$$
$$2\ 3\ 4\ 1\ 2\ 2\ 2\ 2\ 2\ 3\ 1\ 2\ 2\ 4\ 1\ 1\ 2\ 3\ 4\ 1\ 2\ 2\ 2\ 2\ 2\ 3\ 1\ 2\ 2\ 4\ 1\ 1].$$

Let $t_0 = 0$ and $x(0) = \delta_8^1$, when $0 < t \leq \tau$,

$$x(1) = \delta_8^2, \quad x(2) = \delta_8^3, \quad x(3) = \delta_8^4.$$

Now $t_1 = 3$ and $x(3) = \delta_8^4$ is the second sampled point, when $\tau < t \leq 2\tau$,

$$x(4) = \delta_8^2, \quad x(5) = \delta_8^3, \quad x(6) = \delta_8^1.$$

Similarly, $x(6) = \delta_8^1$ is the next sampled point, and the following states are repeated by $\{\delta_8^1, \delta_8^2, \delta_8^3, \delta_8^4, \delta_8^2, \delta_8^3\}$. Thus, the sampled cycle is $C = \{\delta_8^1, \delta_8^2, \delta_8^3, \delta_8^4, \delta_8^2, \delta_8^3\}$.

If $x(0) = \delta_8^3$, when $0 < t \leq \tau$,

$$x(1) = \delta_8^4, \quad x(2) = \delta_8^1, \quad x(3) = \delta_8^2.$$

Later $t_1 = 3$, and $x(3) = \delta_8^2$ is the second sampled point, when $\tau < t \leq 2\tau$,

$$x(4) = \delta_8^3, \quad x(5) = \delta_8^1, \quad x(6) = \delta_8^2.$$

Similarly, $x(6) = \delta_8^2$ is the next sampled point, and the following states are repeated by $\{\delta_8^2, \delta_8^3, \delta_8^1\}$. Thus, $C' = \{\delta_8^2, \delta_8^3, \delta_8^1\}$ is a new sampled cycle, and the trajectory has jumped out the former sampled cycle. Thus, δ_8^3 cannot be regarded as an initial sampled point in the former sampled cycle C.

In this example, it is observed that elements of the sampled cycles C are allowed to be repeated and not any states can be regarded as an initial state. It follows that the states that can be sampled play vital roles in the study of the set stabilization.

Let $\tau = 4$, and

$$L = \delta_8[2\ 3\ 1\ 1\ 2\ 3\ 1\ 2\ 2\ 3\ 1\ 3\ 1\ 2\ 3\ 4],$$
$$E = \delta_2[1\ 2\ 1\ 2\ 1\ 2\ 1\ 2].$$

Using the semi-tensor product, $\widetilde{L}$ is calculated as

$$\widetilde{L} = L \ltimes E = \delta_8[2\ 3\ 1\ 1\ 2\ 3\ 1\ 2\ 2\ 3\ 1\ 3\ 1\ 2\ 3\ 4\ 2\ 3\ 1\ 1\ 2\ 3\ 1\ 2\ 2\ 3\ 1\ 3\ 1\ 2\ 3\ 4$$
$$2\ 3\ 1\ 1\ 2\ 3\ 1\ 2\ 2\ 3\ 1\ 3\ 1\ 2\ 3\ 4\ 2\ 3\ 1\ 1\ 2\ 3\ 1\ 2\ 2\ 3\ 1\ 3\ 1\ 2\ 3\ 4].$$

Let $t_0 = 0$ and $x(0) = \delta_8^1$, when $0 < t \leq \tau$,

$$x(1) = \delta_8^2, \quad x(2) = \delta_8^3, \quad x(3) = \delta_8^1, \quad x(4) = \delta_8^2.$$

Then $t_1 = 4$, and $x(4) = \delta_8^2$ is the second sampled point, so when $\tau < t \le 2\tau$,

$$x(5) = \delta_8^3, \quad x(6) = \delta_8^1, \quad x(7) = \delta_8^2, \quad x(8) = \delta_8^3.$$

Then $t_2 = 8$, and $x(8) = \delta_8^3$ is the third sampled point, so when $2\tau < t \le 3\tau$,

$$x(9) = \delta_8^1, \quad x(10) = \delta_8^2, \quad x(11) = \delta_8^3, \quad x(12) = \delta_8^1.$$

Similarly, $x(12) = \delta_8^1$ is the next sampled point, and the following state is repeated by $\{\delta_8^1, \delta_8^2, \delta_8^3, \delta_8^1, \delta_8^2, \delta_8^3, \delta_8^1, \delta_8^2, \delta_8^3, \delta_8^1, \delta_8^2, \delta_8^3\}$, according to Definition 2.4, the sampled cycle is $\{\delta_8^1, \delta_8^2, \delta_8^3\}$.

The above example shows the necessity of the least positive period. Next, the topological structure of sampled-data controller BCNs is studied

Split $\widetilde{L}$ into 2^n equal blocks as

$$\widetilde{L} = [\widetilde{L}_1 \ \widetilde{L}_2 \ \cdots \ \widetilde{L}_{2^n}], \tag{2.32}$$

where $\widetilde{L}_i \in \mathcal{L}_{2^n \times 2^n}$. The following theorem shows the number of fixed points of system (2.31).

Theorem 2.6 *The point $\delta_{2^n}^i$ is a fixed point of system (2.31), if and only if, the network transition matrix $\widetilde{L}$ satisfies*

$$(\widetilde{L}_i)_{ii} = 1, \tag{2.33}$$

and the number of fixed points, denoted by N_e, equals to

$$N_e = \sum_{i=1}^{2^n} (\widetilde{L}_i)_{ii}. \tag{2.34}$$

Proof (Sufficiency) If $x(0) = \delta_{2^n}^i$ and $(\widetilde{L}_i)_{ii} = 1$, then

$$\begin{aligned} x(1) &= \widetilde{L}x(0)x(0) = \widetilde{L}_i \delta_{2^n}^i = \delta_{2^n}^i, \\ x(2) &= \widetilde{L}x(0)x(1) = \widetilde{L}_i \delta_{2^n}^i = \delta_{2^n}^i, \\ &\vdots \\ x(\tau) &= \widetilde{L}x(0)x(\tau - 1) = \widetilde{L}_i \delta_{2^n}^i = \delta_{2^n}^i, \end{aligned}$$

and $x(\tau)$ is the next sampled point.

From above derivation process, it is noted that if $x(0) = \delta_{2^n}^i$, then $x(t) \equiv \delta_{2^n}^i$ for any $t \ge 0$.

(Necessity) Assume that $\delta_{2^n}^i$ is a fixed point, in terms of Definition 2.3, if $x(0) = \delta_{2^n}^i$, then $x(t) \equiv \delta_{2^n}^i$ for any $t \ge 0$.

If $t = 1$, then

$$x(1) = \delta_{2^n}^i.$$

On the other hand,

$$x(1) = \widetilde{L}x(0)x(0) = (\widetilde{L}\delta_{2^n}^i)\delta_{2^n}^i = \widetilde{L}_i\delta_{2^n}^i.$$

From $\widetilde{L}_i\delta_{2^n}^i = \delta_{2^n}^i$, it is obvious that Eq. (2.33) holds. Further, the number of fixed points can be calculated by

$$N_e = \sum_{i=1}^{2^n}(\widetilde{L}_i)_{ii}.$$

The above theorem finds the number of fixed points for system (2.31). The following construction method is presented to calculate the number of sampled cycles. Owing to the complexity of its expression, the calculation formula for one cycle is omitted.

Before introducing how to find fixed points and sampled cycles for system (2.31), some preliminary results are provided.

Lemma 2.2 ([13]) *Consider the following system*

$$x(t+1) = Lx(t), \tag{2.35}$$

the number of cycles with length s is denoted by N_s, and it can be inductively determined by

$$\begin{cases} N_1 = N_e = tr(L), \\ N_s = \dfrac{tr(L^s) - \sum_{k \in \mathcal{P}(s)} kN_k}{s}, \quad 2 \le s \le 2^n, \end{cases} \tag{2.36}$$

where $\mathcal{P}(s)$ is the set of proper factors of s.

If

$$tr(L^s) - \sum_{k \in \mathcal{P}(s)} kN_k > 0, \tag{2.37}$$

then there exists at least a cycle with length s, and s is called a nontrivial power.

For simple representation, C_s is defined as

$$C_s = \begin{cases} \{\delta_{2^n}^i \mid L_{ii}^s = 1\}, & s \text{ is a nontrivial power} \\ \emptyset, & \text{otherwise} \end{cases} \tag{2.38}$$

for $s = 1, 2, \ldots, 2^n$. In other words, if $x(0) = \delta_{2^n}^i \in C_s$ is an initial state for system (2.35), where s is a nontrivial power, then we have $x(s) = \delta_{2^n}^i$.

Assume that t, s are two nontrivial powers satisfying $t \backslash s$, for any $\delta_{2^n}^i \in C_t$,

$$L^s \delta_{2^n}^i = L^{pt} \delta_{2^n}^i = \underbrace{L^t \cdots L^t}_{p} \delta_{2^n}^i = \delta_{2^n}^i.$$

Therefore, $C_t \subseteq C_s$.

Based on the above analysis, $D_s \subseteq C_s$ can be obtained, where D_s denotes the set of states in cycles with length s for system (2.35). In addition, C_s includes the elements of C_t, $t \in \mathcal{P}(s)$, where $\mathcal{P}(s)$ is the set of proper factors of s.

Therefore, for every nontrivial power s, D_s can be expressed as

$$D_s = C_s \cap_{j \in \mathcal{P}(s)} C_j^c,$$

where C_j^c is the complement of C_j.

When D_s is obtained for every nontrivial power s, for every $\delta_{2^n}^i \in D_s$, a cycle C_i^* with length s for system (2.35) is obtained as follows:

$$C_i^* = \left\{ \delta_{2^n}^i,\ L\delta_{2^n}^i, \ldots,\ (L)^{s-1}\delta_{2^n}^i \right\}.$$

Moreover, $C^* = \{C_i^* \mid i = 1, 2, \ldots, 2^n\}$ is the set of cycles of system (2.35).

Now a construction method is presented to find all fixed points and sampled cycles of system (2.31).

Construct a new logical matrix $L^* \in \mathcal{L}_{2^n \times 2^n}$ as follows:

$$L^* \triangleq \left[(\widetilde{L}_1)^\tau \delta_{2^n}^1\ (\widetilde{L}_2)^\tau \delta_{2^n}^2\ \cdots\ (\widetilde{L}_{2^n})^\tau \delta_{2^n}^{2^n} \right].$$

Meanwhile, a new algebraic system is constructed as follows:

$$x(t+1) = L^* x(t). \tag{2.39}$$

For every cycle C_i^* of system (2.39) and sampling period $\tau > 0$ of system (2.31), corresponding cycle C_i^* is extended as follows:

$$\begin{aligned} \widetilde{C}_i = \{ & \delta_{2^n}^i, (\widetilde{L}\delta_{2^n}^i)\delta_{2^n}^i, \ldots, (\widetilde{L}\delta_{2^n}^i)^{\tau-1}\delta_{2^n}^i, \ldots, \\ & (L^*)^{s-1}\delta_{2^n}^i, \ldots, (\widetilde{L}(L^*)^{s-1}\delta_{2^n}^i)^{\tau-1}(L^*)^{s-1}\delta_{2^n}^i \}, \end{aligned}$$

where $\widetilde{C}_i$ is called an enlarged sampled cycle. The minimum repeating unit of $\widetilde{C}_i$ is called the smallest unit, denoted by C_i. Then the following theorem holds.

Theorem 2.7 *The set of all fixed points and sampled cycles of system (2.31) is* $\Lambda = \{C_i \mid i \in D_s\}$.

Proof Assume $\overline{C}$ is a sampled cycle of system (2.31), there exists a sampled state $\delta_{2^n}^{i_j} \in \overline{C}$. Let $x(0) = \delta_{2^n}^{i_j}$, then $x(\tau), x(2\tau), \ldots$ can be calculated. Owing to the finiteness of Δ_{2^n}, there exists a smallest integer t such that $x(t\tau)$ repeats with $x(0)$. If there exists an integer $l \in [1, t]$ such that $x(t\tau) = x(l\tau)$, it contradicts to the fact that $\delta_{2^n}^{i_j}$ is a sampled state in sampled cycle $\overline{C}$. We have $\delta_{2^n}^{i_j} \in C_t$ and $\delta_{2^n}^{i_j} \notin C_j$, $1 \leq j \leq t-1$. Thus, $\overline{C} \in \Lambda$. From the construction of Λ, any $C_i \in \Lambda$ is apparently a sampled cycle.

A fixed point can be regarded as a cycle with length 1, thus set Λ includes all fixed points. □

Based on the above analysis, Algorithm 3 is presented to find all fixed points and sampled cycles.

Algorithm 3 Find all fixed points and sampled cycles of system (2.31)

1: Construct the new logical transition matrix $L^* \in \mathcal{L}_{2^n \times 2^n}$, where $Col_i(L^*) = (\widetilde{L}_i)^\tau \delta_{2^n}^i$, $i = 1, 2, \ldots, 2^n$.
2: Consider the new BN $x(t+1) = L^* x(t)$, calculate set C_s, $1 \leq s \leq 2^n$. Furthermore, $D_s = C_s \bigcap_{i \in \mathcal{P}(s)} C_i^c$ can be solved.
3: For every $\delta_{2^n}^i \in D_s$, calculate cycle C_i^*, enlarged cycle $\widetilde{C}_i$, and sampled cycle C_i.
4: Get the set $\Lambda = \{C_i \mid i \in D_s\}$.

Then, the minimum number of steps leading all points to limit sets is considered, and it is called the transient period below.

Construct following $r+1$ matrices:

$$(L^*)^0, L^*, (L^*)^2, \ldots, (L^*)^r, \tag{2.40}$$

where $(L^*)^0 = I_{2^n}$ is a $2^n \times 2^n$ matrix and $r = 2^n \times 2^n$. It is found that L^* has only $2^n \times 2^n$ possible independent values, then there must be two equal matrices in the above sequence (2.40).

In the above sequence, denote $r_0 < r$ by the smallest i satisfying the matrix $(L^*)^i$ appears again. More precisely,

$$r_0 = \text{argmin}_{0 \leq i < r} \left\{ (L^*)^i \in \{(L^*)^{i+1}, (L^*)^{i+2}, \ldots, (L^*)^r\} \right\}.$$

Proposition 2.1 *Starting from any initial state, the trajectory of system (2.31) will enter a sampled cycle or fixed point in Λ after $r_0\tau$ iterations.*

In the study of stability, the following equivalent definition of global convergence (stability) is introduced.

Definition 2.7 ([13]) A BN is globally convergent (stable) if the limit set of system (2.31) consists only one fixed point.

Theorem 2.8 *System (2.31) is globally convergent if and only if one of the following equivalent conditions is satisfied.*

(I) $tr(L^*) = 1$ *and there is no other nontrivial power of system (2.39). Furthermore, if* $(L^*)_{ii} = 1$, *then* $(\widetilde{L}_i)_{ii} = 1$.
(II) $tr((L^*)^{2^n}) = 1$. *Furthermore, if* $((L^*)^{2^n})_{ii} = 1$, *then* $(\widetilde{L}_i)_{ii} = 1$.

Proof

(I) (Sufficiency) When $tr(L^*) = 1$, there exists a unique state $\delta_{2^n}^i$ such that $(L^*)\delta_{2^n}^i = \delta_{2^n}^i$. Furthermore, owing to $(L_i)_{ii} = 1$, if $x(0) = \delta_{2^n}^i$, ones have by induction that

$$\begin{aligned} x(1) &= \widetilde{L}x(0)x(0) = \delta_{2^n}^i, \\ x(2) &= \widetilde{L}x(0)x(1) = \delta_{2^n}^i, \\ &\vdots \\ x(\tau) &= \widetilde{L}x(0)x(\tau-1) = \delta_{2^n}^i. \end{aligned}$$

Thus, $\delta_{2^n}^i$ is the only fixed point and there is no other sampled cycle in the limit set Λ. Hence, system (2.31) is globally convergent.

(Necessity) Owing to the definition of global convergence, system (2.31) has only one fixed point, that is, $tr(L^*) = 1$. If $(L^*)_{ii} = 1$, $\delta_{2^n}^i$ is the only one fixed point of system (2.31). Let $x(0) = \delta_{2^n}^i$, it holds that

$$x(1) = \widetilde{L}\delta_{2^n}^i\delta_{2^n}^i = \delta_{2^n}^i \Longleftrightarrow (\widetilde{L}_i)_{ii} = 1.$$

(II) One only needs to prove that $tr((L^*)^{2^n}) = 1$ holds, if and only if $tr(L^*) = 1$ and no other nontrivial power holds at the same time.

When $tr(L^*) = 1$ and there is no other nontrivial power, $tr((L^*)^{2^n}) = 1$ holds.

When $tr((L^*)^{2^n}) = 1$, from Lemma 2.2, we have $tr((L^*)^{2^n}) \geq tr((L^*)^{2^n-1}) \geq \cdots \geq tr(L^*)$, then $tr(L^*) = 1$ or $tr(L^*) = 0$. From (2.36), because N_s is an integer, for any $s > 1$, $tr((L^*)^s) = 0$ or > 1. If $tr(L^*) = 0$, there exists an integer $s_0 > 0$ such that $tr((L^*)^{s_0}) > 1$, which conflicts with $tr((L^*)^{2^n}) = 1$. Thus, the above conditions are equivalent to each other.

□

Remark 2.5 When sampling period is one in system (2.31), all above results can be regarded as a BCN controlled by state feedback controllers. Hence, the result in [13] is a special case of our result.

2.2.3 Set Stabilization for Boolean Control Networks Under Sampled-Data State Feedback Control

Consider system (2.31) with controller (2.30), when $0 \leq t \leq \tau$, it holds that

$$\begin{aligned} x(1) &= LEx(0)x(0), \\ x(2) &= LEx(0)x(1) = (LEx(0))^2x(0), \\ &\vdots \\ x(t) &= LEx(0)x(t-1) = (LEx(0))^t x(t-1). \end{aligned} \tag{2.41}$$

Similarly, when $\tau < t \leq 2\tau$, $x(\tau)$ acts as another initial state, one obtains from (2.41) that

$$x(t) = LEx(\tau)x(t-1) = (LEx(\tau))^{t-\tau}x(\tau). \tag{2.42}$$

By induction, for any integer $t \in (l\tau, (l+1)\tau]$, we have

$$x(t) = (LEx(l\tau))^{t-l\tau}x(l\tau). \tag{2.43}$$

Split L to 2^m matrix as $[L_1\ L_2\ \cdots\ L_{2^m}]$ and assume $E = \delta_{2^m}[p_1\ p_2\ \cdots\ p_{2^n}]$. When $x(t_l) = \delta_{2^n}^i$, where l is an integer and $t_l = l\tau$, for $t_l \leq t < t_{l+1}$,

$$\begin{aligned} x(t) &= x(t_l) = \delta_{2^n}^i, \\ u(t) &= Ex(t_l) = \delta_{2^m}[p_1\ p_2\ \cdots\ p_{2^n}]\delta_{2^n}^i. \end{aligned} \tag{2.44}$$

Definition 2.8 The set $\widetilde{S} \subseteq S$ is said to be an SPCIS of S for BCN (2.28) under SDSFC, if for any $x(t_l) \in \widetilde{S}$, there exists a sampled-data state feedback controller (2.30) such that $x(t_l+1) \in S, x(t_l+2) \in S, \ldots, x(t_l+\tau) = x(t_{l+1}) \in \widetilde{S}$. A set $\widetilde{S}^*$ is called the largest SPCIS of BCN (2.28) under SDSFC, if it contains the largest number of elements among all SPCISs of $\widetilde{S}$.

From the analysis of (2.44), the control u at time t is a feedback on the state at sampled time t_l. Different from conventional state feedback controllers, if a state does not act at a sampled point, it may jump out of S although it has entered the SPCIS. It is easy to explain by the following construction method.

Assume $x(0) = \delta_{2^n}^i \notin \widetilde{S}^*$ and there exists $\delta_{2^m}^j$ such that $Col_i(Blk_j(L)) \in \widetilde{S}^*$ as well as $Col_i([Blk_j(L)]^2) \notin S$. Let $Col_i(E) = \delta_{2^m}^j$, then $x(1) = L\delta_{2^m}^j\delta_{2^n}^i = Col_i(Blk_j(L)) \in \widetilde{S}^*$, but $x(2) = L\delta_{2^m}^jL\delta_{2^m}^j\delta_{2^n}^i = Col_i([Blk_j(L)]^2) \notin S$. It is shown that $x(1)$ is in SPCIS but not sampled, then $x(2)$ still jumps out of S, which is the reason we studied sampled point sets first.

Next, Algorithm 4 is presented to find the sampled point set S^* when S is given. If $S^* = \varnothing$, system (2.31) cannot be S-stabilization via SDSFC.

Algorithm 4 Find the possible sampled point set S^* in a given set S

Initialize: $S^* := \varnothing$.
for e **do**ach state $\delta_{2^n}^{i_a} \in S^*$
  for $\alpha := 1$ to 2^m **do**
    for $k := 1$ to τ **do**
      if $(L_\alpha)^{k-1} Col_{i_a}(L_\alpha) \in S$ **then**
        return $S^* := S^* \cup \{\delta_{2^n}^{i_a}\}$.
      end if
    end for
  end for
end for

Theorem 2.9 *The set S has an SPCIS of BCN (2.31) under SDSFC (2.30) if and only if there exists a set* $\widetilde{S} \subseteq S^*$ *such that for any* $\delta_{2^n}^{i_a} \in \widetilde{S}$*, we can find a control input* $\delta_{2^m}^{\alpha} \in \Delta_{2^m}$ *guaranteeing that the following formula holds:*

$$\begin{aligned} Col_{i_a}(L_\alpha) &\in S, \\ L_\alpha Col_{i_a}(L_\alpha) &\in S, \\ &\vdots \\ (L_\alpha)^{\tau-1} Col_{i_a}(L_\alpha) &\in \widetilde{S}. \end{aligned} \tag{2.45}$$

Proof (Sufficiency) Suppose that (2.45) holds, for any $x(0) = \delta_{2^n}^{i_a} \in \widetilde{S}$, there exists $u(0) = \delta_{2^m}^{\alpha} \in \Delta_{2^m}$ such that

$$\begin{aligned} x(1) &= LEx(0)x(0) = L\delta_{2^m}^{\alpha}\delta_{2^n}^{i_a} = Col_{i_a}(L_\alpha), \\ x(2) &= LEx(0)x(1) = L_\alpha Col_{i_a}(L_\alpha), \\ &\vdots \\ x(\tau) &= LEx(0)x(\tau-1) = {L_\alpha}^{\tau-1} Col_{i_a}(L_\alpha). \end{aligned} \tag{2.46}$$

From (2.46), we have $x(1) \in S, x(2) \in S, \ldots, x(\tau) \in \widetilde{S} \subseteq S$. Thus, the set $\widetilde{S}$ is an SPCIS.

(Necessity) Assume that $\widetilde{S}$ is an SPCIS, then for any $x(0) \in \widetilde{S}$, there exists a control sequence **u** such that $x(1) \in S, x(2) \in S, \ldots, x(\tau) \in \widetilde{S}$. Equation (2.45) holds. □

According to the definition of the sampled point set S^* and SPCIS $\widetilde{S}$, one knows that $\widetilde{S} \subseteq S^*$. Compared with searching for $\widetilde{S^*}$ directly, it is more convenient to delete elements in S^* as follows.

Step 1 Find $\delta_{2^n}^{i_a} \in S^*$, for any input $\delta_{2^n}^{\alpha} \in \Delta_{2^m}$, there exists an integer $j \in [1, \tau - 2]$ such that $(L_\alpha)^j Col_{i_a}(L_\alpha) \notin S$ or $(L_\alpha)^{\tau-1} Col_{i_a}(L_\alpha) \notin S^*$. Let S_1 be the set of all $\delta_{2^n}^{i_a}$, if $S_1 = \varnothing$, for any initial state $x(0) \in S$ and $t > 0$, $x(t) \in S$, so $\widetilde{S^*} = S^*$; otherwise go to Step 2.

Step 2 Let $S_1^* = S^* \backslash S_1$, find $\delta_{2^n}^{i_b} \in S_1^*$, for any input $\delta_{2^m}^{\alpha} \in \Delta_{2^m}$, there exists an integer $j \in [1, \tau - 2]$ such that $(L_\alpha)^j Col_{i_b}(L_\alpha) \notin S$ or $(L_\alpha)^{\tau-1} Col_{i_b}(L_\alpha) \notin S_1^*$. Let S_2 be the set of all such $\delta_{2^n}^{i_b}$, if $S_2 = \varnothing$, $\widetilde{S^*} = S_1^*$; otherwise go to Step 3.

Step 3 Continue a similar process until we find the set S_k^* satisfying for any $\delta_{2^n}^{i_c} \in S_k^*$, there exists $\delta_{2^m}^{\alpha} \in \Delta_{2^m}$ such that $(L_\alpha)^j Col_{i_c}(L_\alpha) \in S,\ j = 1, 2, \ldots, \tau - 2$, and $(L_\alpha)^{\tau-1} Col_{i_c}(L_\alpha) \in S_k^*$. Then $\widetilde{S^*} = S_k^*$ is the largest SPCIS.

Proposition 2.2 *For any $x(k\tau) \in S \backslash \widetilde{S^*}$ and SDSFC (2.30), there exists an integer N such that $x(k\tau + N) \notin S$.*

Proof Let $S^* = S_1 \cup S_2 \cup \cdots \cup S_k \cup \widetilde{S^*}$, we define $S = S^* \cup (S \setminus S^*)$. If $x(k\tau) = \delta_{2^n}^{a_i} \in S \setminus S^*$, according to Algorithm 4, for any $\delta_{2^m}^{\alpha} \in \Delta_{2^m}$, there must exist an integer $j \in [0, \tau - 1]$, such that

$$(L_\alpha)^j Col_{a_i}(L_\alpha) \notin S.$$

Thus, we have $x(k\tau + j) \notin S$.

Otherwise, $x(k\tau) \in S^*$ or $x(k\tau) \in S_t$, for some $t \in [1, k]$. Therefore, there exists an integer $j \in [0, \tau - 2]$ such that $(L_\alpha)^j Col_{a_i}(L_\alpha) \notin S$ or $(L_\alpha)^{\tau-1} Col_{a_i}(L_\alpha) \notin S_{t-1}^*$.

If $(L_\alpha)^j Col_{a_i}(L_\alpha) \notin S$, let $N = j$, then $x(k\tau + j) \notin S$.

On the other hand, $x(k\tau + \tau) \notin S_{t-1}^*$. Then $x(k\tau + \tau) \in S_1, x(k\tau + \tau) \in S_2, \ldots$, or $x(k\tau + \tau) \in S_{t-1}$.

For $x(k\tau + \tau) \in S_m$ holds for some $m \in [1, t - 1]$, if $(L_\alpha)^j Col_{a_i}(L_\alpha) \notin S$, $j \in [0, \tau - 2]$, this theorem holds clearly.

In addition to the above circumstances, it claims that $(L_\alpha)^j Col_{a_i}(L_\alpha) \notin S^*, \ldots$ or $(L_\alpha)^{\tau-1} Col_{a_i}(L_\alpha) \notin S_{t-2}^*$. Because $S^* \supseteq S_1^* \supseteq S_2^* \supseteq \cdots \supseteq \widetilde{S^*}$, $x(k\tau + \tau + \tau) \notin S_{t-2}^*$. Hence, there exists an integer N satisfying $x(k\tau + N - \tau) \notin S^*$, which has been discussed in the first case. □

Next, Algorithm 5 is presented to find the largest SPCIS $\widetilde{S^*}$ in set S^*. The details are as follows:

Denote by $\Omega_l(\widetilde{S^*})$ the set of states that can be steered to $\widetilde{S^*}$ in $l\tau$ steps under some control sequence, that is,

$$\begin{cases} \Omega_1(\widetilde{S^*}) = \{x_0 \in \Delta_{2^n} : (LEx_0)^\tau x_0 \in \widetilde{S^*}\}, \\ \Omega_{l+1}(\widetilde{S^*}) = \{x_0 \in \Delta_{2^n} : (LEx_0)^\tau x_0 \in \Omega_l(\widetilde{S^*})\}. \end{cases} \tag{2.47}$$

Afterwards, the following conclusions can be obtained.

Lemma 2.3

(I)

$$\Omega_1(\widetilde{S^*}) = \{\delta_{2^n}^i \in \Delta_{2^n} : \text{ there exists } \delta_{2^m}^{\alpha} \text{ such that } (L_\alpha)^{\tau-1} Col_i(L_\alpha) \in \widetilde{S^*}\}.$$

Algorithm 5 Determine the largest SPCIS of a given set S. If algorithm returns "$\varnothing$", BCN (2.31) does not have SPCIS in S

1: **function** MAIN
2:     $result := S^*$
3:     **while** ELIG.($result$) is not **True do**
4:         $result := result \setminus$ INELIG.($result$))
5:     **end while**
6:     **return** $result$
7: **end function**
8: **function** ELIG.(S)
9:     **for** each element $\delta_{2^n}^{i_b} \in S$ **do**
10:         **if** $Col_{ib}(L_\alpha) \notin S$ for one $x(0) \in S^*$ $x(t) \in S(t > 0)$ **then**
11:             **return False**
12:         **end if**
13:             **return True**
14:     **end for**
15: **end function**
16: **function** INELIG.(S)
17:     $result := \varnothing$
18:     **for** each subset $\delta_{2^n}^{i_b} \in S$ and $1 < j < \tau$ **do**
19:         **if** $(L_\alpha)^j Col_{i_s}(L_\alpha) \notin S$ or $(L_\alpha)^{\tau-1} Col_{i_s}(L_\alpha) \notin S_1^*$ **then**
20:             $result := result \bigcup \{\delta_{2^n}^{i_b}\}$
21:         **end if**
22:             **return** result
23:     **end for**
24:     $result := S^*$
25: **end function**

(II) $\Omega_{l+1}(\widetilde{S^*}) = \bigcup_{\delta_{2^n}^i \in \Omega_l(\widetilde{S^*})} \{\Omega_1(\delta_{2^n}^i), 1 \leq i \leq 2^n\}$ *for all* $l \geq 1$.

Proof For any $x(k\tau) = \delta_{2^n}^i$, $u(k\tau) = \delta_{2^m}^\alpha$, ones have

$$x((k+1)\tau) = (Lu(k\tau))^\tau x(k\tau) = (L\delta_{2^m}^\alpha)^\tau \delta_{2^n}^i = (L_\alpha)^{\tau-1} Col_i(L_\alpha).$$

Hence, conditions (I) and (II) hold. $\square$

Lemma 2.4

(I) $\widetilde{S^*} \subseteq \Omega_1(\widetilde{S^*})$ *and* $\Omega_l(\widetilde{S^*}) \subseteq \Omega_{l+1}(\widetilde{S^*})$.
(II) *If* $\Omega_1(\widetilde{S^*}) = \widetilde{S^*}$, *then* $\Omega_l(\widetilde{S^*}) = \widetilde{S^*}$, *for all* $l \geq 1$.
(III) *If* $\Omega_{l+1}(\widetilde{S^*}) = \Omega_l(\widetilde{S^*})$ *for some* $l \geq 1$, *then* $\Omega_k(\widetilde{S^*}) = \Omega_l(\widetilde{S^*})$, *for all* $k \geq l$.

Proof

(I) For any $x(k\tau) \in \widetilde{S^*}$, $x((k+1)\tau) \in \widetilde{S^*}$. Thus, $x(k\tau) \in \Omega_1(\widetilde{S^*})$ and $\widetilde{S^*} \subseteq \Omega_1(\widetilde{S^*})$.

From the definition of $\Omega_l(\widetilde{S^*})$, it is noticed that $\Omega_l(\widetilde{S^*}) \subseteq \Omega_{l+1}(\widetilde{S^*})$, thus (II) and (III) follow.

$\square$

Theorem 2.10 *System (2.28) can be globally stabilized to the set S by SDSFC (2.30), if and only if the following two conditions are satisfied simultaneously.*

(I) $\widetilde{S^*} \neq \varnothing$.
(II) there exists an integer T such that $\Omega_T(\widetilde{S^*}) = \Delta_{2^n}$.

Proof (Sufficiency) Assume that $\widetilde{S^*}$ is a nonempty set and there exists an integer T satisfying $\Omega_T(\widetilde{S^*}) = \Delta_{2^n}$, any $x(0) \in \Delta_{2^n}$ can be steered to $x(T\tau) \in \widetilde{S^*}$. Furthermore, when $x(T\tau) \in \widetilde{S^*}$, we have $x(T\tau + N) \in \widetilde{S^*}$ for any integer N. Thus, BCN (2.28) is S-stabilization.

(Necessity) Suppose that BCN (2.28) is S-stabilization by SDSFC (2.30), so for any $x(0) \in \Delta_{2^n}$, there exists an integer T such that when $t \geq T$, $x(t) \in S$.

If $\widetilde{S^*} = \varnothing$, condition (II) is not satisfied. Assume $\Omega_T(\widetilde{S^*}) \neq \Delta_{2^n}$, as a result of the set stabilization, there exists an integer M, when $t \geq M$, $x(t) \in S$. If $x(t) \notin \widetilde{S^*}$, then $x(t) \in S \backslash \widetilde{S^*}$. According to Proposition 2.2, there exists an integer N such that $x(t+N) \notin S$, which contradicts with the definition of the set stabilization. Thus, conditions (I) and (II) are satisfied. □

Observe that when conditions (I) and (II) in Theorem 2.10 are satisfied, Δ_{2^n} can be split into the union of disjoint sets

$$\Delta_{2^n} = \Omega_1(\widetilde{S^*}) \cup \left(\Omega_2(\widetilde{S^*}) \setminus \Omega_1(\widetilde{S^*})\right) \cup \cdots \cup \left(\Omega_T(\widetilde{S^*}) \setminus \Omega_{T-1}(\widetilde{S^*})\right).$$

Hence, for every $1 \leq i \leq 2^n$, there exists a unique integer $1 \leq l_i < T$ such that $\delta_{2^n}^i \in \Omega_{l_i}(\widetilde{S^*}) \setminus \Omega_{l_i-1}(\widetilde{S^*})$ with $\Omega_0(\widetilde{S^*}) = \varnothing$.

If $l_i = 1$ and $\delta_{2^n}^i \in \widetilde{S^*}$, let p_i be the solution such that

$$\begin{array}{c} Col_i(L_{p_i}) \in S, \\ (L_{p_i})Col_i(L_{p_i}) \in S, \\ \vdots \\ (L_{p_i})^{\tau-2}Col_i(L_{p_i}) \in S, \\ (L_{p_i})^{\tau-1}Col_i(L_{p_i}) \in \widetilde{S^*}. \end{array} \tag{2.48}$$

If $l_i = 1$ and $\delta_{2^n}^i \notin \widetilde{S^*}$, let p_i be the solution such that

$$(L_{p_i})^{\tau-1}Col_i(L_{p_i}) \in \widetilde{S^*}. \tag{2.49}$$

If $2 \leq l_i \leq T$, let p_i be the solution such that

$$(L_{p_i})^{\tau-1}Col_i(L_{p_i}) \in \Omega_{l_i}(\widetilde{S^*}) \setminus \Omega_{l_i-1}(\widetilde{S^*}). \tag{2.50}$$

Theorem 2.11 *If there exists SDSFC (2.30) such that system (2.28) can be stabilized to the set S, p_i is the solution of (2.48), (2.49), (2.50). Then the feedback* *law (2.30) with the state feedback matrix E is given as*

$$E = \delta_{2^m}[p_1\ p_2\ \cdots\ p_{2^n}],$$

which globally stabilizes BCN (2.28) to S.

Proof Let $x(0) \in \Delta_{2^n}$, then we have

$$\begin{aligned} x(1) &= Lu(0)x(0) = LEx(0)x(0), \\ x(2) &= Lu(0)x(1) = (LEx(0))^2 x(0), \\ &\vdots \\ x(\tau) &= Lu(0)x(\tau-1) = (LEx(0))^{\tau} x(0). \end{aligned}$$

If $l_i = 1$ and $\delta_{2^n}^i \in \widetilde{S}^*$, then it implies that $x(1) \in S, x(2) \in S, \ldots, x(\tau-1) \in S$ and $x(\tau) \in \widetilde{S}^*$.

If $l_i = 1$ and $\delta_{2^n}^i \notin \widetilde{S}^*$, then $x(\tau) \in \widetilde{S}^*$.

If $l_i \geq 2$, then $\Omega_{l_i - 1}(\widetilde{S}^*)$. □

Remark 2.6 If sampling period is one, this problem can be regarded as state feedback controller to realize global set stabilization in [16]. If the given set S contains a unique element, our result is the case of global stabilization by SDSFC in [19]. When above two cases hold synchronously, global state feedback stabilization is realized, which is the same with results in [5]. Consequently, our result can be regarded as a generation of [5, 16, 19] to some extend.

2.2.4 Example and Simulations

Considering the Boolean model studied in [19], it is a reduced model to describe the lac operon in the bacterium *Escherichia coli*. The following model shows that the lac mRNA and lactose form the core of the lac operon:

$$\begin{cases} x_1(t+1) = \neg u_1(t) \wedge (x_2(t) \vee x_3(t)), \\ x_2(t+1) = \neg u_1(t) \wedge u_2(t) \wedge x_1(t), \\ x_3(t+1) = \neg u_1(t) \wedge (u_2(t) \vee (u_3(t) \wedge x_1(t))), \end{cases} \tag{2.51}$$

where x_1, x_2, x_3 are state variables representing the lac mRNA, the lactose in high concentrations and the lactose in medium concentrations; Control inputs u_1, u_2, u_3 denote extracellular glucose, high extracellular lactose, and medium extracellular lactose, respectively.

$x(t) = \ltimes_{i=1}^3 x_i(t)$ and $u(t) = \ltimes_{i=1}^3 u_i(t)$, using the semi-tensor product, we have

$$x(t+1) = Lu(t)x(t), \tag{2.52}$$

where

$$L = \delta_8[8\ 8\ 8\ 8\ 8\ 8\ 8\ 8\ 8\ 8\ 8\ 8\ 8\ 8\ 8\ 8\ 8\ 8\ 8\ 8\ 8\ 8\ 8\ 8\ 8\ 8\ 8\ 8\ 8\ 8\ 8\ 8\ 8\ 8\ 8\ 8\ 1\ 1\ 1\ 5\ 3\ 3\ 3\ 7\ 1\ 1\ 1\ 5\ 3\ 3\ 3\ 7\ 3\ 3\ 3\ 7\ 4\ 4\ 4\ 8\ 4\ 4\ 4\ 8\ 4\ 4\ 4\ 8].$$

For system (2.51), we consider thc following feedback law:

$$u(t) = Ex(t_l), \quad t_l \le t \le t_{l+1}, \tag{2.53}$$

where the state feedback matrix $E = \widetilde{G}_1 \ltimes_{j=2}^{3} [(I_{2^3} \otimes \widetilde{G}_j)\Phi_3] \in \mathcal{L}_{2^3 \times 2^3}$ and $\widetilde{G}_j,\ j = 1, 2, 3$ are structure matrix of e_j. Suppose that feedback matrix E is presented by

$$E = \delta_8[p_1\ p_2\ \ldots\ p_8].$$

In addition, $\tau = 2$. Now we construct the SDSFC to make system (2.51) globally stabilized to the set $\{\delta_8^1, \delta_8^4, \delta_8^5, \delta_8^7\}$.

Step 1 Find the set of possible sampled points by Algorithm 4.
For $x(t_l) = \delta_8^1$, there exists $u = \delta_8^5$ such that $x(t_l + 1) = \delta_8^1$ and $x(t_l + 2) = \delta_8^1$.
For $x(t_l) = \delta_8^4$, there exists $u = \delta_8^7$ such that $x(t_l + 1) = \delta_8^7$ and $x(t_l + 2) = \delta_8^4$.
For $x(t_l) = \delta_8^5$, there exists $u = \delta_8^7$ such that $x(t_l + 1) = \delta_8^4$ and $x(t_l + 2) = \delta_8^7$.
For $x(t_l) = \delta_8^7$, there exists $u = \delta_8^7$ such that $x(t_l + 1) = \delta_8^4$ and $x(t_l + 2) = \delta_8^7$.
Owing to $\delta_8^1, \delta_8^4, \delta_8^7 \in S$, the sampled point set $S^* = S$.
Step 2 On the basis of Algorithm 5, we find the largest SPCIS $\widetilde{S}^* = S^* = S$.
Step 3 According to Lemma 2.3, it obtains that

$$\Omega_1(\widetilde{S}^*)\backslash\widetilde{S}^* = \{\delta_8^2, \delta_8^3, \delta_8^6\}$$

and

$$\Omega_2(\widetilde{S}^*) \setminus \Omega_1(\widetilde{S}^*) = \delta_8^8.$$

Since $\widetilde{S}^* \neq \varnothing$ and $\Omega_2(\widetilde{S}^*) = \Delta_{2^n}$, system (2.52) can be globally stabilized to set S by SDSFC under Theorem 2.10.
Step 4 Determine $p_i,\ i = 1, 2, \ldots, 8$, as in Theorem 2.11.

For $i = 1, 4, 5, 7$, p_i is calculated to satisfy $Col_i(L_{p_i}) \in S$ and $(L_{p_i})Col_i(L_{p_i}) \in \widetilde{S}^*$. It finds that $p_1 = 5$ or 6, $p_4 = 7$, $p_5 = 7$, and $p_7 = 7$.

For $i = 2, 3, 6$, p_i satisfies $(L_{p_i})Col_i(L_{p_i}) \in \widetilde{S}^*$, then $p_2 = 5$ or 6, $p_3 = 5$ or 6 and $p_6 = 5, 6$ or 7.

For $i = 8$, repeating the same method, $p_8 = 5$ or 6.

Finally, one of feasible feedback laws E is given by

$$E = \delta_8[5\ 6\ 6\ 7\ 7\ 5\ 7\ 5]. \tag{2.54}$$

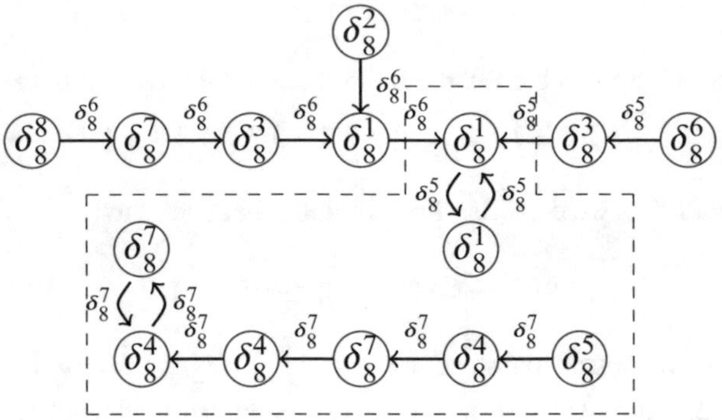

Fig. 2.3 The state trajectory graph of dynamics

Using the method in [13], we can convert the above algebraic form of controller into logical form as follows:

$$\begin{cases} u_1(t) = x_1(t_l) \wedge \neg x_1(t_l), \\ u_2(t) = [x_1(t_l) \wedge (x_2(t_l) \vee x_3(t_l))] \vee [\neg x_1(t_l) \wedge x_3(t_l)], \\ u_3(t) = [x_1(t_l) \wedge (x_2(t_l) \leftrightarrow x_3(t_l))] \vee (\neg x_1(t_l)), \end{cases}$$

under which, system (2.51) can be globally stabilized to the set $\{\delta_8^1, \delta_8^4, \delta_8^5, \delta_8^7\}$. Via controller (2.53), the trajectory of system (2.51) is described in Fig. 2.3 from arbitrary initial state $x(0) \in \Delta_{2^n}$.

In the trajectory graph, every vertex represents the state of system (2.51). The outgoing arcs describe the state transitions associated with the input values u, which has been labeled above the directed arcs. For clear description, the states in blue vertexes are sampled points, while ones in the red vertexes are unsampled points. Simultaneously, the sampled points in the dashed wire frame form the largest SPCIS $\widetilde{S}^*$. For example, let $x(0) = \delta_8^6$ and control $u(0) = \delta_8^5$, then $x(1) = \delta_8^5$ and it is not a sampled point, thus $u(1) = u(0) = \delta_8^5$ keeps unchanged. With the associating of $u(1)$, $x(1)$ evolutes to δ_8^1, i.e., $x(2) = \delta_8^1$. From the trajectory, $x(2)$ belongs to the largest SPCIS and fixes at δ_8^1. Similarly, it is not hard to find that the dynamic (2.51) will globally stabilize to set $\{\delta_8^1, \delta_8^4, \delta_8^7\}$ by designed controller (2.54).

Remark 2.7 Selecting $x(0) = \delta_8^8$ as an initial state, the trajectory of system (2.51) under SDSFC (2.54) is $x(0) : \delta_8^8 \to x(1) : \delta_8^7 \to x(2) : \delta_8^3 \to x(3) : \delta_8^1 \to \cdots \to x(\infty) : \delta_8^1$. The largest SPCIS $\widetilde{S}^*$ is $\{\delta_8^1, \delta_8^4, \delta_8^5, \delta_8^7\}$, and it is noted that state $x(1) : \delta_8^7$ enters $\widetilde{S}^*$ but the next state $x(2) : \delta_8^3 \notin \widetilde{S}^*$, which is different from state feedback control. The main reason is that $x(1) : \delta_8^7$ is not a sampled point, and it is not sampled by controller. When $t = 1$, the controller is still determined by $x(0) : \delta_8^8$. When $t = 4$, state $x(4) : \delta_8^1$ enters $\widetilde{S}^*$ as a sampled point, then trajectory will not jump out of S for $t \geq 4$.

2.3 Summary

In this chapter, we first studied the sampled-date control stabilization problem of BCNs. Necessary and sufficient conditions for the global stabilization by SDSFC have been derived. Different from the normal state feedback control, some new observations have been presented. In terms of the controllability matrix, PCC of BCNs has been discussed as well. We have proved the equivalence of SDSFC and PCC for the control of BCNs. In detail, a BCN can be globally stabilizable by PCC if and only if it is globally stabilizable by SDSFC. Two algorithms have been presented to construct SDSFC.

Without loss of generality, if the sampling period is one, SDSFC is equivalent to regular state feedback control, which has been studied in [5, 13, 16, 17] for the stabilization of BCNs. However, to the best of our knowledge, there is no result on set stabilization of BCNs under SDSFC. Therefore, we then investigated the set stabilization of BCNs under SDSFC. An algorithm has been presented to find all fixed points and sampled cycles in a BCN under given SDSFC. Moreover, the sampled point set and SPCIS have been respectively defined, and they can be found using two algorithms. It has been noted that if a state enters the SPCIS as an unsampled state, then it may run out of the given set, and it is different from conventional BCNs. A necessary and sufficient condition also have been derived for the global set stabilization of BCNs by SDSFC, and sampled-data state feedback controllers also have been designed.

In addition to the results about stabilization of sampled-data BCNs mentioned in this chapter, we have also found some other studies on stabilization of logical control networks. For example, the sampled-data reachability and stabilization of k-valued logical control networks with state and input constraints were investigated in [20]. The necessary and sufficient conditions for constrained sampled-data reachability and constrained sampled-data stabilization of k-valued logical control networks were proposed, respectively. In [21], the set stabilization of switched Boolean control networks under sampled-data feedback control was addressed. Some necessary and sufficient conditions were presented for the set stabilization of switched Boolean control networks by switching signal-dependent sampled-data state feedback control. Furthermore, a constructive procedure was given to design all possible switching signal-dependent sampled-data state feedback controllers.

References

1. Fornasini, E., Valcher, M.E.: On the periodic trajectories of Boolean control networks. Automatica **49**(5), 1506–1509 (2013)
2. Li, R., Yang, M., Chu, T.: State feedback stabilization for probabilistic Boolean networks. Automatica **50**(4), 1272–1278 (2014)
3. Li, H., Wang, Y.: Output feedback stabilization control design for Boolean control networks. Automatica **49**(12), 3641–3645 (2013)

4. Zhao, Y., Cheng, D.: On controllability and stabilizability of probabilistic Boolean control networks. Sci. China Inf. Sci. **57**(1), 1–14 (2014)
5. Li, R., Yang, M., Chu, T.: State feedback stabilization for Boolean control networks. IEEE Trans. Autom. Control **58**(7), 1853–1857 (2013)
6. Li, H., Wang, Y., Liu, Z.: Simultaneous stabilization for a set of Boolean control networks. Syst. Control Lett. **62**(12), 1168–1174 (2013)
7. Nikitin, S.: Piecewise-constant stabilization. SIAM J. Control Optim. **37**(3), 911–933 (1999)
8. Schaller, G.: Open Quantum Systems far from Equilibrium. Springer, Berlin (2014)
9. Voytsekhovsky, D., Hirschorn, R.M.: Stabilization of single-input nonlinear systems using piecewise constant controllers. IEEE Trans. Autom. Control **52**(6), 1150–1154 (2007)
10. Majewski, M.: On an algorithm for construction a piecewise constant control for a continuous Roesser system. In: Proceeding of the 4th International Workshop on Multidimensional Systems, pp. 253–258 (2005)
11. Lefebvre, D.: Finite time control design for contPNs according to piecewise constant control actions. In: Proceeding of the 50th IEEE Conference on Decision and Control and European Control, pp. 5862–5867 (2011)
12. Laschov, D., Margaliot, M.: Controllability of Boolean control networks via the Perron-Frobenius theory. Automatica **48**(6), 1218–1223 (2012)
13. Cheng, D., Qi, H., Li, Z.: Analysis and control of Boolean networks: A semi-tensor product approach. Springer Science & Business Media, New York (2010)
14. Liu, Y., Lu, J., Wu, B.: Some necessary and sufficient conditions for the output controllability of temporal Boolean control networks. ESAIM Control Optim. Calc. Variations **20**(1), 158–173 (2014)
15. Liu, Y., Chen, H., Lu, J., et al.: Controllability of probabilistic Boolean control networks based on transition probability matrices. Automatica **52**, 340–345 (2015)
16. Guo, Y., Wang, P., Gui, W., et al.: Set stability and set stabilization of Boolean control networks based on invariant subsets. Automatica **61**, 106–112 (2015)
17. Li, F., Li, H., Xie, L., et al.: On stabilization and set stabilization of multivalued logical systems. Automatica **80**, 41–47 (2017)
18. Liu, Y., Li, B., Lu, J., et al.: Pinning control for the disturbance decoupling problem of Boolean networks. IEEE Trans. Autom. Control **62**(12), 6595–6601 (2017)
19. Liu, Y., Cao, J., Sun, L., et al.: Sampled-data state feedback stabilization of Boolean control networks. Neural Comput. **28**(4), 778–799 (2016)
20. Li, Y., Li, H., Wang, S.: Constrained sampled-data reachability and stabilization of logical control networks. IEEE Trans. Circuits Syst. II Express Briefs **66**(12), 2002–2006 (2019)
21. Yerudkar, A., Del Vecchio, C., Glielmo, L.: Sampled-data set stabilization of switched Boolean control networks. IFAC-PapersOnLine **53**(2), 6139–6144 (2020)

Chapter 3
Controllability, Observability and Synchronization of Sampled-Data Boolean Control Networks

Abstract In this chapter, we mainly discuss controllability, observability and synchronization of Boolean control networks (BCNs) under periodic sampled-data control.

3.1 Controllability and Observability of Boolean Control Networks via Sampled-Data Control

In this section, the controllability and observability of sampled-data Boolean control networks (BCNs) are investigated. New phenomena are observed in the study of the controllability and observability of sampled-data BCNs. We routinely convert sampled-data BCNs into linear discrete-time systems by the semitensor product of matrices. Necessary and sufficient conditions are derived for the controllability of sampled-data BCNs. After that, we combine two sampled-data BCNs with the same transition matrix into a new sampled-data BCN to study the observability. Using an iterative algorithm, a stable row vector $\mathcal{U}^*$, called the observability row vector, in finite iterations, is obtained. It is proved that a sampled-data BCN is observable, if and only if, $\|\mathcal{U}^*\|_1 = N^2 - N$ with $N := 2^n$, where n is number of state-variables of BNs. Moreover, based on graph theory, a more effective algorithm is given to determine the observability of sampled-data BCNs. Its complexity is not related to the length of the sampling period. In addition, some equivalent necessary and sufficient conditions are put forward for the observability of sampled-data BCNs. Numerical examples are given to demonstrate the effectiveness of the obtained results.

Y. Liu et al., *Sampled-data Control of Logical Networks*,
https://doi.org/10.1007/978-981-19-8261-3_3

3.1.1 Problem Formulation

A sampled-data BCN is described as follows:

$$\begin{cases} x_i(t+1) = f_i(u_1(t), \ldots, u_m(t), x_1(t), \ldots, x_n(t)), \quad i = 1, 2, \ldots, n, \\ y_j(t) = h_j(x_1(t), \ldots, x_n(t)), \quad j = 1, 2, \ldots, p, \\ u_k(t) = u_k(t_l), \quad t_l \le t < t_{l+1}, \ k = 1, 2, \ldots, m, \end{cases} \tag{3.1}$$

where $x_i \in \mathcal{D}, \ i = 1, 2, \ldots, n$ are states, $u_k \in \mathcal{D}, \ k = 1, 2, \ldots, m$ are control inputs, $y_j \in \mathcal{D}, \ j = 1, 2, \ldots, p$ are outputs, $t = 0, 1, 2, \ldots$ is the discrete time, $f_i : \mathcal{D}^{n+m} \to \mathcal{D}, \ i = 1, 2, \ldots, n$ and $h_j : \mathcal{D}^n \to \mathcal{D}, \ j = 1, 2, \ldots, p$ are logical functions. $t_l = l\tau \ge 0$ for $l \in \mathbb{N}$ are sampling instants and $t_{l+1} - t_l = \tau$ denotes the constant sampling period.

Letting $x(t) = \ltimes_{i=1}^{n} x_i(t)$ with $N := 2^n$ and $u(t) = \ltimes_{k=1}^{m} u_k(t)$ with $M := 2^m$, for each logical function $f_i, \ i \in [1, n]$, we can find its structure matrix M_i. Then, the dynamic system (3.1) can be converted into an algebraic form as

$$x_i(t+1) = M_i u(t) x(t), \quad i \in [1, n]. \tag{3.2}$$

Multiplying the equations in (3.2) together yields

$$x(t+1) = L u(t) x(t), \tag{3.3}$$

where $L = *_{i=1}^{n} M_i \in \mathcal{L}_{N \times NM}$ and "*" is Khatri-Rao product.

Similarly, letting $y(t) = \ltimes_{j=1}^{p} y_j(t)$, we have

$$y(t) = H x(t), \tag{3.4}$$

where H is the transition matrix from x to y (calculated in the same way as L). To see more details, please refer to [1].

Then the sampled-data control has the following form:

$$u(t) = u(t_l), \quad t_l \le t < t_{l+1}, t_l = l\tau, l \in \mathbb{N}. \tag{3.5}$$

We denote the sampled-data control sequence with the constant sampling period τ just by

$$\pi^t = \{u(0), u(1), \ldots, u(\tau), \ldots, u(2\tau), \ldots, u(t-1)\}, \quad t > 0.$$

In addition, the set of all sampled-data control sequences

$$\pi^t = \{u(0), u(1), \ldots, u(\tau), \ldots, u(2\tau), \ldots, u(t-1)\}$$

is denoted by $\prod^t,\ t > 0$.

3.1.2 Controllability of Sampled-Data Boolean Control Networks

Now we consider the controllability problem via free sequence control. The following definition is presented.

Definition 3.1 Consider sampled-data BCN (3.1).

1. $x_e \in \Delta_N$ is said to be reachable from $x_0 \in \Delta_N$ at the sth time step if we can find a sampled-data control sequence π^s that steers sampled-data BCN (3.1) from $x(0) = x_0$ to $x(s) = x_e$.
2. Sampled-data BCN (3.1) is said to be controllable at x_0, if for any $x_e \in \Delta_N$, there exists $s > 0$ such that 1) holds. Sampled-data BCN (3.1) is said to be controllable, if for any $x_0,\ x_e \in \Delta_N$, there exists $s > 0$ such that 1) holds.

Since matrix L in (3.3) is an $N \times NM$ logical matrix, we split it into M square blocks as $L = [L_1\ L_2\ \cdots\ L_M]$, where $L_i \in \mathcal{L}_{N\times N},\ i \in [1, M]$. We want to determine the precise logical relationship between $x(t_0)$ and $x(t_0 + t)$ with $t > 0$ (under sampled-data control sequences), where $t_0 = k\tau$ with $k \in \mathbb{N}$ is a sampling instant.

- When $t \le \tau$. It is clear that $x(t_0 + t)$ only depends on $x(t_0)$ and $u(t_0)$. If $u(t_0) = \delta_M^j,\ j \in [1, M]$ and on the basis of Eq. (3.3), one has $x(t_0+t) = (L\delta_M^j)^t x(t_0) = (L_j)^t x(t_0)$, which implies that

$$x(t_0 + t) = {}^tLu(t_0)x(t_0),$$

where ${}^tL := [(L_1)^t\ (L_2)^t\ \cdots\ (L_M)^t]$. Let

$${}^t\mathcal{M} := \sum_{i=1}^{M} {}_{\mathcal{B}}(L_i)^t, \tag{3.6}$$

where ${}^t\mathcal{M} \in \mathcal{B}_{N\times N}$. Notice that if the entry ${}^t\mathcal{M})_{ij} = 1$, it means that there is at least a sampled-data control sequence π^t, which drives $x(t_0) = \delta_N^j$ at the sampling time t_0 to $x(t_0 + t) = \delta_N^i$.

- When $t > \tau$. We have ${}^t\mathcal{M} = {}^{\beta_t}\mathcal{M} \times_{\mathcal{B}} ({}^\tau\mathcal{M})^{(\alpha_t)}$, where $\beta_t = t \mod \tau,\ \alpha_t = \frac{t-\beta_t}{\tau}$. Now, we explore the intrinsic meaning of ${}^t\mathcal{M}$. If $({}^{2\tau}\mathcal{M})_{ij} = \left(({}^\tau\mathcal{M})^{(2)}\right)_{ij} = 1$, then there is at least a number $k \in [1, N]$, such that $({}^\tau\mathcal{M})_{kj} = 1$ and $({}^\tau\mathcal{M})_{ik} = 1$. In accordance with the meaning of $({}^\tau\mathcal{M})_{kj}$ and $({}^\tau\mathcal{M})_{ik}$, there exists at least one sampled-data control sequence $\pi^{2\tau}$ driving $x(t_0) = \delta_N^j$ to $x(t_0 + 2\tau) = \delta_N^i$.

Then by induction, if $({}^{t}\mathcal{M})_{ij} = \left({}^{\beta_t}\mathcal{M} \times_{\mathcal{B}} ({}^{\tau}\mathcal{M})^{(\alpha_t)}\right)_{ij} = 1$, there is at least a number $k \in [1, N]$, such that $({}^{\alpha_t\tau}\mathcal{M})_{kj} = \left(({}^{\tau}\mathcal{M})^{(\alpha_t)}\right)_{kj} = 1$ and $({}^{\beta_t}\mathcal{M})_{ik} = 1$. Similarly, it implies that there exists at least a sampled-data control sequence π^t driving $x(t_0) = \delta_N^j$ to $x(t_0 + t) = \delta_N^i$. For simplicity, let

$$\lfloor {}^{s}\mathcal{M} \rfloor = \begin{cases} \sum_{k=1}^{s} {}_{\mathcal{B}}\, {}^{k}\mathcal{M}, \ s > 0, \\ I, \ s = 0, \end{cases}$$

$$\lfloor ({}^{\tau}\mathcal{M})^{(s)} \rfloor = \begin{cases} \sum_{k=1}^{s} {}_{\mathcal{B}} ({}^{\tau}\mathcal{M})^{(k)}, \ s > 0, \\ I, \ s = 0. \end{cases}$$

Then, one can obtain the following proposition.

Proposition 3.1

(I) For all $s \geq \tau$, it satisfies that

$$\lfloor {}^{s}\mathcal{M} \rfloor = \lfloor {}^{\tau}\mathcal{M} \rfloor \times_{\mathcal{B}} \left(I + \lfloor ({}^{\tau}\mathcal{M})^{(\alpha_s - 1)} \rfloor\right) +_{\mathcal{B}} \lfloor {}^{\beta_s}\mathcal{M} \rfloor \times_{\mathcal{B}} ({}^{\tau}\mathcal{M})^{(\alpha_s)},$$

where and hereafter $\beta_s = s \mod \tau$, $\alpha_s = \frac{s - \beta_s}{\tau}$.
In particular, if $\beta_s = 0$, one has

$$\lfloor {}^{s}\mathcal{M} \rfloor = \lfloor {}^{\tau}\mathcal{M} \rfloor \times_{\mathcal{B}} \left(I +_{\mathcal{B}} \lfloor ({}^{\tau}\mathcal{M})^{(\alpha_s)} \rfloor\right).$$

(II) For any $i \in [1, N]$, there exists $0 \leq s_i^* < N$, such that $Col_i\left(\lfloor ({}^{\tau}\mathcal{M})^{(s)} \rfloor\right) = Col_i\left(\lfloor ({}^{\tau}\mathcal{M})^{(s_i^*)} \rfloor\right)$ holds for all $s \geq s_i^*$.

(III) There exists $0 \leq s^* < N$, such that $\lfloor ({}^{\tau}\mathcal{M})^{(s)} \rfloor = \lfloor ({}^{\tau}\mathcal{M})^{(s^*)} \rfloor$ and $\lfloor {}^{k}\mathcal{M} \rfloor = \lfloor {}^{s^*\tau + \tau - 1}\mathcal{M} \rfloor$ holds for all $s \geq s^*$ and $k \geq s^*\tau + \tau - 1$.

Proof

(I) By the definition of operator "$\lfloor \cdot \rfloor$", it deduces that

$$\lfloor {}^{s}\mathcal{M} \rfloor = \lfloor {}^{\tau}\mathcal{M} \rfloor \times_{\mathcal{B}} \left(I +_{\mathcal{B}} \lfloor ({}^{\tau}\mathcal{M})^{(\alpha_s)} \rfloor\right),$$

on condition $\beta_s = 0$, where

$$
\begin{aligned}
\lfloor {}^s\mathcal{M} \rfloor &= \sum_{k=1}^{s}{}_{\mathcal{B}}\, {}^k\mathcal{M} \\
&= \sum_{k=1}^{\alpha_s\tau}{}_{\mathcal{B}}\, {}^k\mathcal{M} +_{\mathcal{B}} \sum_{k=1}^{\beta_s}{}_{\mathcal{B}}\, {}^{\alpha_s\tau+k}\mathcal{M} \\
&= \left(\sum_{k=1}^{\tau}{}_{\mathcal{B}}\, {}^k\mathcal{M}\right) \times_{\mathcal{B}} \left(I +_{\mathcal{B}} \sum_{k=1}^{\alpha_s-1}{}_{\mathcal{B}}\, {}^{k\tau}\mathcal{M}\right) +_{\mathcal{B}} \left(\sum_{k=1}^{\beta_s}{}_{\mathcal{B}}\, {}^k\mathcal{M}\right) \times_{\mathcal{B}} {}^{\alpha_s\tau}\mathcal{M} \\
&= \lfloor {}^\tau\mathcal{M} \rfloor \times_{\mathcal{B}} \left(I + \lfloor ({}^\tau\mathcal{M})^{(\alpha_s-1)} \rfloor\right) +_{\mathcal{B}} \lfloor {}^{\beta_s}\mathcal{M} \rfloor \times_{\mathcal{B}} ({}^\tau\mathcal{M})^{(\alpha_s)}.
\end{aligned}
$$

(II) It is clear that $Col_i\left(\lfloor ({}^\tau\mathcal{M})^{(s)} \rfloor\right) \le Col_i\left(\lfloor ({}^\tau\mathcal{M})^{(s+1)} \rfloor\right)$ holds for all $s > 0$. So if $Col_i\left(\lfloor ({}^\tau\mathcal{M})^{(s)} \rfloor\right) \neq Col_i\left(\lfloor ({}^\tau\mathcal{M})^{(s+1)} \rfloor\right)$, the norm $\|Col_i\left(\lfloor ({}^\tau\mathcal{M})^{(s+1)} \rfloor\right)\|_1$ (which is the sum of all entries) will increase at least by one. So, at most there are $N-1$ iterations holding $Col_i\left(\lfloor ({}^\tau\mathcal{M})^{(s)} \rfloor\right) \neq Col_i\left(\lfloor ({}^\tau\mathcal{M})^{(s+1)} \rfloor\right)$, and we have $Col_i\left(\lfloor ({}^\tau\mathcal{M})^{(s)} \rfloor\right) = Col_i\left(\lfloor ({}^\tau\mathcal{M})^{(s_i^*)} \rfloor\right)$ for all $s \ge s_i^*$. And finally, (III) is an explicit result from (I) and (II) with $s^* := \max_{i\in[1,N]}\{s_i^*\}$. The proof is completed.

□

Remark 3.1 As mentioned in [2], control problems for BCNs are in general nondeterministic polynomial (NP)-hard. If the considered BCN is a general logical system, the semi-tensor product technique can be a useful algebraic tool for the analysis and control of BNs [1]. In such a sense, the result in (I) of Proposition 3.1 provides an efficient method to calculate $\lfloor {}^s\mathcal{M} \rfloor$, $s \ge \tau$, especially when s is sufficiently large: $s \gg \tau$. Then the multiplication times of calculating $\lfloor {}^s\mathcal{M} \rfloor$ is reduced from $O(s)$ to $O(\max\{\tau, \alpha_s\})$. By the way, there are some exciting and impressive results on conjunctive BNs [3–5] and the tree structure [2] of BNs with polynomial computational complexity. In such a context, the networks are not the general BCNs and thereby, the semi-tensor product technique may lose its usefulness. Here, we focus on the general BCNs with the semi-tensor product technique.

Theorem 3.1 *Consider sampled-data BCN* (3.1), $x(0) = x_0$.

(I) $x_e = \delta_N^i$ *is reachable from* $x_0 = \delta_N^j$ *at the* s*-th time step if and only if* $({}^s\mathcal{M})_{ij} = 1$.
(II) $x_e = \delta_N^i$ *is reachable from* $x_0 = \delta_N^j$ *if and only if* $\lfloor {}^{N\tau}\mathcal{M} \rfloor_{ij} = 1$.
(III) Sampled-data BCN (3.1) *is said to be controllable at* $x_0 = \delta_N^i$, *if and only if* $Col_i\left(\lfloor {}^{N\tau}\mathcal{M} \rfloor\right) > 0$.
(IV) Sampled-data BCN (3.1) *is said to be controllable if and only if* $\lfloor {}^{N\tau}\mathcal{M} \rfloor > 0$.

Proof

(I) According to the above analysis for the intrinsic meaning of ${}^s\mathcal{M}$, it is obvious true of (I).

(II) Assuming there exists a sampled-data control sequence π^s that drives $x_e = \delta_N^j$ to $x_0 = \delta_N^i$, we consider it from two situations.

- $s \mod \tau = 0$ and set $\alpha_s := \frac{s}{\tau}$. There are totally N states of BCNs. Therefore, there must exist a sampled-data control sequence π^s with $\alpha_s \leq N$ within α_s sampling periods such that (II) holds.
- $s \mod \tau \neq 0$ and set $\beta_s := s \mod \tau$, $\alpha_s := \frac{s-\beta_s}{\tau}$. Assume $x(\alpha_s\tau) = \delta_N^p$, $p \neq j$ is the $\alpha_s\tau$th node of the path from $x_0 = \delta_N^j$ to $x_e = \delta_N^i$ under a sampled-data control sequence π^s. From the above arguments, there exists a sampled-data control sequence $\pi^{\alpha_s\tau}$ with $\alpha_s < N$ such that $x(\alpha_s\tau) = \delta_N^p$ is reachable from $x_0 = \delta_N^j$. Then, after another β_s steps and under a sampled-data control sequence π^{β_s}, δ_N^p can be driven to δ_N^i. Therefore, there exists a sampled-data control sequence π^s with $\alpha_s < N$ such that (II) holds.

(III) (III) and (IV) can be directly obtained from (II).

□

Remark 3.2 When $\tau = 1$, the sampled-data BCN can be regarded as a BCN, then the results of the controllability are completely equivalent to the results in [6, 7]. It should be noticed that if $(\mathcal{M}^t)_{ij} = 1$, then $({}^t\mathcal{M})_{ij} = 1$ may not hold. That is to say, $x_0 = \delta_N^j$ can be driven to $x_e = \delta_N^i$ by some free control sequence $\{u_1, u_2, \ldots, u_t\}$ but may not be driven by any sampled-data control sequence π^t. However, the inverse always holds.

Consider the following sampled-data BCN with $\tau = 2$,

$$x(t+1) = Lu(t)x(t),$$

where $L = \delta_4[2\ 3\ 3\ 1\ 3\ 4\ 3\ 1]$, $x(t) \in \Delta_4$ and $u(t) \in \Delta_2$.

As a result, $L_1 = \delta_4[2\ 3\ 3\ 1]$ and $L_2 = \delta_4[3\ 4\ 3\ 1]$. Let $s = 5$, $x_0 = \delta_4^4$, a straightforward computation shows that,

$${}^1\mathcal{M} = \begin{pmatrix} 0\,0\,0\,1 \\ 1\,0\,0\,0 \\ 1\,1\,1\,0 \\ 0\,1\,0\,0 \end{pmatrix}, \ {}^2\mathcal{M} = \begin{pmatrix} 0\,1\,0\,0 \\ 0\,0\,0\,1 \\ 1\,1\,1\,1 \\ 0\,0\,0\,0 \end{pmatrix}, \ldots, {}^5\mathcal{M} = \begin{pmatrix} 0\,0\,0\,0 \\ 0\,0\,0\,1 \\ 1\,1\,1\,1 \\ 0\,0\,0\,0 \end{pmatrix}.$$

Noticing that $({}^5\mathcal{M})_{24} = ({}^5\mathcal{M})_{34} = 1$, by Theorem 3.1, we can conclude that $x_e \in \{\delta_4^2, \delta_4^3\}$ are reachable from $x_0 = \delta_4^4$ at time $t = 5$.

Now, we consider the controllability at $x_0 = \delta_4^2$. We have

$$\lfloor ({}^{\tau}\mathcal{M})^{(2)} \rfloor = \lfloor ({}^{\tau}\mathcal{M})^{(3)} \rfloor = \begin{pmatrix} 0\,1\,0\,1 \\ 0\,0\,0\,1 \\ 1\,1\,1\,1 \\ 0\,0\,0\,0 \end{pmatrix}$$

$$\lfloor {}^{5}\mathcal{M} \rfloor = \begin{pmatrix} 0\,1\,0\,1 \\ 1\,1\,0\,1 \\ 1\,1\,1\,1 \\ 0\,1\,0\,1 \end{pmatrix}$$

From Proposition 3.1, $\lfloor {}^{k}\mathcal{M} \rfloor = \lfloor {}^{2\times2+2-1}\mathcal{M} \rfloor$ for all $k \geq 5$, which implies $\lfloor {}^{2^4\times2+2-1}\mathcal{M} \rfloor = \lfloor {}^{5}\mathcal{M} \rfloor$ and $Col_2\left(\lfloor {}^{33}\mathcal{M} \rfloor\right) > 0$. From Theorem 3.1, it is concluded that the system is controllable at $x_0 = \delta_4^2$. Besides, the system is also controllable at $x_0 = \delta_4^4$ due to $Col_4\left(\lfloor {}^{33}\mathcal{M} \rfloor\right) > 0$. In addition, $\lfloor {}^{33}\mathcal{M} \rfloor \not> 0$, then the system is not controllable.

On the other hand, we calculate matrix $\mathcal{M}^2$ as

$$\mathcal{M}^2 = \begin{pmatrix} 0\,1\,0\,0 \\ 0\,0\,0\,1 \\ 1\,1\,1\,1 \\ 1\,0\,0\,0 \end{pmatrix}.$$

It is observed that $(\mathcal{M}^2)_{41} = 1$ and $({}^{2}\mathcal{M})_{41} = 0$. That is to say, $x_0 = \delta_4^1$ can be driven to $x_e = \delta_4^4$ by some free control sequence $\{u_0, u_1\}$, $u_0, u_1 \in \Delta_2$ (actually, $u_0 = \delta_2^1, u_1 = \delta_2^2$) but not by any sampled-data control sequence π^2 ($\{\delta_2^1, \delta_2^1\}$ or $\{\delta_2^2, \delta_2^2\}$).

Now, we are going to investigate what systems do not lose their controllability regardless of the sampling period. Notably, a BCN can be expressed as a Boolean switched system switching between M possible subsystems (i.e., BNs) by encoding the control inputs as a switching signal [6]. Specifically, for the sampled-data BCNs, such a switching signal sustains in each sampling interval. Moreover, for each subsystem or BN, it may have several cycles, including fixed points. Here, suppose that there are c cycles, named as $C_1, \ldots, C_c$, among the M BNs. Let ω be the least common multiple of the lengthes of all cycles, i.e., $\omega = l.c.m(|C_1|, \ldots, |C_c|)$. Because there are N states in sampled-data BCNs. This implies that, given any initial state and any initial control inputs, if the sampling period $\tau = N$, then $x(\tau)$ must belong to some cycle. Furthermore, if the sampling period $\tau = N + \omega$, then it must hold $x(N) = x(N + \omega)$. In other words, the two sampling periods $\tau = N$ and $\tau = N + \omega$ can result in the same controllability of sampled-data BCNs.

Similarly, any two sampling periods $\tau = N + i$ and $\tau = N + \omega + i$, $i = 1, 2, \ldots, \omega - 1$, can also result in the same controllability of sampled-data BCNs. Accordingly, the following result can be directly obtained.

Theorem 3.2 *The controllability of sampled-data BCN (3.1) is regardless of the sampling period if and only if for any sampling period $\tau \in \{1, 2, \ldots, N + \omega\}$, the controllability of sampled-data BCN (3.1) is invariant. Specifically, sampled-data BCN (3.1) is controllable regardless of the sampling period if and only if for any sampling period $\tau \in \{1, 2, \ldots, N + \omega\}$, sampled-data BCN (3.1) is controllable.*

3.1.3 Observability of Sampled-Data Boolean Control Networks

It is obvious that for a BCN, observability is control dependent. In the following, we investigate the observability of sampled-data BCNs. For convenience, let $x(t; x_0, \pi^t)$ and $y(t; x_0, \pi^t)$ be the state and output of system (3.1) at time t with initial state $x_0 \in \Delta_N$ under the sampled-data control sequence $\pi^t \in \prod^t$. The following definition of observability is presented for sampled-data BCNs.

Definition 3.2 Consider sampled-data BCN (3.1). We call (x, x') the initial distinguishable state pair iff $x \neq x'$ and $Hx \neq Hx'$. Sampled-data BCN (3.1) is observable if for any $x, x' \in \Delta_N$ satisfying $x \neq x'$ and $Hx = Hx'$, there exists a positive integer t and the sampled-data control sequence $\pi^t \in \prod^t$, such that

$$y(t; x, \pi^t) \neq y(t; x', \pi^t).$$

Remark 3.3 Four different definitions were given on the observability of BCNs in the recent literature in [8], which include the one studied by Cheng et al. for BCNs [9]. When constant sampling period $\tau = 1$ in SDBCN (3.1), then Definition 3.2 equals to the standard one in [9].

Let $x'(t + 1) = Lu(t)x'(t)$ be the same system as system (3.1). To facilitate the analysis, let $z(t) = x(t)x'(t)$, which is a bijective map from $\Delta_N \times \Delta_N$ to Δ_{N^2}. Then, we have

$$\begin{aligned} z(t+1) &= x(t+1)x'(t+1) \\ &= Lu(t)x(t)Lu(t)x'(t) \\ &:= Eu(t)z(t), \end{aligned} \tag{3.7}$$

where $E = L * L \in \mathcal{L}_{N^2 \times N^2 M}$.

System (3.7) is an algebraic expression for a new sampled-data BCN, which combines two sampled-data BCNs (3.1) with the same transition matrix. Let

$z(t; z_0, \pi^t)$ be the state of sampled-data BCN (3.7) at time t with initial state $z_0 \in \Delta_{N^2}$ under the sampled-data control sequence $\pi^t \in \prod^t$.

Lemma 3.1 *Consider sampled-data BCN (3.1). Denote B_0 by the set of all initial distinguishable state pairs. Then (δ_N^i, δ_N^j) or $(\delta_N^j, \delta_N^i) \in B_0$, if and only if*

$$(H^T H)_{ij} = 0 = (H^T H)_{ji}.$$

Proof (Necessity) If (δ_N^i, δ_N^j) is an initial distinguishable pair, that is to say the corresponding outputs $H\delta_N^i$ and $H\delta_N^j$ are different, it implies that (δ_N^i, δ_N^j) is also an initial distinguishable pair, and $(H\delta_N^i)^T(H\delta_N^j) = 0 = (H\delta_N^j)^T(H\delta_N^i)$, *i.e.*, $(H^T H)_{ij} = 0 = (H^T H)_{ji}$.

(Sufficiency) If $(H^T H)_{ij} = 0 = (H^T H)_{ji}$, it implies that $(H\delta_N^i)^T(H\delta_N^j) = 0$. That is to say, the output $H\delta_N^i$ of δ_N^i and the output $H\delta_N^j$ of δ_N^j are different. Therefore, (δ_N^i, δ_N^j) is an initial distinguishable pair, i.e., (δ_N^i, δ_N^j), $(\delta_N^i, \delta_N^j) \in B_0$. The lemma is proved. □

By multiplying two elements in each pair in B_0, we can simply express the set B_0 with a new set defined as B. Let $C = \Delta_{N^2} \backslash B \backslash \left\{ \delta_{N^2}^{(i-1)N+i} | i \in [1, N] \right\}$. Then for any $z_0 = \delta_N^i \delta_N^j \in C$, it holds $i \neq j$ and $H\delta_N^j = H\delta_N^i$. And for all $z_0 = \delta_N^i \delta_N^j \in B$, $H\delta_N^j = H\delta_N^i$ does not hold.

Since matrix E in (3.7) is a $N^2 \times N^2 M$ logical matrix, we split it into M square blocks as $E = [E_1 \; E_2 \; \cdots \; E_M]$ (like the method in spitting matrix L before), where $E_j \in \mathcal{L}_{N^2 \times N^2}$, $j \in [1, M]$. Hence, corresponding matrices can be obtained as for matrix L, such as ${}^t\mathcal{M}$.

For convenience and hereafter, we still use these matrices to show the relationship among nodes of combined sampled-data BCN (3.7).

Now, let $\mathcal{U}^0 \in \mathscr{B}_{1 \times N^2}$ be a row vector, satisfying

$$Col_i(\mathcal{U}^0) = \begin{cases} 1, & \delta_{N^2}^i \in B, \\ 0, & \text{otherwise.} \end{cases}$$

Using $\mathcal{U}^0$, we can inductively construct new vectors $\mathcal{U}^k$ and $\mathcal{U}^*$, which will be used to judge the observability as follows:

Notice that at each efficient iteration of "While" in Algorithm 6, the norm $\|\mathcal{U}^k\|_1$ will increase by at least one. So, at most after $N^2 - |B|$ iterations $\mathcal{U}^k \neq \mathcal{U}^{k-1}$, and it holds that $\mathcal{U}^k = \mathcal{U}^{k^*}$, $\forall k \geq k^*$. Consequently, we can always get $\mathcal{U}^*$.

Theorem 3.3 *Consider sampled-data BCN (3.1) (or Eq. (3.7)).*

(I) It is observable, if and only if

$$\|\mathcal{U}^*\|_1 = N^2 - N.$$

Algorithm 6 It is an algorithm for obtaining a stable row vector $\mathcal{U}^*$

$k := 1, \mathcal{Y} := \{i | \delta_{N^2}^i \in B\}$.
for $i \in \mathcal{Y}$ **do**
    $\mathcal{U}^1 := \mathcal{U}^0 +_{\mathcal{B}} Row_i(\lfloor^{\tau-1}\mathcal{M}\rfloor)$
end for
while $\mathcal{U}^k \neq \mathcal{U}^{k-1}$ **do**
    $\mathcal{Y} := \{i | Col_i(\mathcal{U}^k) - Col_i(\mathcal{U}^{k-1}) = 1\}$,
    for $i \in \mathcal{Y}$ **do**
        $\mathcal{U}^{k+1} := \mathcal{U}^k +_{\mathcal{B}} Row_i({}^{\tau}\mathcal{M})$.
    end for
    $k := k + 1$,
end while
$k^* := k - 1, \mathcal{U}^* = \mathcal{U}^{k^*}$.

(II) $(x_0, x_0') \sim x_0 x_0' = \delta_{N^2}^i$ *is an indistinguishable pair, if and only if*

$$Col_i(\mathcal{U}^*) = 0.$$

Proof We prove (I) first.

(Sufficiency) In Algorithm 6, if $Col_i(\mathcal{U}^0) = 0,\ Col_j(\mathcal{U}^0) = 1$ and $\lfloor^{\tau-1}\mathcal{M}\rfloor_{ji} = 1$, then by construction, $\delta_{N^2}^j$ is initially distinguishable and $\delta_{N^2}^i$ is not and there exists at least one sampled-data control sequence $\pi^t,\ t < \tau$, driving $\delta_{N^2}^i$ to $\delta_{N^2}^j$ in less than τ time steps. Since $\delta_{N^2}^j$ is initially distinguishable, we call $\delta_{N^2}^i$ the one-step distinguishable state due to the fact that $\delta_{N^2}^j$ can be determined by only one control from $\delta_{N^2}^i$. So in $\mathcal{U}^1$ we change $Col_i(\mathcal{U}^1) = 0$ into $Col_i(\mathcal{U}^1) = 1$. Repeat this process until $\mathcal{U}^k = \mathcal{U}^{k-1}$. If $\|\mathcal{U}^*\|_1 = N^2 - N$, it is obvious that for all $z_0 = x_0 x_0' \in B \cup C$, (x_0, x_0') is distinguishable. As a result, the sampled-data system is observable.

(Necessity) Let $\Gamma = \{\delta_{N^2}^i | Col_i(\mathcal{U}^*) = 0\}$. Then it is clear that Γ is a control-invariant set. That is, if $\gamma \in \Gamma$, then

$$z(k; \gamma, \pi^k) \in \Gamma, \text{ for any } \pi^k \in \prod^k \text{ and } k > 0.$$

Hence Γ is the set of all indistinguishable states. So, if the sampled-data system is observable, it holds that $\|\mathcal{U}^*\|_1 = N^2 - N$.

From the proof of (I), it is easy to see that (II) is an immediate consequence. □

Remark 3.4 Similar to the controllability, when $\tau > 1$, the observability of sampled-data BCNs is also very different from that of BCNs. Sampled-data BCNs with $\tau > 1$ may not be observable even if the systems are observable with $\tau = 1$, which is shown by the following example.

Consider the following BCN [9]

$$\begin{cases} x_1(t+1) = [u(t) \wedge \neg(x_1(t) \wedge x_2(t) \wedge x_3(t))] \vee \{\neg u(t) \wedge [(x_1(t) \wedge \\ \qquad (x_2(t) \vee \neg x_3(t))) \vee (\neg x_1(t) \wedge x_2(t) \wedge \neg x_3(t))]\}, \\ x_2(t+1) = \{u(t) \wedge [(x_1(t) \wedge x_2(t) \wedge \neg x_3(t)) \vee (\neg x_1(t) \wedge x_2(t) \wedge x_3(t)) \\ \qquad \vee \neg(x_1(t) \vee x_2(t) \vee x_3(t))]\} \vee \{\neg u(t) \wedge [x_1(t) \wedge x_2(t)) \\ \qquad \vee (\neg x_1(t) \wedge x_2(t) \wedge x_3(t))]\}, \\ x_3(t+1) = [u(t) \wedge \neg(x_1(t) \wedge x_2(t))] \vee [\neg u(t) \wedge (\neg x_1(t) \wedge \neg x_2(t) \wedge x_3(t))], \\ y_1(t) = x_1(t) \vee \neg x_2(t) \vee x_3(t), \\ y_2(t) = \neg x_1(t) \vee \neg x_2(t) \wedge \neg x_3(t). \end{cases}$$

Its algebraic form is

$$x(t+1) = Lu(t)x(t),$$
$$y(t) = Hx(t),$$

where

$$L = \delta_8[8\ 1\ 3\ 3\ 2\ 3\ 3\ 1\ 1\ 4\ 5\ 3\ 5\ 3\ 7\ 7],$$
$$H = \delta_4[2\ 1\ 2\ 2\ 2\ 3\ 2\ 1].$$

Assume the sampling period $\tau = 2$, then it is converted to a sampled-data BCN.

Using the variable substitution $z(t) = x(t)x'(t) \in \Delta_{64}$, we have $z(t+1) = Eu(t)z(t)$, where $E = [E_1\ E_2]$,

$$E_1 = \delta_{64}[64\ 57\ 59\ 59\ 58\ 59\ 59\ 57\ 8\ 1\ 3\ 3\ 2\ 3\ 3\ 1\ 24\ 17\ 19\ 19\ 18\ 19\ 19\ 17 \\ 24\ 17\ 19\ 19\ 18\ 19\ 19\ 17\ 16\ 9\ 11\ 11\ 10\ 11\ 11\ 9\ 24\ 17\ 19\ 19\ 18\ 19\ 19\ 17 \\ 24\ 17\ 19\ 19\ 18\ 19\ 19\ 17\ 8\ 1\ 3\ 3\ 2\ 3\ 3\ 1],$$

$$E_2 = \delta_{64}[1\ 4\ 5\ 3\ 5\ 3\ 7\ 7\ 25\ 28\ 29\ 27\ 29\ 27\ 31\ 31\ 33\ 36\ 37\ 35\ 37\ 35\ 39\ 39 \\ 17\ 20\ 21\ 19\ 21\ 19\ 23\ 23\ 33\ 36\ 37\ 35\ 37\ 35\ 39\ 39\ 17\ 20\ 21\ 19\ 21\ 19\ 23 \\ 23\ 49\ 52\ 53\ 51\ 53\ 51\ 55\ 55\ 49\ 52\ 53\ 51\ 53\ 51\ 55\ 55].$$

According to Lemma 3.1, one can easily obtain the initial distinguishable states set

$$B = \delta_{64}\{2, 6, 8, 9, 11, 12, 13, 14, 15, 18, 22, 24, 26, 30, 32, 34, 38, 40, 41, 42, 43, \\ 44, 45, 47, 48, 50, 54, 56, 57, 59, 60, 61, 62, 63\}, \quad |B| = 34.$$

It is easy to figure out $\lfloor {}^1\mathcal{M} \rfloor$ and ${}^2\mathcal{M}$ by E_1 and E_2.

$$\lfloor {}^1\mathcal{M} \rfloor = {}^1\mathcal{M} = E_1 +_{\mathscr{B}} E_2, {}^2\mathcal{M} = E_1^2 +_{\mathscr{B}} E_2^2,$$

Using set B, we can obtain $\mathcal{U}^0$ and $\mathcal{Y}$, where $\|\mathcal{U}^0\|_1 = 34$ and $\mathcal{Y} = \{i | \delta_{64}^i \in B\}$.
Firstly, according to Algorithm 6, using $\lfloor {}^1\mathcal{M} \rfloor$ and $\mathcal{U}^0$, we have

$$\begin{aligned}
\mathcal{U}^1 &= \mathcal{U}^0 +_{\mathscr{B}} Row_i(\lfloor {}^1\mathcal{M} \rfloor) \\
&= [0\,1\,1\,1\,0\,1\,1\,1\,1\,0\,1\,1\,1\,1\,1\,0\,1\,1\,0\,0\,1\,1\,0 \\
&\quad\; 1\,1\,1\,0\,0\,1\,1\,0\,1\,0\,1\,1\,1\,0\,1\,1\,1\,1\,1\,1\,1\,1\,0 \\
&\quad\; 1\,1\,1\,1\,0\,0\,1\,1\,0\,1\,1\,0\,1\,1\,1\,1\,1\,0], \quad i \in \mathcal{Y}, \\
\|\mathcal{U}^1\|_1 &= 46, \\
\mathcal{Y} :&= \{i | Col_i(\mathcal{U}^1) - Col_i(\mathcal{U}^0) = 1\} \\
&= \{3, 4, 7, 17, 21, 25, 29, 35, 36, 39, 49, 53\}.
\end{aligned}$$

Next, on the basis of Algorithm 6, it follows that

$$\begin{aligned}
\mathcal{U}^2 &= \mathcal{U}^1 +_{\mathscr{B}} Row_i({}^2\mathcal{M}) \\
&= [0\,1\,1\,1\,0\,1\,1\,1\,1\,0\,1\,1\,1\,1\,1\,0\,1\,1\,0\,0\,1\,1\,1 \\
&\quad\; 1\,1\,1\,0\,0\,1\,1\,1\,1\,0\,1\,1\,1\,0\,1\,1\,1\,1\,1\,1\,1\,1\,0 \\
&\quad\; 1\,1\,1\,1\,1\,1\,1\,1\,0\,1\,1\,0\,1\,1\,1\,1\,1\,0], \quad i \in \mathcal{Y}, \\
\|\mathcal{U}^2\|_1 &= 50, \\
\mathcal{Y} :&= \{i | Col_i(\mathcal{U}^1) - Col_i(\mathcal{U}^0) = 1\} \\
&= \{23, 31, 51, 52\}.
\end{aligned}$$

Similarly, we have

$$\begin{aligned}
\mathcal{U}^3 &= \mathcal{U}^2 +_{\mathscr{B}} Row_i({}^2\mathcal{M}) \\
&= [0\,1\,1\,1\,0\,1\,1\,1\,1\,0\,1\,1\,1\,1\,1\,1\,1\,1\,0\,0\,1\,1\,1 \\
&\quad\; 1\,1\,1\,0\,0\,1\,1\,1\,1\,0\,1\,1\,1\,0\,1\,1\,1\,1\,1\,1\,1\,1\,0 \\
&\quad\; 1\,1\,1\,1\,1\,1\,1\,1\,0\,1\,1\,1\,1\,1\,1\,1\,1\,0], \quad i \in \mathcal{Y}, \\
\|\mathcal{U}^3\|_1 &= 52, \\
\mathcal{Y} :&= \{i | Col_i(\mathcal{U}^1) - Col_i(\mathcal{U}^0) = 1\} \\
&= \{16, 58\}.
\end{aligned}$$

As a result, we have $\mathcal{U}^* = \mathcal{U}^3 = \mathcal{U}^4 = \mathcal{U}^3 +_{\mathscr{B}} Row_i({}^2\mathcal{M}),\ i \in \{16, 58\}$ and $\|\mathcal{U}^*\|_1 = 52 < 2^6 - 2^3$.

It is concluded that the considered sampled-data BCN is not observable from Theorem 3.3, while the system is observable with $\tau = 1$ by Cheng et al. [9].

3.1.4 More Effective Algorithm for Observability

A BCN can be expressed as a Boolean switched system by encoding the control inputs as a switching signal [6]. Therefore, a sampled-data BCN can be also expressed as a sampled-data Boolean switched system by encoding the control inputs as a switching signal. Then it results in a switched sampled-data Boolean networks

$$z(t+1) = E_{\sigma(t)} z(t), \tag{3.8}$$

where $\sigma : \mathbb{N} \to [1, M]$ is the switching signal, $\sigma(t) = \sigma(t_l),\ t_l \le t < t_{l+1},\ t_l = l\tau, l = 0, 1, \ldots$, and $E_{\sigma(t)} = Eu(t_l\tau)$. Notably, in each sampling period $t_l \le t < t_{l+1}$ of sampled-data BCNs (3.5), once the value of control at time instance t_l is determined, the system can be regarded as a typical BN and $E_{\sigma(t_l)}$ is the structure matrix of this BN. Therefore, in each sampling period, we only need to focus on some BN, which is significant for the following observability analysis. Before giving the main results, we first introduce some properties of BNs.

For a BN, we can spilt the state-space graph (or dynamic graph) of BNs into some directed subgraphs. Let k be the time for node z_0 getting into the cycle. We call this k the transient time. Corresponding to k, there is a path outside the cycle, called the transient path. Hence, the trajectory of state z_0 consists of the following three parts:

(I) the transient path.
(II) the limit cycle (if it exists).
(III) z_k—the entry of the cycle starting from z_0.

If the trajectory of state z_0 has no transient path, i.e., $k = 0$, then it is called a pure periodic trajectory, else we call z_k a switching node of the limit cycle. For any state z_0' in the trajectory of state z_0, z_0' either belongs to the transient path or the limit cycle of z_0. The next example shows the directed subgraphs of BNs.

Consider the following BN,

$$z(t+1) = \delta_8[2\ 3\ 4\ 2\ 3\ 6\ 5\ 5]z(t),$$

where $x(t) \in \Delta_8$. Then, we can obtain two directed subgraphs of the considered BN, which are shown in Fig. 3.1.

Some important notations are given for system (3.8).

Fig. 3.1 Subgraphs of the considered BN. For simplicity, number i in each circle denotes the node δ_8^i

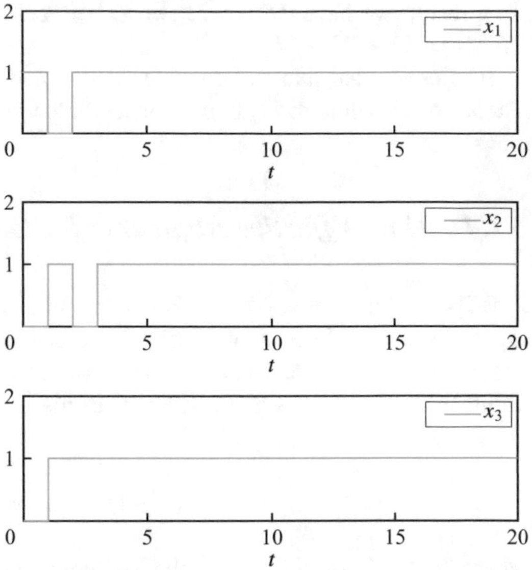

- $\Omega \subseteq \Delta_{N^2}$—for any $z \in \Omega$, node z can reach some initial distinguishable state $z_0 \in B$ by some sampled-data control sequence $\pi^s = \{u(0), u(1), ..., u(s-1)\}$ with $u(0) = \cdots = u(s-1)$, $0 < s \le \tau$.
- $D_i^\tau \subseteq \Delta_{N^2}$—for any $z \in D_i^\tau$, node z can be driven to the terminal state $\delta_{N^2}^i$ by some sampled-data control sequence π^τ (exactly a sampling period);
- $S_l \subseteq \Delta_{N^2}$—the set of states that can reach some initial distinguishable state $z_0 \in B$ in the time interval $[l\tau + 1, (l+1)\tau]$ and can not reach any initial distinguishable state $z_0 \in B$ in the time interval $[1, l\tau]$. That is to say, for a state $z_0' \in S_l$, there exists some sampled-data control sequence $\pi^s = \{u(0), u(1), ..., u(s-1)\}, l\tau \le s < (l+1)\tau$, such that $z(s) = z_0$ with $z(0) = z_0'$, but not by any sampled-data control sequence π^s with $0 < s < l\tau$;
- $\Lambda \subseteq \Delta_{N^2}$—for any $z \in \Omega$, node z can reach some initial distinguishable state $z_0 \in B$ by some sampled-data control sequence $\pi^s = \{u(0), u(1), ..., u(s-1)\}$, $s > 0$. Notably, it holds $\Lambda = \bigcup_{n=0}^{\infty} S_n$.
- $\mathcal{S}_C \subseteq \Delta_{N^2}$—the switching nodes set of some cycle C;
- $Pre_j^i \subseteq \Delta_{N^2}$, $j \in [1, N], i \in [1, M]$—the set of states that can reach node $\delta_{N^2}^j$ in exactly one step in the i-th BN $z(t+1) = E_i z(t) = E\delta_M^i z(t)$;
- $vis[i] \in \mathcal{D}$—if $vis[i] = True$, it means $\delta_{N^2}^i \in \Lambda \cup B$ and otherwise.
- Set Q is called a queue (which is discussed in any Data Structure book), if Q can be added an element in the back of Q and removed an element from the front of Q. In other words, a queue Q is a first in first out (FIFO) data structure. Furthermore, $Q.empty()$ denotes a Boolean value (True or False) with true for the fact Q is empty and vice versa. $Q.front()$ and $Q.pop()$ respectively represent an action for returning and deleting the first element of queue Q, while $Q.push(\cdot)$ signifies that a element is added in the end of queue Q.

Now, considering the system (3.8) and based on the constructed subgraphs of each BN $z(t + 1) = E_i z(t) = E\delta^i_M z(t)$, $i \in [1, M]$, the pseudocode for obtaining the set Ω is displayed in Algorithm 7.

Algorithm 7 It is an algorithm for obtaining the set Ω

1: Initialize: $\Omega := \varnothing$.
2: **for** $i := 1$ to M **do**
3:     Construct subgraphs of BN $z(t + 1) = E\delta^i_M z(t)$.
4:     Calculate set Pre^i_j, $j \in [1, N]$.
5:     **for** each directed subgraph $G = \{V, \mathcal{E}\}$ **do**
6:         $\mathcal{A} := B \cap V$.
7:         $step[j] := 0$, $j \in [1, N^2]$, which means that $\delta^j_{N^2}$ can reach some state
8:         $z_0 \in \mathcal{A}$ in exactly $step[j]$ time steps.
9:         **for** each node $\delta^\alpha_{N^2} \in \mathcal{A}$ **do**
10:             $Q := \{\alpha\}$.
11:             **while** $Q.empty() = False$ **do**
12:                 $k := Q.front()$.
13:                 **for** $\delta^j_{N^2} \in Pre^i_k$ **do**
14:                     **if** $\delta^j_{N^2} \notin \mathcal{A}$ and $step[k] < \tau$ **then**
15:                         $Q.push(j)$.
16:                         $\Omega := \Omega \cup \{\delta^j_{N^2}\}$.
17:                         $step[j] := step[k] + 1$.
18:                   **end if**
19:                 **end for**
20:                 $Q.pop()$.
21:              **end while**
22:         **end for**
23:     **end for**
24: **end for**

To better understand the process of Algorithm 7, the following example shows the detailed process.

Consider the following directed graph $G = \{V, \mathcal{E}\}$ shown in Fig. 3.2. Suppose $\tau = 2$, while the node set $V = \Delta_{11}$ and set $\mathcal{A} = \{\delta^2_{11}, \delta^4_{11}, \delta^6_{11}\}$. As for the edge set $\mathcal{E}$, please refer to Fig. 3.2. Notably, though the graph in Fig. 3.2 cannot be a state-space graph of an BN (actually a mixed-value logical network [1]), it does not affect our analysis in this example. According to Algorithm 7, for each node $\delta^i_{11} \in \mathcal{A}$, one can construct a directed graph of each node δ^i_{11} within two steps shown in Fig. 3.3. Therefore, $\Omega = \{\delta^1_{11}, \delta^3_{11}, \delta^5_{11}, \delta^7_{11}, \delta^8_{11}, \delta^9_{11}, \delta^{10}_{11}\} = V \backslash \mathcal{A} \backslash \{\delta^{11}_{11}\}$. That is to say, except for node δ^{11}_{11}, node $\delta^i_{11} \in \Omega$ with $i \neq 11$ can reach some node $\delta^j_{11} \in \mathcal{A}$ within $\tau = 2$ steps.

In the following, we propose an algorithm to figure out set D^τ_i with computation complexity $O(N^2 M)$, having nothing to do with the length of the sampling period τ.

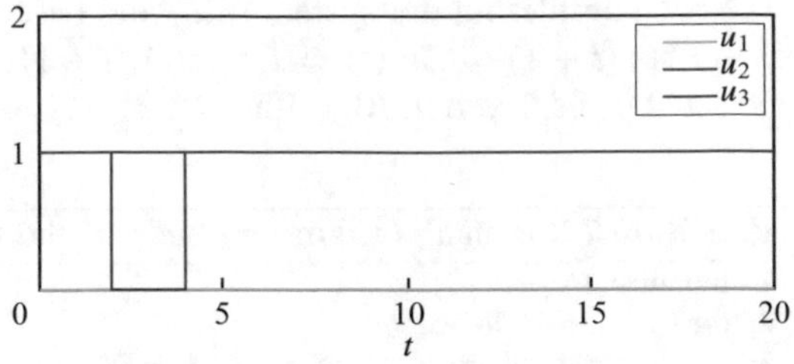

Fig. 3.2 The directed graph $\mathcal{G} = \{\mathcal{V}, \mathcal{E}\}$, double circles denote some node $\delta_{11}^i \in \mathcal{A}$, and single circles otherwise

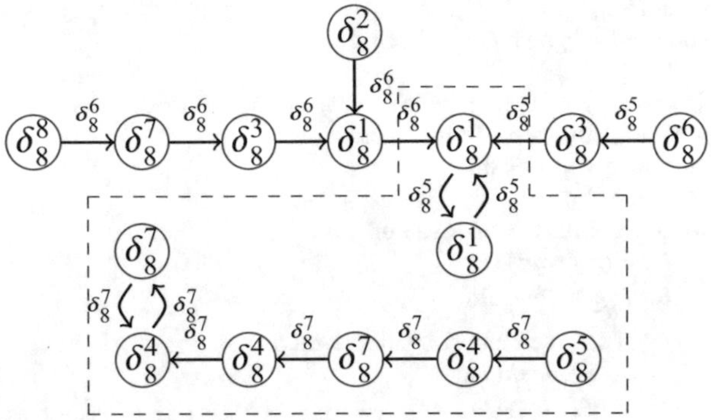

Fig. 3.3 Three directed graphs of each node $i \in \mathcal{A}$. Double circles denote some node $i \in \mathcal{A}$ with $\tau = 2$, and single circles otherwise

The following example shows how Algorithm 8 works.

Consider the graph shown in Fig. 3.4. Suppose $\tau = 100$ and the limit cycle $C^* = (\delta_6^{\alpha_0}, \delta_6^{\alpha_1}, \delta_6^{\alpha_2}) = (\delta_6^2, \delta_6^3, \delta_6^1)$ (which is regarded as the cycle C^* in Algorithm 8). Notably, though the graph in Fig. 3.4 cannot be a state-space graph of an BN (actually a mixed-value logical network [1]), one can still use this example to show the process of Algorithm 8. We just set the variable i equaling 1 in the first "for" loop Algorithm 8.

Then we have

$$Pre_1^1 = \{\delta_6^3\}, \ Pre_2^1 = \{\delta_6^1\},$$

$$Pre_3^1 = \{\delta_6^2, \delta_6^4\}, \ Pre_4^1 = \{\delta_6^5, \delta_6^6\}$$

$$Pre_5^1 = Pre_6^1 = \varnothing.$$

Due to δ_6^3 being the only switching node and according to the 14-th line in Algorithm 8, we reset Pre_3^1 as $Pre_3^1 \backslash \{\delta_6^2\} = \{\delta_6^4\}$.

We then consider this example from two steps.

Step 1. Consider the cycle C^*. According to the 8-th line in Algorithm 8, it is easily concluded that $D_1^{100} := \{3\}$, $D_2^{100} := \{1\}$, and $D_3^{100} := \{2\}$.

Algorithm 8 It is an algorithm for obtaining each states set D_i^τ, $i \in [1, N^2]$

1: Initialize: $D_i^\tau := \varnothing$, $i \in [1, N^2]$.
2: **for** $i := 1$ to M **do**
3:     Construct subgraphs of BN $z(t+1) = E\delta_M^i z(t)$.
4:     Calculate set Pre_j^i, $j \in [1, N^2]$.
5:     **for** each directed subgraph $G = \{V, \mathcal{E}\}$ **do**
6:         Obtain limit cycle $C^* = \left(\delta_{N^2}^{\alpha_0}, \delta_{N^2}^{\alpha_1}, \cdots, \delta_{N^2}^{\alpha_{w-1}}\right)$, w is the length of cycle C^*.
7:         **for** $j := 0$ to $w - 1$ **do**
8:             $p := j + \tau \mod w$, $D_p^\tau := D_p^\tau \cup \{\delta_{N^2}^{\alpha_j}\}$.
9:         **end for**
10:         Obtain set S_{C^*}, which denotes the switching nodes set of cycle C^*.
11:         $D_i := D_i \backslash C^*$.
12:         **for** each node $\delta_{N^2}^{\alpha_l} \in S_{C^*}$ **do**
13:             $top := 0$, $stack[top] := \alpha_l$.
14:             Reset $Pre_{\alpha_l}^i$ as $Pre_{\alpha_l}^i \backslash \{\delta_{N^2}^{\alpha_{l-1}}\}$.
15:             DFS(l, α_l, i).
16:         **end for**
17:     **end for**
18: **end for**
19: **function** DFS(l, k, i)
20:     **for** $\delta_{N^2}^j \in Pre_k^i$ **do**
21:         $top := top + 1$, $stack[top] = j$.
22:         **if** $top \geq \tau$ **then**
23:             $p := stack[top - \tau]$, $D_p^\tau := D_p^\tau \cup \{\delta_{N^2}^j\}$.
24:         **else**
25:             $p := \tau - top$, $h := (l + p) \mod w$.
26:             $q := \alpha_h$, $D_q^\tau := D_q^\tau \cup \{\delta_{N^2}^j\}$.
27:         **end if**
28:         DFS(l, j, i).
29:         $top := top - 1$.
30:     **end for**
31: **end function**

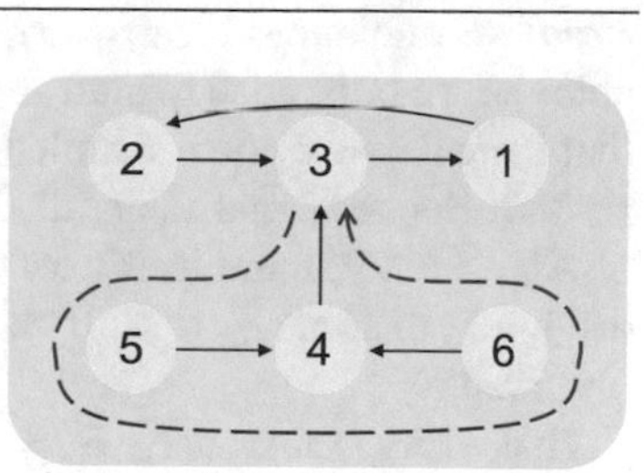

Fig. 3.4 A directed graph, can be regarded as a subgraph of BNs. The dotted line is the processing of function $DFS(3)$

Step 2. Consider parts outside cycle C^*. Invoke function depth first search $(1, 3, 1)$ and simulate the specific process of array $stack$ and number top as

1. $stack[0] = 3$, $top = 0$,
2. $stack[1] = 4$, $top = 1$, $D_3^{100} := D_3^{100} \cup \{\delta_6^4\} = \{\delta_6^2, \delta_6^4\}$,
3. $stack[2] = 5$, $top = 2$, $D_2^{100} := D_2^{100} \cup \{\delta_6^5\} = \{\delta_6^1, \delta_6^5\}$,
4. $top = 1$,

5. $stack[2] = 6, top = 2, D_2^{100} := D_2^{100} \cup \{\delta_6^6\} = \{\delta_6^1, \delta_6^5, \delta_6^6\}$,
6. $top = 1$,
7. $top = 0$.

Using the sets Ω and D_i^τ obtained through Algorithms 7 and 8, the pseudocode for getting the set Λ is displayed in Algorithm 9. Consequently, the following result for the observability of sampled-data BCNs is presented.

Algorithm 9 It is an algorithm for obtaining set Λ

Initialize: $\Lambda := \Omega,\ S_0 = \Omega,\ Q := \{i \mid \delta_{N^2}^i \in \Omega\};\ vis[i] := True$, for $\delta_{N^2}^i \in B \cup \Omega$; $vis[j] := False$, for $\delta_{N^2}^j \in C \backslash \Omega$; $step[k] := 0$, where $\delta_{N^2}^k \in C \cup B$, which means that $\delta_{N^2}^k \in S_{step[j]}$.
while $Q.empty() = False$ **do**
  $i := Q.front()$.
  for $\delta_{N^2}^j \in \mathcal{D}_i^\tau$ **do**
    if $\delta_{N^2}^j \in C$ and $vis[j] = False$ **then**
      $vis[j] = True, step[j] := step[i] + 1$.
      $S_{step[j]} := S_{step[j]} \cup \{\delta_{N^2}^j\}$.
      $\Lambda := \Lambda \cup \{\delta_{N^2}^j\}$.
      $Q.push(j)$.
    end if
  end for
  $Q.pop()$.
end while

Theorem 3.4 *Consider sampled-data BCN (3.1). It is observable if and only if* $\Lambda = C$.

Proof (Sufficiency) If $z_0 = x_0 x_0' \in B$, then the pair (x_0, x_0') is already distinguishable. So we only need to study $z_0 \in \Lambda$. If $z_0 \in \Omega$, based on Algorithm 8, it is clear that there is $t < \tau$ and a sampled-data control sequence π^t, driving z_0 to some node $z_0' \in B$. Hence x_0 and x_0' are distinguishable. If $z_0 \in \Lambda \backslash \Omega$, according to the process of Algorithm 9, there is at least one sampled-data control sequence π^t with $t > \tau$ driving z_0 to some $z_0' \in B$. If $\Lambda = C$, it is obvious that the sampled-data system is observable.

(Necessity) Obviously, $B \neq \varnothing,\ \Lambda \subseteq C$. If $C = \varnothing$, it is clear that $\Lambda = \varnothing$. Hereafter, Λ can be obtained using Algorithm 9. If $\Lambda \subsetneq C$, there exists at least one node $z_0 = x_0 x_0' \in C \backslash \Lambda$, such that for any sampled-data control sequence π^t with $t > 0$, $Hx_0 = Hx_0'$, which contradicts with the fact that the sampled-data system is observable. If the sampled-data system is observable, then it holds that $\Lambda = C$.

The proof is completed. □

Remark 3.5 It is clear that the computational complexity of Algorithm 7 for obtaining the states set Ω is $O(N^2 M)$ because each node in each BN $z(t+1) = E_i z(t)$ is traversed exactly at one time and there are N^2 nodes and M BNs. Similarly, for each BN $z(t+1) = E_i z(t)$, any node no matter belonging to its limit cycle or not is also traversed exactly at one time in Algorithm 8, resulting in the sum of $|D_i^\tau|$ at most N^2M and the computational complexity is $O(N^2M)$. In the worst case, each node of D_i^τ in Algorithm 9 can be traversed at most one time. Therefore, the computational complexity of Algorithm 9 is $O(N^2M)$ as well.

Remark 3.6 It is observed that the complexity of Algorithm 7, Algorithm 8 or Algorithm 9 has noting to do with τ. That is to say, the does not increase with the growth of sampling period length τ accordingly. Therefore, the observability of sampled-data BCNs in the sense of small number of Boolean variables is easily computable.

Remark 3.7 Scalability of the proposed algorithm is still a serious issue with the increase in the number of the Boolean states. n Boolean states have 2^n number of possible states and it has been proved that the computational complexity of BN-related problems of interest is NP-hard [10]. Therefore, the analysis and control problems for large-scale Boolean networks are still challenging at present.

The following example is given to illustrate the effectiveness of Algorithm 9 in determining the observability of sampled-data BCNs.

Consider the following sampled-data BCN

$$\begin{aligned} x(t+1) &= Lu(t)x(t), \\ y(t) &= Hx(t), \\ u(t) &= u(t_l), \quad t_l \le t < t_{l+1}, \end{aligned}$$

where $x(t) \in \Delta_{16}$, $u(t) \in \Delta_2$, $\tau = 100$, $t_l = l\tau \ge 0$, $l \in \mathbb{N}$, $t_{l+1} - t_l = \tau$, and

$$\begin{aligned} L &= \delta_{16}[3\ 4\ 5\ 6\ 1\ 2\ 1\ 2\ 7\ 8\ 7\ 8\ 9\ 10\ 3\ 4\ 1\ 8\ 3\ 2\ 9\ 10\ 2\ 8\ 10\ 1\ 5\ 8\ 1\ 1\ 3\ 15], \\ H &= \delta_2[1\ 1\ 1\ 1\ 1\ 1\ 1\ 1\ 1\ 2\ 1\ 2\ 2\ 1\ 2\ 1]. \end{aligned}$$

Using the variable substitution $z(t) = x(t)x'(t) \in \Delta_{256}$, then we have $z(t+1) = Eu(t)z(t)$, where $E = [E_1\ E_2]$,

$$\begin{aligned} E_1 &= \delta_{256}[35\ 36\ 37\ 38\ 33\ \cdots\ 55\ 56\ 57\ 58\ 51\ 52], \\ E_2 &= \delta_{256}[1\ 8\ 3\ 2\ 9\ \cdots\ 229\ 232\ 225\ 225\ 227\ 239]. \end{aligned}$$

According to Lemma 3.1, one can easily obtain the initial distinguishable states set

$$B = \{\delta_{256}^{10}, \delta_{256}^{12}, \delta_{256}^{15}, \ldots, \delta_{256}^{250}, \delta_{256}^{252}, \delta_{256}^{253}, \delta_{256}^{255}\}, \quad |B| = 96.$$

Based on Algorithm 7, a straightforward computation shows that

$$\Omega = \{\delta_{256}^{5}, \delta_{256}^{6}, \delta_{256}^{9}, \ldots, \delta_{256}^{247}, \delta_{256}^{248}, \delta_{256}^{251}, \delta_{256}^{254}\}, \quad |\Omega| = 96.$$

Using Algorithm 8, one can calculate the states set D_i^{100}, for instance

$$\begin{aligned}
D_1^{100} &= \{\delta_{256}^1, \delta_{256}^5, \delta_{256}^6, \ldots, \delta_{256}^{219}, \delta_{256}^{221}, \delta_{256}^{222}\},\\
D_2^{100} &= \{\delta_{256}^{70}, \delta_{256}^{72}, \delta_{256}^{102}, \delta_{256}^{104}\},\\
&\vdots\\
D_{256}^{100} &= \varnothing.
\end{aligned}$$

Using Algorithm 9, we have

$$\begin{aligned}
S_0 &= \Omega,\\
S_1 &= \{\delta_{256}^3, \delta_{256}^4, \delta_{256}^{19}, \delta_{256}^{20}, \delta_{256}^{33}, \delta_{256}^{34}, \delta_{256}^{36}, \delta_{256}^{39}, \delta_{256}^{40}, \delta_{256}^{49}, \delta_{256}^{50}, \delta_{256}^{51}, \delta_{256}^{55}, \delta_{256}^{56},\\
&\quad \delta_{256}^{89}, \delta_{256}^{99}, \delta_{256}^{100}, \delta_{256}^{115}, \delta_{256}^{116}, \delta_{256}^{134}, \delta_{256}^{144}, \delta_{256}^{157}, \delta_{256}^{159}, \delta_{256}^{189}, \delta_{256}^{191}, \delta_{256}^{202}, \delta_{256}^{204}, \delta_{256}^{234},\\
&\quad \delta_{256}^{236}, \delta_{256}^{249}\},\\
S_2 &= \{\delta_{256}^2, \delta_{256}^7, \delta_{256}^8, \delta_{256}^{17}, \delta_{256}^{23}, \delta_{256}^{24}, \delta_{256}^{96}, \delta_{256}^{97}, \delta_{256}^{98}, \delta_{256}^{113}, \delta_{256}^{114}, \delta_{256}^{207}, \delta_{256}^{237}, \delta_{256}^{246}\},\\
S_3 &= \{\delta_{256}^{104}, \delta_{256}^{119}, \delta_{256}^{156}, \delta_{256}^{186}\}.
\end{aligned}$$

Hence $\Lambda := S_0 \cup S_1 \cup S_2 \cup S_3 = C$, signifying that the considered sampled-data BCN is observable.

3.2 Sampled-Data Control for the Synchronization of Boolean Control Networks

In this section, we investigate the SDSFC for the synchronization of BCNs under the configuration of drive-response coupling. Necessary and sufficient conditions for the complete synchronization of BCNs are obtained by the algebraic representations of logical dynamics. Here, the relationship between the cycles of systems and sampling periods is discussed to analyze the existence of SDSFC. To achieve the synchronization, how to choose a suitable sampling period is presented based on the states. Based on the analysis of the sampling periods, we establish an algorithm to guarantee the synchronization of drive-response coupled BCNs by SDSFC. An example is given to illustrate the significance of the obtained results.

3.2.1 Problem Formulation

Consider a drive BN as

$$\begin{cases} x_1(t+1) = f_1(x_1(t), \ldots, x_n(t)), \\ \quad\vdots \\ x_n(t+1) = f_n(x_n(t), \ldots, x_n(t)), \end{cases} \tag{3.9}$$

for $t \geq 0$, where $f_i : \mathcal{D}^n \to \mathcal{D},\ i = 1, 2, \ldots, n$ are logical functions. Letting $x(t) = \ltimes_{i=1}^{n} x_i(t)$, the drive BN (3.9) can further be converted into the following discrete-time system:

$$x(t+1) = Mx(t), \quad t \geq 0, \tag{3.10}$$

where $M = M_1 \ltimes_{i=2}^{n} [(I_{2^n} \otimes M_i)\Phi_n] \in \mathcal{L}_{2^n \times 2^n}$ and M_i is the structure matrix of f_i.

A response BCN expressed by

$$\begin{cases} y_1(t+1) = g_1(x_1(t), \ldots, x_n(t), y_1(t), \ldots, y_n(t), u_1(t), \ldots, u_m(t)), \\ \quad\vdots \\ y_n(t+1) = g_n(x_1(t), \ldots, x_n(t), y_1(t), \ldots, y_n(t), u_1(t), \ldots, u_m(t)), \end{cases} \tag{3.11}$$

for $t \geq 0$, where $g_i : \mathcal{D}^{2n+m} \to \mathcal{D},\ i = 1, 2, \ldots, n$ are logical functions, and $u_j \in \mathcal{D},\ j = 1, 2, \ldots, m$ are control inputs. Similarly, let $y(t) = \ltimes_{i=1}^{n} y_i(t)$ and $u(t) = \ltimes_{j=1}^{m} u_j(t)$, then we have

$$y(t+1) = Hu(t)x(t)y(t), \quad t \geq 0, \tag{3.12}$$

where $H = H_1 \ltimes_{i=2}^{n} [(I_{2^{2n+m}} \otimes H_i)\Phi_{m+2n}] \in \mathcal{L}_{2^n \times 2^{m+2n}}$ and H_i is the structure matrix of g_i. The sampled-data state feedback law to be determined for system (3.11) is in the following from,

$$u_j(t) = e_j(x_1(t_l), \ldots, x_n(t_l), y_1(t_l), \ldots, y_n(t_l)), \tag{3.13}$$

for $t_l \leq t < t_{l+1}$, where $e_j : \mathcal{D}^{2n} \to \mathcal{D},\ 1 \leq j \leq m$ are logical functions. Similar to the above analysis, (3.13) is equivalent to the following algebraic form

$$u(t) = Ex(t_l)y(t_l) = Ez(t_l), \tag{3.14}$$

for $t_l \leq t < t_{l+1}$, where $z(t) = x(t) \ltimes y(t),\ E = E_1 \ltimes_{j=2}^{m} [(I_{2^{2n}} \otimes E_j)\Phi_{2n}] \in \mathcal{L}_{2^m \times 2^{2n}}$, E_j is the structure matrix of e_j, and the range for $t_l \leq t < t_{l+1}$ represents the time from the state $z(t_l)$ to $z(t_{l+1})$ with the controller $u(t)$. Meanwhile, t_l are the sampling instants for $l = 0, 1, 2, \dots$ and $t_{l+1} - t_l = \tau_l$ is sampling period which with the change of states $z(t)$ of iteration.

Definition 3.3 ([11]) The drive BN (3.9) and the response BCN (3.11) achieve complete synchronization for any initial states $x(0)$ and $y(0)$, if there exist a logical control sequence $U = \{u(t), t = 0, 1, 2, \dots\}$ and an integer $k \geq 0$, such that for all $t \geq k$, $x(t) = y(t)$.

In the following, the completely synchronization problem of $x(t)$ in system (3.9) and $y(t)$ in system (3.11) is analyzed under SDSFC.

From the above analysis, we have

$$\begin{aligned} z(t+1) =& x(t+1)y(t+1) \\ =& Mx(t)Hu(t)x(t)y(t) \\ =& M(I_{2^n} \otimes H)x(t)u(t)x(t)y(t) \\ =& M(I_{2^n} \otimes H)W_{[2^m,2^n]}u(t)x(t)x(t)y(t) \\ =& M(I_{2^n} \otimes H)W_{[2^m,2^n]}(I_{2^m} \otimes \Phi_n)u(t)z(t) \\ \triangleq& Lu(t)z(t), t \geq 0, \end{aligned} \tag{3.15}$$

where $L = M(I_{2^n} \otimes H)W_{[2^m,2^n]}(I_{2^m} \otimes \Phi_n) \in \mathcal{L}_{2^{2n} \times 2^{m+2n}}$.

First, we consider that $t_l = l\tau$ in (3.14) with some $\tau > 0$. According to drive BN (3.9), the response BCN (3.11) and the SDSFC (3.13), for $0 \leq t \leq \tau$ we have

$$\begin{aligned} z(1) =& LEz(0)z(0) = LEW_{[2^{2n}]}z(0)z(0) \\ =& LEW_{[4^n]}\Phi_{2n}z(0) \\ z(2) =& LEz(0)z(1) = LEW_{[4^n]}z(1)z(0) \\ =& (LEW_{[4^n]})^2\Phi_{2n}^2 z(0) \\ & \vdots \\ z(t) =& (LEW_{[4^n]})^t\Phi_{2n}^t z(0). \end{aligned} \tag{3.16}$$

Similarly, regard $z(\tau)$ as an initial state for $\tau < t \leq 2\tau$, it is learned from (3.16) that

$$\begin{aligned} z(t) =& (LEW_{[4^n]})^{t-\tau}\Phi_{2n}^{t-\tau}z(\tau) \\ =& (LEW_{[4^n]})^{t-\tau}\Phi_{2n}^{t-\tau}(LEW_{[4^n]})^{\tau}\Phi_{2n}^{\tau}z(0). \end{aligned} \tag{3.17}$$

Consequently, for $k\tau < t \leq (k+1)\tau$, we have by the induction that

$$z(t) = (LEW_{[4^n]})^{t-k\tau}\Phi_{2n}^{t-k\tau}((LEW_{[4^n]})^{\tau}\Phi_{2n}^{\tau})^k z(0). \tag{3.18}$$

Remark 3.8 The complete synchronization occurs between the drive BN (3.9) and the response BCN (3.11) for any initial states $x(0)$ and $y(0)$, if there exists an integer $d \geq 0$, such that for all $t \geq d$, $x(t) = y(t) = \delta_{2^n}^i$, $i \in [1, 2^n]$, where i depends on t. Moreover, $z(t) = \delta_{2^{2n}}^{(i-1)2^n+i} \in Col\left\{(LEW_{[4^n]})^t \Phi_{2n}^t\right\}$.

3.2.2 Main Results

Fixed points can be regarded as special limit cycles with period 1, so we only consider limit cycles in general. Let the set of all limit cycles of system (3.9) be $C_x = \{C_x^1, C_x^2, \ldots, C_x^h\}$, where $C_x^i = \{q_1^i, q_2^i, \ldots, q_{k_i}^i\}$, $1 \leq i \leq h$. Denote $q_d^i = \delta_{2^n}^{b_{id}}$, $b_{id} \in \Omega_n$ for $1 \leq d \leq k_i$, $1 \leq i \leq h$.

Remark 3.9 With respect to the investigation of cycles and fixed points of system (3.9), a set of methods have been presented to acquire the cycles and fixed points, in Chapter 5 of [1] with Definition 5.4, Theorem 5.2 and Theorem 5.3. The above two theorems show the number of fixed points and cycles of different lengths. Using this approach, one can quickly and effectively find the cycles of BNs.

Lemma 3.2 ([12]) *Systems (3.9) and (3.11) are completely synchronized if and only if (3.15) has the set of cycles* $C_z = \{C_z^1, C_z^2, \ldots, C_z^h\}$ *with* $C_z^i = \{q_1^i q_1^i, \ldots, q_{k_i}^i q_{k_i}^i\}$, *where* $q_d^i q_d^i = \delta_{2^{2n}}^{(b_{id}-1)2^{2n}+b_{id}}$ *and* $1 \leq d \leq k_i$, $1 \leq i \leq h$.

Next, we will study the synchronization of drive system $x(t)$ and response system $y(t)$ by system $z(t)$. Without loss of generality, the system $z(t)$ will enter the cycles (or fixed points) in the iterative process, then the sampling periods will be determined by the states update, which consists of sampling periods inside the cycle and outside the cycle, respectively.

Let the bth limit cycle for $z(t)$ be $C_z^b = \{q_1^b q_1^b, q_2^b q_2^b, \ldots, q_{k_b}^b q_{k_b}^b\}$. Use $\delta_{2^{2n}}^{r_{bi}}$ to represent a certain element in C_z^b. Let

$$L = \delta_{2^{2n}}[\alpha_1\ \alpha_2\ \cdots\ \alpha_{2^{m+2n}}],$$

where $\alpha_1, \alpha_2, \ldots, \alpha_{2^{m+2n}} \in \Omega_{2n}$, and

$$E = \delta_{2^m}[p_1\ p_2\ \cdots\ p_{2^{2n}}],$$

where $p_1, p_2, \ldots, p_{2^{2n}} \in \Omega_m$. Define

$$S_k^\tau(C_z^b) = S_k^\tau(i) = \{z(0) \in \Delta_{2^{2n}} : ((LEW_{[4^n]})^\tau \Phi_{2n}^\tau)^k z(0) = \delta_{2^{2n}}^i \in C_z^b\},$$

where $k \geq 1$, $1 \leq i \leq 2^{2n}$ and $1 \leq b \leq h$.

Lemma 3.3 *$S_1^\tau(r_{b1})$ can be rewritten by a set $\{\delta_{2^{2n}}^i : 1 \leq i \leq 2^{2n}, w_i^\tau = r_{b1},$ for some $1 \leq p_i \leq 2^m\}$, where $\delta_{2^{2n}}^{r_{b1}}$ is the first point which enters into the bth limit cycle, and*

$$\begin{cases} w_i^1 = \alpha_{(p_i-1)2^{2n}+i}, \\ w_i^{l+1} = \alpha_{(p_i-1)2^{2n}+w_i^l}, l \geq 1. \end{cases} \tag{3.19}$$

Proof Assuming $\delta_{2^{2n}}^i \in S_1^\tau(r_{b1})$, we have from (3.15) that

$$\begin{aligned} \delta_{2^{2n}}^{r_{b1}} =&(LEW_{[4^n]})^\tau \Phi_{2n}^\tau \delta_{2^{2n}}^i \\ =&(LEW_{[4^n]})^\tau (\delta_{2^{2n}}^i)^{\tau+1} \\ =&(LEW_{[4^n]})^{\tau-1} LEW_{[4^n]} (\delta_{2^{2n}}^i)^2 (\delta_{2^{2n}}^i)^{\tau-1} \\ =&(LEW_{[4^n]})^{\tau-1} [\delta_{2^{2n}}^{\alpha_{(p_i-1)2^{2n}+i}}] (\delta_{2^{2n}}^i)^{\tau-1} \\ =&(LEW_{[4^n]})^{\tau-1} (\delta_{2^{2n}}^{w_i^1}) (\delta_{2^{2n}}^i)^{\tau-1}. \end{aligned}$$

Moreover, for $1 \leq j \leq \tau - 1$,

$$\begin{aligned} &(LEW_{[4^n]})^{\tau-j} (\delta_{2^{2n}}^{w_i^j}) (\delta_{2^{2n}}^i)^{\tau-j} \\ =&(LEW_{[4^n]})^{\tau-(j+1)} (LEW_{[4^n]} \delta_{2^{2n}}^{w_i^j} \delta_{2^{2n}}^i) (\delta_{2^{2n}}^i)^{\tau-(j+1)} \\ =&(LEW_{[4^n]})^{\tau-(j+1)} (\delta_{2^{2n}}^{w_i^{j+1}}) (\delta_{2^{2n}}^i)^{\tau-(j+1)}. \end{aligned}$$

By induction, we have $\delta_{2^{2n}}^{r_{b1}} = \delta_{2^{2n}}^{w_i^\tau}$ for $j = \tau - 1$. Therefore, $w_i^\tau = r_{b1}$ and the proof is completed. □

The Relationship Between the Sampling Periods and Cycles

If the complete synchronization occurs between the drive BN (3.9) and the response BCN (3.11), then the drive-response system will eventually synchronize into the limit cycles. Let the sampling period be τ as assumed previously, and

$$l = \lambda\tau + \eta, \quad \eta < \tau,$$

where l is the length of limit cycle C_z^b for some b, where $1 \leq b \leq h$. η and λ are non-negative integers. For the relationship between l and τ, following three cases will be discussed.

Case a: $\lambda = 0,\ \eta > 0$, where $l = \eta,\ \eta < \tau$. In this case, the length of the cycle is smaller than τ. Then, we consider the sampling period to be l rather than τ, which will reduce the complexity of control designing. We will analyze the reason why the sampling period should be adjusted to the length of the cycle later. Assume $\delta_{2^{2n}}^{r_{b1}}$ to be the first point enters into the cycle C_z^b, then we have

$$\begin{cases} r_{b(i+1)} = \alpha_{(p_{r_{b1}}-1)2^{2n}+r_{bi}}, \\ r_{b1} = \alpha_{(p_{r_{b1}}-1)2^{2n}+r_{bk_b}}, \quad 1 \leq i \leq k_b, 1 \leq b \leq h. \end{cases} \tag{3.20}$$

The above formula guarantees that any state in the bth cycle will not jump out of C_z^b.

Case b: $\lambda > 0,\ \eta = 0$, with $l = \lambda\tau$. This case shows that the sampling period can divide the length of the cycle. Then letting sampling period be τ, one has λ sampling intervals in the cycle from the first state of the cycle. For the first sampling period, we have

$$\begin{cases} r_{b2} = \alpha_{(p_{r_{b1}}-1)2^{2n}+r_{b1}}, \\ r_{b(\tau+1)} = \alpha_{(p_{r_{b1}}-1)2^{2n}+r_{b\iota}}. \end{cases}$$

For the ith sampling period with $1 < i \leq \lambda$,

$$\begin{cases} r_{b((i-1)\tau+2)} = \alpha_{(p_{r_{b((i-1)\tau+1)}}-1)2^{2n}+r_{b((i-1)\tau+1)}}, \\ r_{b(i\tau+1)} = \alpha_{(p_{r_{b((i-1)\tau+1)}}-1)2^{2n}+r_{bi\tau}}. \end{cases}$$

Based on thc above analysis, we have for $1 \leq i \leq \lambda,\ 1 \leq b \leq h$,

$$\begin{cases} r_{b((i-1)\tau+2)} = \alpha_{(p_{r_{b((i-1)\tau+1)}}-1)2^{2n}+r_{b((i-1)\tau+1)}}, \\ r_{b(i\tau+1)} = \alpha_{(p_{r_{b(i-1)\tau}}+1)2^{2n}+r_{bi\tau}}, \\ r_{b1} = \alpha_{(p_{r_{b(\lambda-1)\tau}}+1)2^{2n}+r_{b\lambda\tau}}. \end{cases} \tag{3.21}$$

which guarantees that the bth cycle is closed.

Case c: $\lambda > 0,\ \eta > 0$, with $l = \lambda\tau + \eta$. This case means that the sampling period cannot divide the length of cycle C_z^b. The same as Case b, we need to analyze the state in the cycle according to the sampling periods. For the first λ sampling period, one get the same relationships for the states in the cycle as Case b. In detail, they can be determined by

$$\begin{cases} r_{b2} = \alpha_{(p_{r_{b1}}-1)2^{2n}+r_{b1}}, \\ r_{b(\tau+1)} = \alpha_{(p_{r_{b1}}-1)2^{2n}+r_{b\tau}}, \\ \vdots \\ r_{b(\lambda\tau+1)} = \alpha_{(p_{r_{b(\lambda-1)\tau}}+1)2^{2n}+r_{b\lambda\tau}}. \end{cases}$$

For the last step, the sampling period τ needs to be replaced by η, which is similar with Case a. Therefore,

$$\begin{cases} r_{b(\lambda\tau+2)} = \alpha_{(p_{r_{b(\lambda\tau+1)}}-1)2^{2n}+r_{b(\lambda\tau+1)}}, \\ r_{b1} = \alpha_{(p_{r_{b(\lambda\tau+1)}}-1)2^{2n}+r_{b(\lambda\tau+\eta)}}. \end{cases}$$

As a result, we have for $1 \leq i \leq \lambda,\ 1 \leq b \leq h$,

$$\begin{cases} r_{b(i-1)\tau+2} = \alpha_{(p_{r_{b((i-1)\tau+1)}}-1)2^{2n}+r_{b((i-1)\tau+1)}}, \\ r_{bi\tau+1} = \alpha_{(p_{r_{b((i-1)\tau+1)}}-1)2^{2n}+r_{bi\tau}}, \\ r_{b1} = \alpha_{(p_{r_{b(\lambda\tau+1)}}-1)2^{2n}+r_{b(\lambda\tau+\eta)}}. \end{cases} \tag{3.22}$$

which guarantees the closure of C_z^b.

The sampled-data control in a cycle has been analyzed in the above, which presents the relationship between the sampling period and the length of the cycle. These three cases cover all the possibilities in a cycle.

Remark 3.10 For a given sampling period, if it cannot make the cycle closed, then the sampling period should be altered such that the cycle closed. It gives an approach to design the sampling period and the SDSFC to achieve synchronization.

Design of Sampling Periods and SDSFC

Since (3.15) has the set of cycles, assumed by $C_z = \{C_z^1, C_z^2, \ldots, C_z^h\}$ with $C_z^b = \{\delta_{2^{2n}}^{r_{b1}}, \delta_{2^{2n}}^{r_{b2}}, \ldots, \delta_{2^{2n}}^{r_{bk_b}}\}$, for $1 \leq k_b \leq 2^{2n},\ 1 \leq b \leq h$. Therefore, to design the SDSFC for the synchronization, it is necessary to design a suitable sampling period for each cycle and make one of previous three cases satisfied. The main purpose of this part is to choose an appropriate sampling period for each cycle such that the sampled-data control can be simple and cost less.

Let l_b represents the length of the bth cycle, τ_b means the sampling period of the bth cycle which has l_b choices for $\tau_b^j \in [1, l_b]$ and $\tau_b^1 > \tau_b^2 > \cdots > \tau_b^{l_b}$. Then we have $l_b = \lambda_b^j \tau_b^j + \eta_b^j$, for $1 \leq b \leq h, 1 \leq j \leq l_b, 1 \leq \tau_b^j \leq l_b, \eta_b^j \leq \tau_b^j$, where λ_b^j is a corresponding factor of τ_b^j, and η_b^j is a remainder.

In each cycle, we do not select a larger value than the length of the cycle for sampling period. The reasons are given as follows.

Without loss of generality, take $C_z^1 = \{\delta_{2^{2n}}^{i_1}, \delta_{2^{2n}}^{i_2}, \ldots, \delta_{2^{2n}}^{i_p}\}$ for example, where $i_1, \ldots, i_p \in \Omega_{2^n}$. Assume that there are two sampling periods τ_1 and τ_2, with $\tau_1 = l_1,\ \tau_2 = l_1 + 1$.

When sampling period is equal to τ_1, it follows from (3.20) that

$$\begin{cases} i_2 = \alpha_{(p_{i_1}-1)2^{2n}+i_1}, \\ i_3 = \alpha_{(p_{i_1}-1)2^{2n}+i_2}, \\ \vdots \\ i_p = \alpha_{(p_{i_1}-1)2^{2n}+i_{l_1-1}}, \\ i_1 = \alpha_{(p_{i_1}-1)2^{2n}+i_{l_1}}. \end{cases} \tag{3.23}$$

It can be seen from the above formula, solving a such p_{i_1} needs to satisfy p equations. While $p_{i_2}, p_{i_3}, \ldots, p_{i_p}$ can take any value which belongs to Ω_{2^n}.

When sampling period is equal to τ_2, we have

$$\begin{cases} i_2 = \alpha_{(p_{i_1}-1)2^{2n}+i_1}, \\ i_3 = \alpha_{(p_{i_1}-1)2^{2n}+i_2}, \\ \vdots \\ i_p = \alpha_{(p_{i_1}-1)2^{2n}+i_{l_1-1}}, \\ i_1 = \alpha_{(p_{i_1}-1)2^{2n}+i_{l_1}}, \\ i_2 = \alpha_{(p_{i_1}-1)2^{2n}+i_1}, \\ i_3 = \alpha_{(p_{i_2}-1)2^{2n}+i_2}, \\ \vdots \end{cases} \tag{3.24}$$

From (3.24), there are also p equations to meet such that p_{i_1} can be solved. Meanwhile, we need to solve the other p_{i_j}, for $2 \leq j \leq p$. That is, compared with sampling period is equal to τ_1, we need to solve p^2 equations for C_z^1, which will be more complicated and restrictive. Therefore, the maximum value of the sampling period will be the length of the cycle, which can not only reduce the complexity of the study, but also be easier to get the solutions.

For C_z^b, from the above analyses, we have $l_b = \lambda_b^j \tau_b^j + \eta_b^j$, where $b = 1, \ldots, h$. There are l_b kinds of potential sampling periods, such as $\tau_b^1 = l_b$, $\tau_b^2 = l_b - 1$, $\tau_b^3 = l_b - 2, \ldots, \tau_b^{l_b} = 1$. In the following, we give the Algorithm 10 to obtain suitable sampling periods for these cycles.

If the sampling period for each cycle has been well found, then we are in the position to obtain the sampled-data controller for the purpose of synchronization

---

Algorithm 10 This algorithm is to find the maximum sampling period of each cycle

---

Require: System (3.15) and the set of cycles C_z.
Ensure: The biggest sampling period for each cycle of system (3.15).

1: **for** $b = 1$ to h **do**
2: choose the biggest τ_b^i for C_z^b as l_b, for $i = 1, \ldots, l_b$.
3: **if** $\delta_{2^{2n}}^{r_{b1}}$ be the first point enters into the cycle, then the application of Case b,
4: leads to one solution of $p_{r_{b1}}$ in logical matrix E **then**
5: go to 7
6: **else**
7: let $\delta_{2^{2n}}^{r_{b2}}$ be the first point in this cycle, and go to 3
8: **if** taking over all points in the cycle, one can not find the solution,
9: changing sampling period l_b to $\tau_b^2 = l_b - 1$ **then**
10: go to 3, applying Case c, one can potentially get the solution of $p_{r_{b1}}$ and
11: $p_{r_{bl_b}}$.
12: **else**
13: go to 19
14: **if** there is no such solution for such $p_{r_{bi}}$ **then**
15: let $\delta_{2^{2n}}^{r_{b2}}$ be the first point in this cycle, and go to 3
16: **else**
17: go to 7
18: **end if**
19: **if** taking over all points in the cycle, there is no solution **then**
20: choosing a smaller sampling period between $[1, l_b - 2]$, and go to 3
21: **else**
22: we find the biggest sampling period for bth cycle and all the solution
23: of $p_{r_{bi}}$, $i = 1, 2, \ldots, l_b$.
24: **end if**
25: **end if**
26: **end if**
27: **end for**

---

in Algorithm 11. In Algorithm 11, define the set $R = \{\delta_{2^{2n}}^{j} \mid (\delta_{2^{2n}}^{j} \in C_z^b) \cup S_k^{\tau_b^{l}}(r_{bq}), 1 \le j \le 2^{2n}, k \ge 1\}$.

Given the drive and response BCNs, as well as the initial states, we are in the position to design the SDSFC based on algorithms obtained. The initial state $z(0)$ will be divided into two types: in a cycle and outside cycles, respectively.

When the initial state is in the ith cycle, and $z^i(t)$ represents the state $z(t)$ is in the ith cycle, then we have

$$u(t) = Ex(t_l)y(t_l) = Ez(t_l) = Ez^i(t_l), \quad t_l \le t < t_{l+1}, \tag{3.25}$$

for $l = 0, 1, 2, \ldots, t_0 = 0$, where $t_l = l\tau_i$, and τ_i is the sampling period of the ith cycle obtained by Algorithm 10. Assume that $\delta_{2^{2n}}^{r_{i1}}$ represents the first state enters into the ith cycle by $C_z^i = \{\delta_{2^{2n}}^{r_{i1}}, \delta_{2^{2n}}^{r_{i2}}, \ldots, \delta_{2^{2n}}^{r_{il_b}}\}$, as in Algorithm 10. Meanwhile, the sampling points have been identified as $\delta_{2^{2n}}^{r_{i1}}$, $\delta_{2^{2n}}^{r_{i(\tau_i+1)}}$, $\delta_{2^{2n}}^{r_{i(2\tau_i+1)}}$, ..., and denote $\Lambda = \{\delta_{2^{2n}}^{r_{i1}}, \delta_{2^{2n}}^{r_{i(\tau_i+1)}}, \delta_{2^{2n}}^{r_{i(2\tau_i+1)}}, \ldots\}$. If the initial state $z^i(0) = \delta_{2^{2n}}^{r_{i(k\tau_i+p)}} \notin \Lambda$, where

---

Algorithm 11 This algorithm is to design the SDSFC for system (3.11)

1: **for** $b = 1$ to h, $i = 1$ to l_b **do**
2: to determine $p_{r_{bi}}$; for a C_z^b, one has an appropriate sampling period from
3: Algorithm 10 and $p_{r_{bi}}$ can be determined.
4: **end for**
5: assume that the first point by $\delta_{2^{2n}}^{r_{bq}}$ in the bth cycle. Let τ_b' be the sampling period before states entering the bth cycle, and make τ_b' equal to τ_b^j for $\tau_b' \in [1, \tau_b^j]$.
6: solving $w_j^{\tau_b'} = r_{iq}$ for $f_j = 1$ with $j \neq r_{bq}$, to get one solution of p_j for each C_z^b.
7: by the definition of $S_k^{\tau}(C_z)$, for every j, there exists a unique integer f_j such that $\delta_{2^{2n}}^j \in S_{f_j}^{\tau_b'}(C_z^b) \backslash S_{f_j-1}^{\tau_b'}(C_z^b)$; to determine the rest of p_j.
8: **if** one can regard $\delta_{2^{2n}}^j \in S_{f_j-1}^{\tau_b'}(C_z^b) \backslash S_{f_j-2}^{\tau_b'}(C_z^b)$ to be $\delta_{2^{2n}}^{r_{bq}}$ for $f_j = 1$ and solve $\delta_{2^{2n}}^{w_j^{\tau_b'}} \in S_{f_j-1}^{\tau_b'}(C_z^b) \backslash S_{f_j-2}^{\tau_b'}(C_z^b)$ for $f_j \geq 2$ to get one solution of each p_j **then**
9: go to 14
10: **else**
11: take $\delta_{2^{2n}}^{r_{bq'}}$ to replace $\delta_{2^{2n}}^{r_{bq}}$ as the first point in the bth cycle, and go to 4
12: **if** taking over all states in the cycle, there is no solution of such p_j **then**
13: E does not exist in terms of τ_b', one should adjust τ_b' as $\tau_b^j - 1, \tau_b^j - 2, \dots, 1$.
14: **else**
15: **if** $|R| < 2^{2n}$ **then**
16: make one of the τ_b' more smaller, and return to 3
17: **if** $|R| < 2^{2n}$ as well **then**
18: change the sampling period as $\tau_1' - 2, \tau_1' - 3, \dots, 1$, return to 4
19: **else**
20: go to 27
21: **end if**
22: **if** change all of the sampling periods for τ_1', cannot get the rest of p_j **then**
23: change another sampling period such as τ_2' for C_z^2 and return to 4
24: **else**
25: go to 27
26: **end if**
27: **else**
28: we get $u(t)$ with $E = \delta_{2^m}[p_1 \ p_2 \ \dots \ p_{2^{2n}}]$.
29: **end if**
30: **end if**
31: **end if**

$1 \leq k \leq \lambda_i$ and $1 < p \leq \tau_i$, which is not a sampling point, then we need to find the position of this $z^i(0)$ corresponding to the ith cycle, and according to the analysis of sampling points, the value for sampling should be $\delta_{2^{2n}}^{r_{i(k\tau_i+1)}}$. So taking $\delta_{2^{2n}}^{r_{i(k\tau_i+1)}}$ as the sampling point into (3.25), we have

$$u(t) = E \ltimes \delta_{2^{2n}}^{r_{i(k\tau_i+1)}}, \quad 0 \leq t < \tau_i + 1 - p.$$

When the initial states is on the trajectory into the ith cycle, then

$$u(t) = Ex(t_l)y(t_l) = Ez(t_l) = Ez^{i'}(t_l), \quad t_l \le t < t_{l+1}, \tag{3.26}$$

for $l = 0, 1, 2, \ldots$. Typically, $t_l = l\tau_i'$, where τ_i' is the sampling period outside the ith cycle, and $z^{i'}(t)$ represents the state $z(t)$ which before entering the ith cycle.

Theorem 3.5 *Assumed that an SDSFC is obtained from Algorithm 10 and 11, then the set of all cycles for (3.15) can be described by C_z if and only if one of the above three cases and the following condition is satisfied:*

Case d: For any initial state $z(0) \in \Delta_{2^{2n}}$, there exist b, $(1 \le b \le l)$ and an integer $M \ge 0$, such that $z(M) = \delta_{2^{2n}}^{r_{b1}}$.

Proof (Necessity) As for the definition of cycles, any initial state of (3.15) will enter one of the cycles at some time instant. Therefore, Case d obviously holds. After the analysis of Case a, Case b and Case c, and according to Algorithm 10 and 11, we get the SDSFC. So, the necessity is evident.

(Sufficiency) Conditions in the Case a, Case b and Case c guarantee that any state in the bth cycle will not jump out of C_z^b. Therefore, C_z consists of all the cycles of (3.15). Case d shows that any state will enter into a certain cycle in C_z. □

3.2.3 Example and Its Simulations

Consider the following Boolean control network, which is a reduced model of the lac operon in the Escherichia coli [13].

$$\begin{cases} x_1(t+1) = \neg u_1(t) \wedge (x_2(t) \vee x_3(t)), \\ x_2(t+1) = \neg u_1(t) \wedge u_2(t) \wedge x_1(t) \wedge \xi(t), \\ x_3(t+1) = \neg u_1(t) \wedge (u_2(t) \vee (u_3(t) \wedge x_1(t))), \end{cases} \tag{3.27}$$

where x_1, x_2 and x_3 are state variables denoting the lac mRNA, the lactose in high concentrations, and the lactose in medium concentrations, respectively; u_1, u_2 and u_3 are control inputs which represent the extracellular glucose, the high extracellular lactose, and the medium extracellular lactose, respectively; and ξ is an external disturbance.

Letting $\xi = 0$, $u_1(t) = 0$, $u_2(t) = x_1(t)$, $u_3(t) = 1$, we have the drive BN as follows

$$\begin{cases} x_1(t+1) = x_2(t) \vee x_3(t), \\ x_2(t+1) = 0, \\ x_3(t+1) = x_1(t), \end{cases} \tag{3.28}$$

Letting $\xi = 1$, the response BN is given by

$$\begin{cases} y_1(t+1) = \neg u_1(t) \wedge (y_2(t) \vee y_3(t)), \\ y_2(t+1) = \neg u_1(t) \wedge u_2(t) \wedge y_1(t), \\ y_3(t+1) = \neg u_1(t) \wedge (u_2(t) \vee (u_3(t) \wedge y_1(t))). \end{cases} \tag{3.29}$$

Using the vector form of logical variables, and setting $x(t) = \ltimes_{i=1}^{3} x_i(t),\ y(t) = \ltimes_{i=1}^{3} y_i(t)$ and $u(t) = \ltimes_{j=1}^{3} u_j(t)$, we have

$$M = \delta_8[3\ 3\ 3\ 7\ 4\ 4\ 4\ 8]$$

$$H = \delta_8[8\ 8\ 8\ 8\ 8\ 8\ 8\ 8\ 8\ 8\ 8\ 8\ 8\ 8\ 8\ 8\ 8\ 8\ 8\ 8\ 8\ 8\ 8\ 8\ 8\ 8\ 8\ 8\ 8\ 8\ 8\ 8 \\ 1\ 1\ 1\ 5\ 3\ 3\ 3\ 7\ 1\ 1\ 1\ 5\ 3\ 3\ 3\ 7\ 3\ 3\ 3\ 7\ 4\ 4\ 4\ 8\ 4\ 4\ 4\ 8\ 4\ 4\ 4\ 8].$$

Let $L = \delta_{64}[\alpha_1\ \alpha_2\ \cdots\ \alpha_{512}]$, where $\alpha_1, \alpha_2, \ldots, \alpha_{512} \in \Omega_6$. Assume the sampled-data state feedback matrix E is expressed by $E = \delta_8[p_1\ p_2\ \cdots\ p_{64}]$.

According to the drive BN, the limit cycle of (3.28) can be computed by $C_x = \{C_x^1, C_x^2, C_x^3\}$ with $C_x^1 = \{\delta_8^3\}$, $C_x^2 = \{\delta_8^4, \delta_8^7\}$ and $C_x^3 = \{\delta_8^8\}$. Then we have $C_z = \{C_z^1, C_z^2, C_z^3\}$ with $C_z^1 = \{\delta_{64}^{19}\}$, $C_z^2 = \{\delta_{64}^{28}, \delta_{64}^{55}\}$ and $C_z^3 = \{\delta_{64}^{64}\}$. We can see, $l_1 = 1, l_2 = 2, l_3 = 1$ and by the analysis from Algorithm 10 and 11, we have $\tau_2 = 2$. Then we choose the sampling period $\tau_1' = \tau_2' = \tau_3' = 2$ for C_z.

Step 1: Determine $p_{19},\ p_{28},\ p_{55},\ p_{64}$.
If $\alpha_{(p_{19}-1)2^6+19} = 19,\ (p_{19}-1)2^6 + 19 = 261, 262, 263, \cdots, 402, 403, 404$, then $p_{19} = 7$. If $\alpha_{(p_{28}-1)2^6+28} = 55, \alpha_{(p_{28}-1)2^6+55} = 28, (p_{28}-1)2^6 + 28 = 404$ *and* $(p_{28}-1)2^6 + 55 = 439$ *or* 503, then $p_{28} = 7$. If $\alpha_{(p_{64}-1)2^6+64} = 64$, then $p_{64} = 1$ *or* 2 *or* 3 *or* 4 *or* 7 *or* 8. As a result, we get $p_{19} = 7,\ p_{28} = 7,\ p_{55} = 7$ *or* $8,\ p_{64} = 1$ *or* 2 *or* 3 *or* 4 *or* 7 *or* 8. And p_{55} is arbitrary, with $p_{55} \in \Delta_8$.

Step 2: For C_z^1, since $w_j^2 = 19$, using the equations in (3.19) from Lemma 3.3. We have $w_j^1 = \alpha_{(p_j-1)2^6+j},\ w_j^2 = \alpha_{(p_j-1)2^6+w_j^1} = 19$. Then we get $j = 4, 16$, and $p_4 = p_{16} = 5$. For C_z^2, since $w_j^2 = 28$, using the equations in (3.19), and solving it. We can get $p_{32} = 5$ *or* 6, $p_{25} = p_{26} = p_{27} = 6$, $p_{29} = p_{30} = p_{31} = 5$. Then for C_z^3, we have $w_j^2 = 64$, solving the following equations $w_j^1 = \alpha_{(p_j-1)2^6+j}$, $w_j^2 = \alpha_{(p_j-1)2^6+w_j^1} = 64$. We obtain $p_6 = p_7 = p_8 = p_{10} = p_{11} = p_{12} = 6$ and $p_{57} = p_{58} = p_{59} = 6$.

Step 3: To determine the rest of p_j, where

$$j \in \Omega_6 \setminus \{4, 6, 7, 8, 10, 11, 12, 16, 19, 25, 26, 27, 28, 29, 30, 31, 32, 55, 57, 58, 59, 64\}.$$

Regard the rest of $p_j \in S^2_{f_j-1}(C^i_z) \setminus S^2_{f_j-2}(C^i_z),\ i = 1, 2, 3$. Assume

$$w_j^2 = 4\ or\ 6\ or\ 7 \ldots\ or\ 59,$$

which remove the point with in the cycles. And solving the equations in (3.19). We get the rest p_j, they are $p_1 = p_2 = p_3 = p_5 = 1,\ p_{13} = 6,\ p_{14} = p_{15} = p_{17} = p_{18} = 1$ and $p_{20} = 6 \ldots$.

Finally, we get all the vales of p_j for $j \in \Omega_6$. And one of the constructed values of E is given by

$$E = \delta_8[1\ 1\ 1\ 5\ 1\ 6\ 6\ 6\ 6\ 6\ 6\ 6\ 6\ 6\ 1\ 1\ 5\ 1\ 1\ 7\ 6\ 1\ 1\ 1\ 6\ 6\ 6\ 6\ 6\ 7\ 6\ 6\ 6\ 6$$
$$1\ 1\ 1\ 1\ 1\ 1\ 1\ 1\ 1\ 1\ 1\ 1\ 1\ 1\ 1\ 1\ 1\ 1\ 1\ 1\ 1\ 1\ 1\ 1\ 1\ 1\ 6\ 6\ 6\ 6\ 6\ 6\ 6\ 6].$$

Now, we know that whether the states in the cycle or outside the cycle, the sampling period of the system is 2. Therefore, we can get the SDSFC as

$$u(t) = Ex(t_l)y(t_l) = Ez(t_l), \quad t_l \le t < t_{l+1},$$

for $l = 0, 1, 2, \ldots,\ t_0 = 0$, where $t_{l+1} - t_l = 2$ and $t_l = 2l$. Of course, $z(t)$ represents the sampling point of the system, described by $z(t) \in \{\Delta_{64} \backslash \delta_{64}^{55}\}$. Simulations for the complete synchronization of (3.28) and (3.29) with some initial states, as well as the designed controller are given by the following Figs. 3.5 and 3.6.

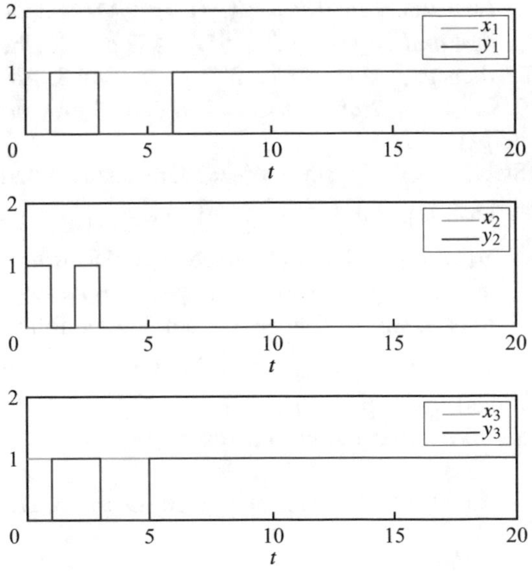

Fig. 3.5 The complete synchronization of (3.28) and (3.29) with initial states $x_1(0) = \delta_2^1, x_2(0) = \delta_2^2, x_3(0) = \delta_2^1$ and $y_1(0) = \delta_2^2, y_2(0) = \delta_2^1, y_3(0) = \delta_2^2$

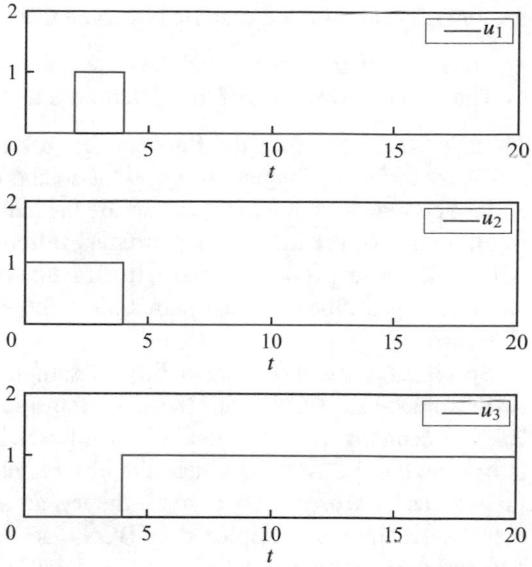

Fig. 3.6 An illustration of the designed $u(t)$ to achieve the complete synchronization of (3.28) and (3.29)

3.3 Summary

In this chapter, the controllability and observability of sampled-data BCNs have been studied. Based on the properties of the semi-tensor product, necessary and sufficient conditions have been given for controllability and observability of sampled-data BCNs. Some new phenomena have been observed that an SDBCN may lose the controllability or observability compared to the system with the sampling period $\tau = 1$. Here, we would like to explain such a phenomena as far as possible from the admissible control and system switching views:

- Intuitively, when $\tau > 1$, the control input should be constant in each sampling period, implying that the control strategy is not flexible to control the system. In other words, the space of the admissible control sequence with the sampling period $\tau > 1$ is much smaller than the one with the sampling period $\tau = 1$. Such constrained admissible control can contribute to the loss of controllability and observability.
- Because the states of control inputs are in the finite field. A BCN can be regarded as a Boolean switched system switching between M possible subsystems [6]. From this veiw, slow switching rate can contribute to the loss of the controllability and observability. It should be pointing out that for a continuous-time linear system $\dot{x} = Ax$, there may exist a pathological sampling such that the converted discrete-time system (difference equation) loses the observability and/or controllability [14, 15].

As for BCNs or sampled-data BCNs, some facts we may concern:

- The states of the systems and the states of control inputs are in the finite field.
- The intrinsic dynamics of the systems are highly nonlinear (logical operations).

Though we can translate the Boolean network dynamics into an continuous form (ordinary differential equation) using the method described in Wittmann et al. [16] and the continuous dynamics can capture the main properties (such as stable points) of original BNs, the transformed ordinary differential equation is highly nonlinear. All in all, these properties make it difficult to track the pathological sampling analysis (eigenvalue analysis) path akin to the continuous-time linear system case (Fig. 3.6).

Specifically, for the observability of sampled-data BCNs, we have combined two sampled-data BCNs with the same transition matrix into a new sampled-data BCN. According to initial distinguishable set, a stable observability row vector is obtained, and based on which, the observability of sampled-data BCNs can be judged easily. Moreover, via graph theory, an algorithm is presented for judging the observability of sampled-data BCNs. It is much more effectively with the computational complexity $O(N^2M)$, which is also found to be independent of the sampling period τ. It is worth pointing out that we actually did not directly study the observability but to transfer the observability problem to the reachability problem to be investigated. Numerical examples have been given to illustrate the obtained results.

Overall, it provided a general framework for using system and control techniques to analyze and manipulate sampled-data BCNs. Moreover, all obtained results can be further extended to the aperiodic sampling periods [17] case and some akin to sampled-data control, such as, event-triggered or time-triggered control.

Moreover, we also studied the SDSFC for the synchronization of Boolean control networks. Necessary and sufficient conditions for the synchronization was derived. An algorithm was given to construct the sampled-data state feedback controllers. In detail, we studied the relationship between the sampling periods and the length of the cycles.

In addition to the controllable research on sampled-data BCNs mentioned in this chapter, sampled-data controllability and stabilizability of BCNs were also considered in [18]. Here, based on a simplified intelligent traffic control system, the BCNs with nonuniform sampling periods were presented. Several necessary and sufficient conditions were obtained to determine nonuniform sampled-data controllability and stabilizability of BCNs.

References

1. Cheng, D., Qi, H., Li, Z.: Analysis and control of Boolean networks: A semi-tensor product approach. Springer Science & Business Media, New York (2011)
2. Akutsu, T., Hayashida, M., Ching, W.K., et al.: Control of Boolean networks: hardness results and algorithms for tree structured networks. J. Theor. Biol. **244**(4), 670–679 (2007)

3. Gao, Z., Chen, X., Başar, T.: Stability structures of conjunctive Boolean networks. Automatica **89**, 8–20 (2018)
4. Gao, Z., Chen, X., Başar, T.: Controllability of conjunctive Boolean networks with application to gene regulation. IEEE Trans. Control Netw. Syst. **5**(2), 770–781 (2017)
5. Weiss, E., Margaliot, M.: A polynomial-time algorithm for solving the minimal observability problem in conjunctive Boolean networks. IEEE Trans. Autom. Control **64**(7), 2727–2736 (2018)
6. Laschov, D., Margaliot, M.: Controllability of Boolean control networks via the Perron-Frobenius theory. Automatica **48**(6), 1218–1223 (2012)
7. Zhao, Y., Qi, H., Cheng, D.: Input-state incidence matrix of Boolean control networks and its applications. Syst. Control Lett. **59**(12), 767–774 (2010)
8. Zhang, K., Zhang, L.: Observability of Boolean control networks: a unified approach based on finite automata. IEEE Trans. Autom. Control **61**(9), 2733–2738 (2015)
9. Cheng, D., Qi, H., Liu, T., et al.: A note on observability of Boolean control networks. Syst. Control Lett. **87**, 76–82 (2016)
10. Zhao, Q.: A remark on “Scalar equations for synchronous Boolean networks with biological Applications” by C. Farrow, J. Heidel, J. Maloney, and J. Rogers. IEEE Trans. Neural Netw. **16**(6), 1715–1716 (2005)
11. Naghshtabrizi, P., Hespanha, J.P., Teel, A.R.: On the robust stability and stabilization of sampled-data systems: A hybrid system approach. In: Proceedings of the 45th IEEE Conference on Decision and Control, pp. 4873–4878 (2006)
12. Liu, Y., Sun, L., Lu, J., et al.: Feedback controller design for the synchronization of Boolean control networks. IEEE Trans. Neural Netw. Learn. Syst. **27**(9), 1991–1996 (2015)
13. Li, H., Wang, Y., Liu, Z.: Simultaneous stabilization for a set of Boolean control networks. Syst. Control Lett. **62**(12), 1168–1174 (2013)
14. Franklin, G.F., Ragazzini, J.R.: Sampled-Data Control Systems. Me Graw Hill, New York (1958)
15. Kalman, J.: Controllability of linear dynamical systems. Contribut. Differ. Equ. **1**, 189–213 (1963)
16. Wittmann, D.M., Krumsiek, J., Saez-Rodriguez, J., et al.: Transforming Boolean models to continuous models: Methodology and application to T-cell receptor signaling. BMC Syst. Biol. **3**(1), 1–21 (2009)
17. Lu, J., Sun, L., Liu, Y., et al.: Stabilization of Boolean control networks under aperiodic sampled-data control. SIAM J. Control Optim. **56**(6), 4385–4404 (2018)
18. Yu, Y., Feng, J., Wang, B., et al.: Sampled-data controllability and stabilizability of Boolean control networks: nonuniform sampling. J. Franklin Inst. **355**(12), 5324–5335 (2018)

Chapter 4
Stabilization of Probabilistic Boolean Control Networks Under Sampled-Data Control

Abstract In this chapter, we address the stabilization of probabilistic Boolean control networks (PBCNs) via a sampled-data state feedback controller (SDSFC).

4.1 Sampled-Data Stabilization of Probabilistic Boolean Control Networks

Probabilistic Boolean networks (PBNs) are discrete-time systems composed of a family of BNs, between which the PBN switches in a stochastic fashion. In this section, we address the stabilization of probabilistic Boolean control networks (PBCNs) via a SDSFC. Based on the algebraic representation of logical functions, a necessary and sufficient condition is derived for the existence of SDSFCs for the global stabilization of PBCNs, and the controller is further designed. A biological example is presented to illustrate the effectiveness of the obtained results.

4.1.1 An Algebraic Form of a Probabilistic Boolean Control Network

A PBN with a set of Boolean variables $x_1, x_2, \ldots, x_n$ can be described as

$$\begin{cases} x_1(t+1) = f_1(x_1(t), x_2(t), \ldots, x_n(t)) \\ x_2(t+1) = f_2(x_1(t), x_2(t), \ldots, x_n(t)) \\ \quad\vdots \\ x_n(t+1) = f_n(x_1(t), x_2(t), \ldots, x_n(t)) \end{cases} \tag{4.1}$$

where $x_i \in D$, $i = 1, 2, \ldots, n$ are logical variables and for each time step, $f_i(t)$ is chosen randomly from a given finite set of Boolean functions $\mathcal{F}_i = \{f_i^1, f_i^2, \ldots, f_i^{l_i}\}$. There are totally $\sigma = \prod_{i=1}^n l_i$ models. Denote by $\Sigma_\lambda = \{f_1^{\lambda_1}, f_2^{\lambda_2}, \ldots, f_n^{\lambda_n}\}$ the λth model, where $\lambda = \Sigma_{i=1}^{n-1}(\lambda_i - 1)\prod_{j=1}^{n-1} l_{j+1} + \lambda_n$ and

Y. Liu et al., *Sampled-data Control of Logical Networks*,
https://doi.org/10.1007/978-981-19-8261-3_4

$1 \leq \lambda_i \leq l_i$. We assume that the probability of f_i being f_i^j is

$$\mathbf{P}\left(f_i = f_i^j\right) = p_i^j, \quad j = 1, 2, \ldots, l_i,$$

where $\Sigma_{j=1}^{l_i} p_i^j = 1$. And the probability of Σ_λ to be active is

$$P_\lambda = \mathbf{P}\,(\text{network } \Sigma_\lambda \text{ is selected}) = \prod_{i=1}^{n} p_i^{\lambda_i}.$$

In order to convert system (4.1) into an algebraic form, we define $x(t) = \ltimes_{i=1}^{n} x_i(t)$. Now we can obtain the algebraic form of (4.1) as follows:

$$\begin{cases} x_1(t+1) = M_1 x(t) \\ x_2(t+1) = M_2 x(t) \\ \quad\vdots \\ x_n(t+1) = M_n x(t) \end{cases} \tag{4.2}$$

where $M_i,\ i = 1, 2, \cdots, n$ are logical matric and can be chosen from a corresponding matric set $\{M_i^1, \ldots, M_i^{l_i}\}$. Let $\mathbb{E}x$ denote the overall expected value of x, then we have $\mathbb{E}x_i(t+1) = \hat{M}_i \mathbb{E}x(t),\ i = 1, 2, \ldots, n$, where $\hat{M}_i = \Sigma_{j=1}^{l_i} p_i^j M_i^j$, which leads to the following equation:

$$\mathbb{E}x(t+1) = M\mathbb{E}x(t), \tag{4.3}$$

where $M = \hat{M}_1 * \hat{M}_2 * \cdots * \hat{M}_n$.

Definition 4.1 ([1]) For PBN (4.1), a state $x^* \in \{0, 1\}^n$ is said to be globally stable with probability one, if for every initial state $X(0) := x_0,\ x_0 \in \{0, 1\}^n$, there exists a positive integer T, such that $\mathbf{P}(X(t) = x^* | X(0) = x_0) = 1$ for $t \geq T$.

4.1.2 Sampled-Data State Feedback Control for Probabilistic Boolean Control Networks

A PBCN with a set of Boolean variables $x_1, \ldots, x_n$ and controllers $u_1, \ldots, u_m$ can be described as

$$\begin{cases} x_1(t+1) = f_1(x_1(t), \ldots, x_n(t), u_1(t), \ldots, u_m(t)) \\ x_2(t+1) = f_2(x_1(t), \ldots, x_n(t), u_1(t), \ldots, u_m(t)) \\ \quad\vdots \\ x_n(t+1) = f_n(x_1(t), \ldots, x_n(t), u_1(t), \ldots, u_m(t)) \end{cases} \tag{4.4}$$

where $x_i \in \mathcal{D},\ i = 1, \ldots, n$ and $u_j \in \mathcal{D},\ j = 1, \ldots, m$ are logical variables and for each time, $f_i : \mathcal{D}^{m+n} \to \mathcal{D}^n$ is chosen randomly from a given finite set of Boolean functions $\mathcal{F}_i = \{f_i^1, f_i^2, \ldots, f_i^{l_i}\}$.

Here, we shall deal with a state feedback stabilization problem for PBCN (4.4). More precisely, we are concerned with the design of SDSFCs such that the system can be stabilized. The feedback law to be determined for system (4.4) is in the following form,

$$\begin{cases} u_1(t) = h_1(x_1(t_l), \ldots, x_n(t_l)), \\ \quad\vdots \qquad\qquad\qquad\qquad\qquad\qquad t_l \le t < t_{l+1} \\ u_m(t) = h_m(x_1(t_l), \ldots, x_n(t_l)), \end{cases}$$

where $h_j : \mathcal{D}^n \to \mathcal{D},\ 1 \le j \le m$ are Boolean functions, $t_l = l\tau \ge 0$ for $l = 0, 1, \ldots$ are sampling instants and $t_{l+1} - t_l = \tau$ denotes the constant sampling period.

Let $u(t) = \ltimes_{i=1}^m u_i(t)$, with similar transformation of (4.3), PBCN (4.4) can be represented by

$$\mathbb{E}x(t+1) = Lu(t)\mathbb{E}x(t), \tag{4.5}$$

where $L \in \Upsilon_{2^n \times 2^{m+n}}$ and the controller $u(t)$ can be converted to the following algebraic form,

$$u(t) = Hx(t_l), \quad t_l \le t < t_{l+1}. \tag{4.6}$$

Denote by

$$L = [\alpha_1 \ \cdots \ \alpha_{2^n} \ \alpha_{2^n+1} \ \cdots \ \alpha_{(2^m-1)2^n+1} \ \cdots \ \alpha_{2^{m+n}}]$$

where $\alpha_s \in \Upsilon_{2^n}$ for $s = 1, 2, \ldots, 2^{m+n}$. Splitting it into 2^m equal blocks as follows:

$$L = [Blk_1(L) \ \cdots \ Blk_{2^m}(L)],$$

where $Blk_j(L) \in \Upsilon_{2^n \times 2^n}$ for $j = 1, 2, \ldots, 2^m$. Denote by $H = \delta_{2^m}[p_1 \ p_2 \ \cdots \ p_{2^n}]$, where $p_j \in \{1, 2, \ldots, 2^m\}$ for $j = 1, 2, \ldots, 2^n$.

For every $x^* = \delta_{2^n}^r \in \mathcal{D}^n$, we define a sequence of sets $\{S_k(r)\}$ as follows:

$$S_1(r) = \{\delta_{2^n}^i \in \Delta_{2^n} | (Blk_{p_i}(L))^\tau \delta_{2^n}^i = \delta_{2^n}^r\},$$
$$S_{k+1}(r) = \{\delta_{2^n}^i | \left[(Blk_{p_i}(L))^\tau \delta_{2^n}^i\right] \circ (\mathbf{1}_{2^n} - \Sigma_{a \in S_k(r)} a) = \mathbf{0}_{2^n}\}.$$

Lemma 4.1 *If $\delta_{2^n}^r \in S_1(r)$, then $S_k(r) \subseteq S_{k+1}(r)$ for $k \ge 1$.*

Proof We shall use induction on k. When $k = 1$, suppose that $\delta_{2^n}^s \in S_1(r)$, which implies that $(Blk_{p_s}(L))^\tau \delta_{2^n}^s = \delta_{2^n}^r$. Then we have

$$\left[(Blk_{p_s}(L))^\tau \delta_{2^n}^s\right] \circ (\mathbf{1}_{2^n} - \Sigma_{a\in S_1(r)} a) = \delta_{2^n}^r \circ (\mathbf{1}_{2^n} - \Sigma_{a\in S_1(r)} a).$$

Since $\delta_{2^n}^r \in S_1(r)$, we have $\delta_{2^n}^r \circ (\mathbf{1}_{2^n} - \Sigma_{a\in S_1(r)} a) = \mathbf{0}_{2^n}$, which implies that $\delta_{2^n}^s \in S_2(r)$.

Now let $k > 1$ and assume by induction that $S_{k-1}(r) \subseteq S_k(r)$. Suppose that $\delta_{2^n}^s \in S_k(r)$, then $\left[(Blk_{p_s}(L))^\tau \delta_{2^n}^s\right] \circ (\mathbf{1}_{2^n} - \Sigma_{a\in S_{k-1}(r)} a) = \mathbf{0}_{2^n}$. Since $S_{k-1}(r) \subseteq S_k(r)$, we have

$$\left[(Blk_{p_s}(L))^\tau \delta_{2^n}^s\right] \circ (\mathbf{1}_{2^n} - \Sigma_{a\in S_k(r)} a) = \mathbf{0}_{2^n}.$$

Hence $\delta_{2^n}^s \in S_{k+1}(r)$, so that $S_k(r) \subseteq S_{k+1}(r)$. □

Lemma 4.2

(I) If $S_1(r) = \{\delta_{2^n}^r\}$, then $S_k(r) = \{\delta_{2^n}^r\}$ holds for all $k \geq 1$.
(II) If $S_{j+1}(r) = S_j(r)$ for some $j \geq 1$, then $S_k(r) = S_j(r)$ holds for all $k \geq j$.

Proof We first prove part (I), and the proof of part (II) is quite similar. The case of $k = 1$ is trivial, so we take $k > 1$ and assume that the result holds for $k - 1$. By lemma 4.1, $S_k(r) \neq \emptyset$. Now let $\delta_{2^n}^s \in S_k(r)$, then $\left[(Blk_{p_s}(L))^\tau \delta_{2^n}^s\right] \circ (\mathbf{1}_{2^n} - \delta_{2^n}^r) = \mathbf{0}_{2^n}$, and we have $(Blk_{p_s}(L))^\tau \delta_{2^n}^s = \delta_{2^n}^r$, which implies that $\delta_{2^n}^s \in S_1(r)$. Consequently $\delta_{2^n}^s = \delta_{2^n}^r$, which leads to $S_k(r) = S_1(r) = \{\delta_{2^n}^r\}$. □

Theorem 4.1 *Consider PBCN (4.4), and let $x^* = \delta_{2^n}^r$. If there is a SDSFC law (4.6) such that PBCN (4.4) can be globally stabilized to x^* with probability one, then*

(I) $\delta_{2^n}^r \in S_1(r)$.
(II) There exists a positive integer $N \leq 2^n - 1$, such that $S_N(r) = \Delta_{2^n}$.

Proof By Definition 4.1, there exist $k > 0$ and $T = k\tau$ such that for any initial state $\delta_{2^n}^\alpha$, $x(t) \equiv \delta_{2^n}^r$ for all $t \geq T$. Then

$$\begin{aligned} x((k+1)\tau) &= LHx(k\tau)x((k+1)\tau - 1) \\ &= (LH\delta_{2^n}^r)^\tau \delta_{2^n}^r \\ &= (Blk_{p_r}(L))^\tau \delta_{2^n}^r = \delta_{2^n}^r, \end{aligned}$$

which implies that $\delta_{2^n}^r \in S_1(r)$, proving (I).

Since $\delta_{2^n}^r$ is globally stable state, then for any initial state $\delta_{2^n}^j$, $j = 1, 2, \ldots, 2^n$, there is a positive integer N_j, such that $\delta_{2^n}^j$ can be steered to $\delta_{2^n}^r$ in $N_j\tau$ steps. That is to say if $P[x(N_j\tau) = \delta_{2^n}^r | x(0) = \delta_{2^n}^j, U] = 1$, then $\delta_{2^n}^j \in S_{N_j}(r)$. We still show it by induction.

When $N_j = 1$, then $x(\tau) = (Blk_{p_j}(L))^\tau \delta_{2^n}^j = \delta_{2^n}^r$, which implies $\delta_{2^n}^j \in S_1(r)$. Suppose that $N_j = k - 1$, and $P[x((k-1)\tau) = \delta_{2^n}^r | x(0) = \delta_{2^n}^j, U] = 1$ implies

$\delta_{2^n}^j \in S_{k-1}(r)$. Then for $N_j = k$,

$$x(k\tau) = (LHx((k-1)\tau))^\tau \cdots (LH\delta_{2^n}^j)^\tau \delta_{2^n}^j.$$

Without loss of generality, let

$$(LH\delta_{2^n}^j)^\tau \delta_{2^n}^j = (m_1 \cdots m_k \, 0 \cdots 0)^{\mathrm{T}},$$

where $\Sigma_{t=1}^k m_t = 1,\ m_j > 0$. Since

$$1 = P[x(k\tau) = \delta_{2^n}^r | x(0) = \delta_{2^n}^j, U] = \Sigma_{s=1}^k m_s P[x((k-1)\tau) = \delta_{2^n}^r | x(0) = \delta_{2^n}^s, U],$$

then we have $P[x((k-1)\tau) = \delta_{2^n}^r | x(0) = \delta_{2^n}^s, U] = 1$, which implies that $\delta_{2^n}^s \in S_{k-1}(r)$ for $t = 1, \ldots, k$. Then

$$[(Blk_{p_j}(L))^\tau \delta_{2^n}^j] \circ (\mathbf{1}_{2^n} - \Sigma_{a \in S_{k-1}(r)} a) = \mathbf{0}_{2^n},$$

and we have $\delta_{2^n}^j \in S_k(r)$.

Let $N = \arg\min\{N_j | j = 1, \ldots, 2^n\}$ such that $S_N(r) = \Delta_{2^n}$. Then we show that $N \leq 2^n - 1$. It is enough to show that $|S_k(r)| \geq k+1$ for every $1 \leq k \leq N$. When $k = 1$, if $S_1(r) < 2$, since we have proved that $\delta_{2^n}^r \in S_1(r)$, then we have $S_1(r) = \{\delta_{2^n}^r\}$. It implies that $S_N(r) = \{\delta_{2^n}^r\}$, which is contrary to $S_N(r) = \Delta_{2^n}$. Hence $S_1(r) \geq 2$. Now let $1 < k \leq N$ and assume by induction that $|S_{k-1}(r)| \geq k$. Since $\delta_{2^n}^r \in S_1(r)$, Lemma 4.1 shows that $S_{k-1}(r) \subseteq S_k(r)$, thus $|S_k(r)| \geq |S_{k-1}(r)| \geq k$. If $|S_k(r)| < k+1$, then $|S_k(r)| = k$, which implies that $S_k(r) = S_{k-1}(r)$. So that $S_{k-1}(r) = S_N(r) = \Delta_{2^n}$ by Lemma 4.2. This contradicts the minimality of N. Thus $|S_k(r)| \geq k+1$.

Observe that if the above conditions (I) and (II) are satisfied, then Δ_{2^n} can be represented as the union of disjoint sets

$$\Delta_{2^n} = S_1(r) \cup (S_2(r) \backslash S_1(r)) \cup \cdots \cup (S_N(r) \cup S_{N-1}(r)).$$

Hence for each $1 \leq i \leq 2^n$, there exists a unique integer $1 \leq l_i \leq N$, such that $\delta_{2^n}^i \in S_{l_i}(r) \backslash S_{l_i - 1}(r)$, where $S_0(r) = \varnothing$. □

Theorem 4.2 *Consider PBCN (4.4), and let* $x^* = \delta_{2^n}^r$. *If there is a SDSFC law (4.6) such that PBCN (4.4) can be globally stabilized to* x^* *with probability one, then for every initial state* $\delta_{2^n}^i$, $i = 1, \ldots, 2^n$, *there exists a unique integer* $1 \leq l_i \leq N$, *such that* $\delta_{2^n}^i \in S_{l_i}(r) \backslash S_{l_i-1}(r)$ *with* $S_0(r) = \varnothing$. *Moreover let* p_i *be the solution of*

$$\begin{cases} \alpha_{(p_r-1)2^n+r} = \delta_{2^n}^r, \\ (Blk_{p_i}(L))^\tau \delta_{2^n}^i = \delta_{2^n}^r, \quad l_i = 1, \\ \left[(Blk_{p_i}(L))^\tau \delta_{2^n}^i\right] \circ (\mathbf{1}_{2^n} - \Sigma_{a \in S_{l_i-1}(r)} a) = \mathbf{0}_{2^n}, \quad l_i \geq 2. \end{cases}$$

then SDSFC can be determined by $H = \delta_{2^m}[p_1 \ \cdots \ p_{2^n}]$.

Proof Since $\delta_{2^n}^r$ is a stable state, then $P[x(k\tau+1)=\delta_{2^n}^r|x(k\tau)=\delta_{2^n}^r, u(k\tau)=H\delta_{2^n}^r,]=1$ implies that $Blk_{p_r}(L)\delta_{2^n}^r=\alpha_{(p_r-1)2^n+r}=\delta_{2^n}^r$.

For $l_i=1$, $i\neq r$, $\delta_{2^n}^i\in S_1(r)$, we need to find the corresponding p_i such that $(Blk_{p_i}(L))^\tau\delta_{2^n}^i=\delta_{2^n}^r$, then $P[x(\tau)=\delta_{2^n}^r|x(0)=\delta_{2^n}^i, u(t)=H\delta_{2^n}^i, 0\leq t<\tau]=1$.

For $l_i\geq 2$, regard $S_{l_i-1}(r)$ as the stable states set such that $(Blk_{p_i}(L))^\tau\delta_{2^n}^i$ can be represented by the elements in $S_{l_i-1}(r)$, then it can be stabilized to $\delta_{2^n}^r$. We show it by induction.

When $l_i=2$, $\delta_{2^n}^i\in S_2(r)\backslash S_1(r)$, if we find the corresponding p_i, such that $(Blk_{p_i}(L))^\tau\delta_{2^n}^i=\Sigma_{j=1}^k m_j\delta_{2^n}^j$, where $\delta_{2^n}^j\in S_1(r)$ and $\Sigma_{j=1}^k m_j=1$. Then $x(2\tau)=(LHx(\tau))^\tau(LHx(0))^\tau x(0)=\Sigma_{j=1}^k m_j(LH\delta_{2^n}^j)^\tau\delta_{2^n}^j=\Sigma_{j=1}^k m_j\delta_{2^n}^r=\delta_{2^n}^r$.

Suppose that for $l_i=k$, $\delta_{2^n}^i\in S_k(r)\backslash S_{k-1}(r)$, we have found the corresponding p_i, such that $(Blk_{p_i})^\tau\delta_{2^n}^i$ can be represented by the elements in $S_{k-1}(r)$, and $P[x(k\tau)=\delta_{2^n}^r|x(0)=\delta_{2^n}^i, u(t)=Hx(t_l), t_l\leq t<t_{l+1}]=1$. Then for $l_i=k+1$, $\delta_{2^n}^i\in S_{k+1}(r)\backslash S_k(r)$. If there exists p_i such that $(Blk_{p_i}(L))^\tau\delta_{2^n}^i$ can be represented by the elements in $S_k(r)$, then we have $x((k+1)\tau)=(LHx(k\tau))^\tau\cdots(Blk_{p_i}(L))^\tau\delta_{2^n}^i$. Denote by $(Blk_{p_i})^\tau\delta_{2^n}^i=\Sigma_{j=1}^k m_j\delta_{2^n}^j$, where $\Sigma_{j=1}^k m_j=1$ and $\delta_{2^n}^j$ are elements in $S_k(r)$. Then $P[x((k+1)\tau)=\delta_{2^n}^r|x(0)=\delta_{2^n}^i, u(t)=Hx(t_l), t_l\leq t<t_{l+1}]=\Sigma_{j=1}^k m_jP[x((k)\tau)=\delta_{2^n}^r|x(0)=\delta_{2^n}^j, u(t)=Hx(t_l), t_l\leq t<t_{l+1}]=1$. □

Remark 4.1 When $\tau=1$, the SDSFC reduces to state feedback controller. Then the condition $(Blk_{p_i}(L))\delta_{2^n}^i=\delta_{2^n}^r$ holds for $l_i=1$ implies that $Col_{(p_i-1)2^n+i}(L)=\delta_{2^n}^r$. Since $\delta_{2^n}^r\in S_1(r)$, $\alpha_{(p_r-1)2^n+r}=\delta_{2^n}^r$ can be obtained. When $\tau>1$, this argument is not enough, which means that $\alpha_{(p_r-1)2^n+r}=\delta_{2^n}^r$ cannot be guaranteed. The following example illustrates this point.

Let $\tau=2$, $m=1$, $n=2$, $r=1$ and

$$L=\begin{bmatrix}0&0&1&1&0&\frac{1}{3}&0&1\\0&0&0&0&1&0&1&0\\\frac{1}{2}&1&0&0&0&\frac{2}{3}&0&0\\\frac{1}{2}&0&0&0&0&0&0&0\end{bmatrix}.$$

It can be calculated that

$$Blk_1(L)^2=\begin{bmatrix}1&1&0&0\\0&0&0&0\\0&0&\frac{1}{2}&\frac{1}{2}\\0&0&\frac{1}{2}&\frac{1}{2}\end{bmatrix},$$

then we have $(Blk_1(L))^2\delta_4^1=\delta_4^1$. That is to say $x(2k)=\delta_4^1$, $k=0,1,\ldots$, while $x(2k+1)\neq\delta_4^1$. It implies that $\alpha_{(p_r-1)2^n+r}=\delta_{2^n}^r$ is not necessarily guaranteed by $(Blk_{p_i}(L))^\tau\delta_{2^n}^i=\delta_{2^n}^r$ for $\tau>1$.

Remark 4.2 From Theorem 4.2, for a given state δ_{2n}^r to be stabilized and sampling period τ, the process of designing the controller $H = [p_1\ p_2\ \cdots\ p_{2^n}]$ can be summarized as follows. First split L into 2^m equal blocks, and calculate $(Blk_i(L))^\tau$ for $i = 1, 2, \ldots, 2^m$. Then find the possible p_r set $N_r = \{p_r | Col_{(p_r-1)2^n+r}(L) = \delta_{2^n}^r\}$.

Next for $\delta_{2^n}^i \in S_1 \backslash \{\delta_{2^n}^r\}$, find the corresponding p_i set

$$N_i = \{p_i | (Blk_{p_i}(L))^\tau \delta_{2^n}^i = \delta_{2^n}^r\}.$$

For $\delta_{2^n}^i \in S_k(r) \backslash S_{k-1}(r),\ k \geq 2$, denote by

$$N_i = \left\{p_i | ((Blk_{p_i}(L))^\tau \delta_{2^n}^i) \circ (1_{2^n} - \Sigma_{a \in S_{k-1}(r)} a) = 0_{2^n}\right\}.$$

Based on the procedure for the SDSFC design, the computational complexity of it can be calculated. It needs 2^m times to find the fixed point, and then for each block of L, τ steps will be needed to get $(Blk_i(L))^\tau$. In order to find the sets $S_1(r), S_2(r), \ldots$, it needs $2^m(2^n - 1) + 2^m(2^n - 2) + \cdots + 2^m$. So the total complexity will be $2^m(2^{2n-1} + 2^{n-1} + \tau + 1)$. Let $Z = |N_1| \times |N_2| \times \cdots \times |N_{2^n}|$, then Z is the number of the least steps that the system can be stabilized.

Remark 4.3 The process of designing the controller above degenerates the method in [2] since the probabilistic matric can be degraded into a determined one. Besides, compared with the state feedback controller in [1], the SDSFC may reduce the controller steps in some case. The following example can illustrate it.

Remark 4.4 The results of the output feedback control has been well considered for BCN in [3]. According to some methods presented in [3], the results of Theorem 4.2 above can also be extended to the case of a sampled-data output feedback controller. As a result, we can consider a PBCN as follows,

$$\begin{cases} Ex(t+1) = Lu(t)Ex(t) \\ Ey(t) = HEx(t) \end{cases} \tag{4.7}$$

and the sampled-data output feedback controller is described as

$$u(t) = Ky(t), \quad t_l \leq t < t_{l+1}, \tag{4.8}$$

where $x(t) \in \Delta_{2^n},\ y(t) \in \Delta_{2^p},\ u(t) \in \Delta_{2^m},\ L \in \Upsilon_{2^n \times 2^{m+n}}$, and $H_{2^p \times 2^n},\ K_{2^m \times 2^p}$ are logical matric.

Assume $H = \delta_{2^p}[h_1\ h_2\ \cdots\ h_{2^n}]$ and $K = \delta_{2^m}[k_1\ k_2\ \cdots\ k_{2^p}]$, then (4.8) can be rewritten as

$$\begin{aligned} u(t) &= \delta_{2^m}[k_1\ k_2\ \cdots\ k_{2^p}]\delta_{2^p}[h_1\ h_2\ \cdots\ h_{2^n}]x(t), \\ &= \delta_{2^m}[k_{h_1}\ k_{h_2}\ \cdots\ k_{h_{2n}}]x(t). \end{aligned}$$

Let $x^* = \delta_{2n}^r$ be the desired stabilized state. If there is a sampled-data output feedback controller law (4.8) such that PBCN (4.7) can be globally stabilized to

x^* with probability one, then for every initial state $\delta_{2^n}^i$, $i = 1, \ldots, 2^n$, there exists a unique integer $1 \leq l_i \leq N$, such that $\delta_{2^n}^i \in S_{l_i}(r) \backslash S_{l_i-1}(r)$ with $S_0(r) = \varnothing$. Moreover, let k_i be the solution of

$$\begin{cases} \alpha_{(k_{h_r}-1)2^n+r} = \delta_{2^n}^r, \\ (Blk_{k_{h_i}}(L))^\tau \delta_{2^n}^i = \delta_{2^n}^r, \quad l_i = 1, \\ \left[(Blk_{k_{h_i}}(L))^\tau \delta_{2^n}^i\right] \circ (\mathbf{1}_{2^n} - \Sigma_{a \in S_{l_i-1}(r)} a) = \mathbf{0}_{2^n}, \quad l_i \geq 2, \end{cases}$$

then sampled-data output feedback controller can be determined by

$$K = \delta_{2^m}[k_1 \ \cdots \ k_{2^p}].$$

4.1.3 Examples

Next we consider a 7-gene network containing the genes WNT5A, pirin, S100P, RET1, MART1, HADHB and STC2 [4]. There are four figures in [4], which represents the four highly probable Boolean networks that are used as the constituent Boolean networks in the PBN, with their selection probabilities based on their Bayesian scores. What's more in [4], the pirin gene has been chosen as a control gene to the designed PBN. Assume $WNT5A = x_1$, $pirin = u$, $S100P = x_2$, $RET1 = x_3$, $MART1 = x_4$, $HADHB = x_5$, $STC2 = x_6$, here we consider a PBN constructed by two of the figures with probability p_1 and p_2, satisfying $p_1 + p_2 = 1$.

The Boolean network [4].

$$\begin{aligned} x_1^+ &= x_5 \vee (x_4 \neg x_5 \wedge \neg x_6) \\ x_2^+ &= u \rightarrow (x_4 \wedge x_6 \wedge u) \\ x_3^+ &= (x_5 \wedge x_6 \wedge \neg x_7) \vee (x_5 \wedge \neg x_6) \\ x_4^+ &= [x_1 \wedge (x_6 \vee \neg x_6 \wedge x_7)] \vee \\ &\quad \neg x_1 \wedge [(x_6 \wedge x_7) \vee (\neg x_6 \wedge \neg x_7)] \\ x_5^+ &= x_1 \rightarrow [x_1 \wedge (u \rightarrow x_7 \wedge u)] \\ x_6^+ &= [x_1 \rightarrow (x_1 \wedge \neg x_2)] \vee (x_1 \wedge x_5 \wedge u) \end{aligned} \tag{4.9}$$

The Boolean network [4].

$$\begin{aligned} x_1^+ &= x_5 \\ x_2^+ &= u \rightarrow (x_4 \wedge \neg x_6 \wedge u) \\ x_3^+ &= x_5 \wedge (x_6 \rightarrow x_6 \wedge \neg x_7) \\ x_4^+ &= (x_6 \wedge x_7) \vee (\neg x_6 \wedge \neg x_7) \\ x_5^+ &= [x_1 \rightarrow (x_1 \wedge \neg u)] \vee (x_1 \wedge x_7 \wedge u) \\ x_6^+ &= u \rightarrow \{[x_5 \vee (\neg x_5 \wedge \neg x_6)] \wedge u\} \end{aligned} \tag{4.10}$$

Then we are going to design the state feedback control pirin gene u with sampling period $\tau = 3$ to stabilize the PBN to state δ_{64}^{25} with probability one. For the PBN system constructed by (4.9) and (4.10), their transition matrices L_2 and L_3 can be calculated as follows,

$$L_2 = \delta_{64}[\underbrace{25\ 19\ 1\ 7\ 58\ 60\ 14\ 16\ 25\ 19\ 17\ 23\ 58\ 60\ 58\ 64\ \cdots}_{32}$$
$$\underbrace{25\ 21\ 5\ 1\ 57\ 61\ 13\ 9\ 25\ 21\ 21\ 17\ 57\ 61\ 29\ 25\ \ldots}_{32}$$
$$\underbrace{9\ 1\ 1\ 5\ 41\ 41\ 9\ 13\ \cdots}_{32}\ \underbrace{9\ 5\ 5\ 1\ 41\ 45\ 13\ 9\ \cdots}_{32}].$$

$$L_3 = \delta_{64}[\underbrace{25\ 19\ 1\ 7\ 57\ 59\ 41\ 47\ 25\ 19\ 17\ 23\ 58\ 60\ 57\ 59\ \cdots}_{32}$$
$$\underbrace{9\ 7\ 5\ 3\ 42\ 48\ 45\ 43\ 9\ 23\ 5\ 19\ 42\ 64\ 45\ 59\ \cdots}_{32}$$
$$\underbrace{25\ 3\ 1\ 7\ 57\ 59\ 57\ 63\ \cdots}_{32}\ \underbrace{25\ 7\ 5\ 3\ 57\ 63\ 61\ 59\ \cdots}_{32}].$$

Thus we have

$$Ex(t+1) = (p_1 L_2 + p_2 L_3)u(t)Ex(t). \tag{4.11}$$

Let $\widetilde{L}_2 = [(Blk_1(L_2))^3\ (Blk_2(L_2))^3]$ and $\widetilde{L}_3 = [(Blk_1(L_3))^3\ (Blk_2(L_3))^3]$, then $\widetilde{L}_2$ and $\widetilde{L}_3$ can be obtained respectively as follows,

$$\widetilde{L}_2 = \delta_{64}[\underbrace{25\ 25\ 25\ 60\ 58\ 25\ 17\ 25\ 25\ 25\ 25\ 60\ 58\ 25\ 58\ 25\ \cdots}_{32}$$
$$\underbrace{25\ 21\ 21\ 25\ 25\ 25\ 21\ 25\ \cdots}_{32}\underbrace{9\ 9\ \cdots\ 9\ 9}_{64}].$$

$$\widetilde{L}_3 = \delta_{64}[\underbrace{25\ 25\ 25\ 9\ 25\ 57\ 25\ 42\ 25\ 25\ 25\ 9\ 41\ 1\ 25\ 57\ \cdots}_{32}$$
$$\underbrace{25\ 9\ 9\ 25\ 41\ 5\ 23\ 57\ \cdots}_{32}\ \underbrace{25\ 25\ 25\ 25\ 25\ 57\ 25\ 57\ \cdots}_{40}$$
$$25\ 25\ 25\ 25\ 25\ 57\ 25\ 1\ \underbrace{25\ 25\ 25\ 25\ 25\ 57\ 25\ 57\ \cdots}_{16}].$$

Based on $p_1\widetilde{L}_2 + p_2\widetilde{L}_3$ we can get

$$\begin{aligned} S_0(25) &= \{\ 25\} \\ S_1(25) &= \{\ 1, 2, 3, 9, 10, 11, 17, 18, 19, 26, 27, 33, 36, 41, 44, 49, 52, 57, 60\} \\ S_2(25) &= \{\ 4, 5, 6, 7, 8, 12, 13, 14, 15, 16, 20, 21, 22, 23, 24, 28, 29, 30, 31, 32, \\ &\quad 34, 35, 37, 38, 39, 40, 42, 43, 45, 46, 47, 48, 50, 51, 53, 54, 55, 56, 58, \\ &\quad 59, 61, 62, 63, 64\}. \end{aligned} \tag{4.12}$$

Meanwhile, the control u can be design as follows:

$$u(t) = \delta_2[\underbrace{1\ 1\ 1\ 1\ 2\ 2\ 2\ \cdots\ 1\ 1\ 1\ 2\ 2\ 2\ 2}_{32}\underbrace{1\ 2\ 2\ 1\ 2\ 2\ 2\ 2\ \cdots\ 1\ 2\ 2\ 1\ 2\ 2\ 2\ 2}_{32}]x(t), \tag{4.13}$$

the logical relationship is

$$u^+ = (x_1 \wedge x_4) \vee \{(\neg x_1 \wedge x_4) \wedge [(x_5 \wedge x_6) \vee (\neg x_5 \wedge \neg x_6)]\}.$$

4.2 Sampled-Data Partial Stabilization of Probabilistic Boolean Control Networks

In this section, we investigate the partial stabilization problem of PBCNs under SDSFC with a control Lyapunov function approach. First, the probability structure matrix of the considered PBCN is represented by a Boolean matrix, based on which, a new algebraic form of the system is obtained. Second, we convert the partial stabilization problem of PBCNs into the global set stabilization one. Third, we define control Lyapunov function and its structural matrix under SDSFC. It is found that the existence of a control Lyapunov function is equivalent to that of SDSFC. Then, a necessary and sufficient condition is obtained for the existence of control Lyapunov function under SDSFC, based on which, all possible sample-data state feedback controllers and corresponding structural matrices of control Lyapunov function are designed by two different methods.

4.2.1 Problem Formulation

Consider a PBCN with n nodes, m control inputs and s sub-networks as

$$\begin{cases} x_1(t+1) = f_1(x_1(t), \ldots, x_n(t), u_1(t), \ldots, u_m(t)), \\ x_2(t+1) = f_2(x_1(t), \ldots, x_n(t), u_1(t), \ldots, u_m(t)), \\ \vdots \\ x_n(t+1) = f_n(x_1(t), \ldots, x_n(t), u_1(t), \ldots, u_m(t)), \end{cases} \tag{4.14}$$

where $x_i \in \mathcal{D}$, $i = 1, 2, \ldots, n$ are logical states, $u_i \in \mathcal{D}$, $i = 1, 2, \ldots, m$ are control inputs, respectively. As a logical function, f_i represents the chosen one from a known finite set of Boolean functions $F_i = \{f_i^1, f_i^2, \ldots, f_i^{m_i}\}$. Assume $f_i^j : \mathcal{D}^{m+n} \to \mathcal{D}$, $i = 1, \ldots, n$, $j = 1, \ldots, m_i$ is chosen with probability $\mathbf{P}(f_i = f_i^j) = \lambda_i^j$, $j = 1, 2, \ldots, m_i$, where $\sum_{j=1}^{m_i} \lambda_i^j = 1$. Obviously, there are total $s = \prod_{i=1}^{n} m_i$ sub-systems. Let $\upsilon_\alpha = \{f_1^{\alpha_1}, f_2^{\alpha_2}, \ldots, f_n^{\alpha_n}\}$ denotes the α-th sub-system, then probability of choosing the α-th sub-system is $P_\alpha = \prod_{j=1}^{n} \lambda_j^{\alpha_j}$. Let $x(t) = \ltimes_{i=1}^{n} x_i(t)$ and $u(t) = \ltimes_{i=1}^{m} u_i(t)$, the algebraic form of system (4.14) is expressed as

$$\begin{cases} x_1(t+1) = L_1 u(t) x(t), \\ x_2(t+1) = L_2 u(t) x(t), \\ \vdots \\ x_n(t+1) = L_n u(t) x(t), \end{cases} \tag{4.15}$$

where L_i, $i \in [1, n]$ are logical matrices and can be chosen from a corresponding matrix set $\{L_i^1, L_i^2, \ldots, L_i^{m_i}\}$.

When the value of $x_i(t)$, $i = 1, 2, \ldots, n$ has been decided, the value of $x(t) = \ltimes_{i=1}^{n} x_i(t)$ is confirmed. Define $x^i(t)$, $i = 1, 2, \ldots, 2^n$ as $x(t)$ with different values. If $i \neq j$, $x^i(t) \neq x^j(t)$.

Let $\mathbb{E}x$ represents the overall expectation of x. Multiply the equations in (4.15) together and generates the following algebraic form eventually as

$$\mathbb{E}x(t+1) = \bar{L} u(t) \mathbb{E}x(t), \tag{4.16}$$

where $\bar{L_i} = \sum_{j=1}^{m_i} \lambda_i^j L_i^j$ and $\bar{L} = \bar{L}_1 * \bar{L}_2 * \cdots * \bar{L}_n$ is a matrix when considering the overall expected value of x.

The sampled-data state feedback law for PBCN (4.14) is determined as follows:

$$\begin{cases} u_1(t) = e_1(x_1(t_l), \ldots, x_n(t_l)), \\ u_2(t) = e_2(x_1(t_l), \ldots, x_n(t_l)), \\ \vdots \\ u_m(t) = e_m(x_1(t_l), \ldots, x_n(t_l)), \end{cases} \quad t_l \leq t < t_{l+1}, \tag{4.17}$$

where $e_j : \mathcal{D}^n \to \mathcal{D},\ j = 1, 2, \ldots, m$ are the Boolean functions, constant sampling period $\tau := t_{l+1} - t_l \in \mathbb{Z}_+$, sampling instants $t_l := l\tau \geq 0,\ l = 0, 1, \ldots$. Similarly, the algebraic form of control (4.17) is expressed as

$$u(t) = Hx(t_l), \quad t_l \leq t < t_{l+1}, \tag{4.18}$$

where $H = H_1 \ltimes_{j=2}^{m} [(I_{2^n} \otimes H_j)\Phi_n] \in \mathcal{L}_{2^m \times 2^n}$.

It should be noted that when $\tau = 1$, controller (4.18) can be seen as a conventional state feedback one.

In PBCN (4.14) with SDSFC (4.18), we define $X(t) := (x_1(t), x_2(t), \ldots, x_n(t))$. The state of the first r $(1 \leq r \leq n)$ elements for X can be expressed by $X_r(t) := (x_1(t), x_2(t), \ldots, x_r(t))$, and the rest of elements of X can be denoted as $X_{n-r}(t) := (x_{r+1}(t), x_{r+2}(t), \ldots, x_n(t))$. Thus $X = (X_r, X_{n-r})$, where $X_r \in \mathcal{D}^r$ and $X_{n-r} \in \mathcal{D}^{n-r}$. Let $X(t; X_0, u)$ denotes the trajectory of PBCN (4.14) for a certain initial state $X_0 \in \mathcal{D}_n$ under a control u. Similarly, $X_r(t; X_0, u)$ represents the trajectory of the first r nodes of PBCN (4.14) for $X_0 \in \mathcal{D}_n$ under a control u.

Definition 4.2 For a given state $x_r = \ltimes_{i=1}^{r} x_i(t) \in \Delta_{2^r}$, the PBCN is said to be partially stabilized to x_r with probability 1, if for the given initial state $x(0) = \ltimes_{i=1}^{n} x_i(0) \in \Delta_{2^n}$, there exist an integer T and a control sequence u such that $\mathbf{P}(X_r(t) = x_r | x(0), u) = 1,\ \forall t \geq T$.

Definition 4.3 A set $S^* \subseteq S$ is called the sampled point set of PBCN (4.14) with sample period τ, if for any $x(0) \in S^*$, there exists a SDSFC (4.18) such that $x(1) \in S, x(2) \in S, \ldots, x(\tau) \in S$. It is worth noticing that $x(i)$ may hold more than one value with a certain probability for any $i \in [1, \tau]$.

Definition 4.4 A set $\widetilde{S} \subseteq S^*$ is called the sampled point control-invariant set with sample period τ, if for any $x(t_l) \in S^*$, there exists SDSFC (4.18) such that $x(t_l + 1) \in S, x(t_l + 2) \in S, \ldots, x(t_{l+1}) \in S^*$. A set $\widetilde{S^*}$ is called the largest sampled point control-invariant set if it contains the largest number of elements of $\widetilde{S}$.

Definition 4.5 PBCN (4.14) is called to be S-stabilization with probability 1 for a given set $S \subseteq \Delta_{2^n}$, if for any initial state $x(0) \in \Delta_{2^n}$, there exist a positive integer $T \geq 0$ and a control sequence u such that $\mathbf{P}(x(t) \in S | x(0), u) = 1,\ \forall t \geq T$.

If PBCN (4.14) is partially stabilized to $x_r = \delta_{2^r}^{\lambda}$ under SDSFC (4.18), the partial stabilization problem can be converted into set stabilization problem and the stabilization set is $S = \{\delta_{2^r}^{\lambda} \ltimes \delta_2^{i_1} \ltimes \cdots \ltimes \delta_2^{i_{n-r}}\}$, where $i_1, i_2, \ldots, i_{n-r} \in \{1, 2\}$.

Remark 4.5 Without loss of generality, we consider the partial stabilization with respect to the first r nodes. In fact, through a certain coordinate transformation $z = Ax$, the r nodes can be converted into the first r positions. Therefore, without loss of generality, we only study the first r nodes for the partial stabilization problem.

4.2.2 Main Results

Considering PBCN (4.16) with SDSFC (4.18), the following system is obtained,

$$\mathbb{E}x(t+1) = \bar{L}u(t)\mathbb{E}x(t) = \bar{L}H\mathbb{E}x(t_l)\mathbb{E}x(t), \quad t_l \le t < t_{l+1}, \tag{4.19}$$

where $t_{l+1} - t_l = \tau$ and $\mathbb{E}x(t) \in S$ means the overall expectation of x belongs to S. Foremost, an algorithm is given to discover the sample point control invariant set of S.

Algorithm 18 Search for the sampled point set S^* of set S

Input: $S^* := S$.
for each state $\delta_{2^n}^i \in S$ **do**
    for cach integer $1 \le j \le 2^m$ **do** $u := \delta_{2^m}^j$
        $Flag := 0$
        for each integer $1 \le k \le \tau - 1$ and an integer τ **do**
        $(Lu)^k x := x_k$ and $(Lu)^\tau x := x^*$
            if $x_k \in S$ or $x^* \in S^*$ **then**
                $Flag := 1$
            end if
        end for
        if Flag:=1 **then**
            return $S^* := S \backslash \{\delta_{2^n}^i\}$.
        end if
    end for
end for

For PBCNs, consider the S-stabilization with probability 1, for any $x(0) \in \Delta_{2^n}$ only when there exists an integer k such that an initial state enters $\widetilde{S}^*$ in $k\tau$ steps, it can always stay in S after $k\tau$ steps from Definition 4.4.

Considering S-stabilization problem of PBCN (4.19) with $\rho = 1$, $\bar{L}$ can be substituted for L, which replaces all the nonzero elements of $\bar{L}$ into 1. Actually, for any $x(0) \in \Delta_{2^n}$, a maximum integer k satisfies that $\mathbb{E}x(t) = (\bar{L}u)^{k\tau}x(0) = \alpha_1 x^1(t) + \alpha_2 x^2(t) + \cdots + \alpha_k x^k(t)$, $1 \le k \le 2^n$ with $\sum_{i=1}^k \alpha_i = 1$, $0 \le \alpha_i \le 1$, which means $x(0)$ can reach $x^1(t)$ with probability α_1 in $k\tau$ steps under the control u. As considering the S-stabilization problem of PBCN (4.19) with $\rho = 1$, a necessary and sufficient condition of S-stabilization is obtained, that is $x^i(t) \in \widetilde{S}^*$, $i = 1, 2, \ldots, k$, which means that one can replace L by $\bar{L}$ regardless of probability α_i.

Obviously, through the substitution above, the calculation process can be simplified greatly, but the expression (4.19) of system (4.14) should be changed at the same time. Next, define $[x(t)] = x^1(t) +_{\mathcal{B}} \cdots +_{\mathcal{B}} x^k(t)$ as a column including the BN addition of all states belong to $M = \{x^i(t), i \in [1, k]\}$. Besides, for any state $x \in M$, $x \wedge [x(t)] = x$. Evidently, $[x(t)]$ can be split to a set $\{x^i(t), i \in [1, k]\}$. The algebraic form of system (4.19) here is presented as

$$[x(t+1)] = Lu(t)[x(t)], \quad t_l \leq t < t_{l+1}, \tag{4.20}$$

for $\forall x^i \in M$, we have $Lu(t)x^i \wedge [x(t+1)] = Lu(t)x^i$, define $Lu(t)x^i = [x^i(t+1)]$, thus, $[x^i(t+1)] \wedge [x(t+1)] = [x^i(t+1)]$.

Next, we account for the relation of system (4.20) and system (4.19). For any initial state $x(0) \in \Delta_{2^n}$, we assume $\mathbb{E}x(1) = (Lu)^\tau x(0) = \sum_{i=1}^{k} \alpha_i x^i(1)$, according to the construction of L from the probability transition matrix $\bar{L}$, it is evident that $[x(1)] = x^1(1) \vee \cdots \vee x^k(1)$. System (4.19) is S-stabilized if and only if for any $x(0) \in \Delta_{2^n}$, a maximum integer l exists such that $\mathbb{E}x(l) = (Lu)^{l\tau} x(0) = \sum_{i=1}^{m} \beta_i x^i(l)$ with $x^i(l) \in \widetilde{S}^*$, $i = 1, 2, \ldots, m$. Equivalently, $[x(l)] = x^1(l) \vee \cdots \vee x^m(l)$. Thus, $x^i(l) \in \widetilde{S}^*$ is equivalent to $[x(l)] \wedge [\widetilde{S}^*] = [x(l)]$, where $[\widetilde{S}^*]$ denotes the BN addition of all states belong to $\widetilde{S}^*$. Obviously, we can also split $[x(l)]$ into the set $\{x^i(l),\ i = 1, 2, \ldots, m\}$, that is, $[x(l)]$ and $\{x^i(l),\ i = 1, 2, \ldots, m\}$ can translate each other.

Then, we give the definition of control Lyapunov function of PBCNs.

Definition 4.6 For a given set S and corresponding $\widetilde{S}^*$, if there exists a mapping $\Psi(x) : \Delta_{2^n} \to \sum_i R_i$ satisfying.

(I) For $\forall x^0 \in \widetilde{S}^*$ and $x^0 \wedge [x(t)] = x^0$, there exist a sequence of states $\{x_i, i \in [1, l]\} \in \widetilde{S}^*$, a sequence of corresponding states $\{x_{k_j}, k \in [1, \tau - 1], j \in [1, n]\} \in S$, and a control u^* such that

$$\Psi((Lu^*)^\tau x^0) - \Psi([x^0(t+1)]) = 0$$

and

$$\Psi((Lu^*)^k x_0) - \Psi([x_k^0]) = 0$$

for any $k \in [1, \tau)$. Here, $[x^0(t+1)] \wedge [x(t+1)] = [x^0(t+1)]$, $[x_k^0] \wedge [x_k] = [x_k^0]$ and $\Psi((Lu^*)^\tau [x(t)]) - \Psi([x(t+1)]) = 0$, $\Psi((Lu^*)^k [x(t)]) - \Psi([x_k]) = 0$;

(II) For $\forall x^i \notin \widetilde{S}^*$, $x^i \wedge [x(t)] = x^i$, there exists a control u such that

$$\Psi((Lu)^\tau x^i) - \Psi(x^i) > 0.$$

Then, the mapping $\Psi(x) : \Delta_{2^n} \to \sum_i R_i$ is called a control Lyapunov function of PBCN (4.20).

Definition 4.6 is motivated by a control Lyapunov function design to state-feedback control of BCNs [5].

Especially, when the PBCN reduces to BCN in system (4.14), the definition of control Lyapunov function of BCNs can be rewritten as follows.

Definition 4.7 (Control Lyapunov Function of BCNs) For the known set S and corresponding set $\widetilde{S}^*$, if there exists a mapping $\Psi(x) : \Delta_{2^n} \to R$ satisfying:

(I) For any $x_0 \in \widetilde{S}^*$, there exists a state $x^* \in \widetilde{S}^*$, a sequence of corresponding states $\{x_k, k \in [1, \tau - 1]\} \in S$, and a control u^* such that $\Psi((Lu^*)^\tau x_0) - \Psi(x^*) = 0$ and $\Psi((Lu^*)^k x_0) - \Psi(x_k) = 0$ for any $k \in [1, \tau)$;

(II) For any state $x \notin \widetilde{S}^*$, there exists a control u such that

$$\Psi((Lu)^\tau x) - \Psi(x) > 0.$$

The mapping $\Psi(x) : \Delta_{2^n} \to \sum_i R_i$ is called a control Lyapunov function of BCN.

Next, we are intended to show the great importance of control Lyapunov function for such a class of problems about controllers, such as stabilization problem and control design. First, considering PBCN (4.20) with SDSFC (4.18), when $1 \leq t \leq \tau$, we can get

$$\begin{aligned}
[x(1)] &= LHx(0)x(0) = LHW_{[2^n]}x(0)x(0) \\
&= LHW_{[2^n]}\Phi_n x(0), \\
[x(2)] &= LHx(0)x(1) = LHW_{[2^n]}x(1)x(0) \\
&= (LHW_{[2^n]})^2\Phi_n^2 x(0), \\
&\vdots \\
[x(t)] &= (LHW_{[2^n]})^t\Phi_n^t x(0).
\end{aligned} \tag{4.21}$$

In the same way, when $\tau \leq t \leq 2\tau$, we can get

$$\begin{aligned}
[x(\tau)] &= LHx(\tau)x(\tau) = LHW_{[2^n]}x(\tau)x(\tau) \\
&= LHW_{[2^n]}\Phi_n x(\tau), \\
[x(\tau+1)] &= LHx(\tau)x(\tau+1) = LHW_{[2^n]}x(\tau+1)x(\tau) \\
&= (LHW_{[2^n]})^2\Phi_n^2 x(\tau), \\
&\vdots \\
[x(t)] &= (LHW_{[2^n]})^{t-\tau}\Phi_n^{t-\tau} x(\tau) \\
&= (LHW_{[2^n]})^{t-\tau}\Phi_n^{t-\tau}(LHW_{[2^n]})^\tau\Phi_n^\tau x(0).
\end{aligned} \tag{4.22}$$

When $k\tau \leq t \leq (k+1)\tau$, we can get

$$\begin{aligned}
[x(t)] &= (LHW_{[2^n]})^{t-\tau}\Phi_n^{t-\tau} x(k\tau) \\
&= (LHW_{[2^n]})^{t-k\tau}\Phi_n^{t-k\tau}((LHW_{[2^n]})^\tau\Phi_n^\tau)^k x(0).
\end{aligned} \tag{4.23}$$

Lemma 4.3 *For any $x^i(t) \wedge [x(t)] = x^i(t)$, if there exists a control u, such that $\Psi((Lu)^\tau x^i(t)) - \Psi(x^i(t)) > 0$. Then, $\Psi((Lu)^\tau [x(t)]) - \Psi([x(t)]) > 0$.*

Base on the definition of control Lyapunov function, we assume $\Psi(x) = Gx$, where $G = [\lambda_1\ \lambda_2\ \cdots\ \lambda_{2^n}]$ represents the structural matrix of mapping.

Theorem 4.3 *System (4.20) is globally stabilized to a set S by SDSFC (4.18), if and only if there exists a control Lyapunov function: $\Psi(x) = Gx$.*

Proof (Sufficiency) First, we show that $\Psi\left((Lu^*)^\tau x^0\right) - \Psi\left([x^0(t+1)]\right) = 0$ is equivalent to $(Lu^*)^\tau x^0 = [x^0(t+1)]$. If $(Lu^*)^\tau x^0 = [x^0(t+1)]$, thus $\Psi\left((Lu^*)^\tau x^0\right) - \Psi\left([x^0(t+1)]\right) = 0$ is apparent. When $\Psi\left((Lu^*)^\tau x^0\right) - \Psi\left([x^0(t+1)]\right) = 0$, we assume $(Lu^*)^\tau x^0 \neq [x^0(t+1)]$, let $[y_1] = (Lu^*)^\tau x^0$, then $\Psi\left((Lu^*)^\tau x^0\right) - \Psi\left([y_1]\right) = 0$. For $y_1 \wedge [y_1] = y_1$, there exists a control u_1 s.t. $(Lu_1)^\tau y_1 = [y_2]$, then we have $\Psi\left((Lu_1)^\tau y_1\right) - \Psi\left([y_2]\right) = 0$. In the same way, we have $\Psi\left((Lu_2)^\tau y_2\right) - \Psi\left([y_3]\right) = 0$, $\Psi\left((Lu_3)^\tau y_3\right) - \Psi\left([y_4]\right) = 0, \ldots, \Psi\left((Lu_n)^\tau y_n\right) - \Psi\left([y_{(n+1)}]\right) = 0, \ldots$. According to the second condition of Definition 4.6 and Lemma 4.3, if $x^i \notin \widetilde{S}^*$, we have $\Psi((Lu)^\tau x^i) - \Psi(x^i) > 0$, then $\Psi((Lu)^\tau [x]) - \Psi([x]) > 0$. As the elements of $\widetilde{S}^*$ are limited, there exist states $y_{i_k} \notin \widetilde{S}^*$, $y_{i_k} \wedge [y_i] = y_{i_k}$, $\Psi([y_i]) < \Psi([y_{i+1}]) < \cdots < \Psi([y_n]) < \cdots$, which is incompatible with that the system only has 2^n states. Therefore, $\Psi\left((Lu^*)^\tau x_0\right) - \Psi\left([x^0(t+1)]\right) = 0$ is equivalent to $(Lu^*)^\tau x_0 = [x^0(t+1)]$. Similarly, $\Psi((Lu_1)^k x_0) - \Psi([x_k^0]) = 0$ for any $k \in [1, \tau)$ is equivalent to $(Lu_1)^k x_0 = [x_k^0]$ for any $k \in [1, \tau)$. In other words, when $x(t\tau) \in \widetilde{S}^*$, we have $x((t+1)\tau) \in \widetilde{S}^*$ and $x(t\tau + k) \in S$ for any $k \in [1, \tau)$. That is, for any $x_0 \in \widetilde{S}^*$, there exists a homologous control u, such that $x(t; x_0, u) \in S$, for any $t \geq 1$.

Second, when a state $x_0 \notin \widetilde{S}^*$, we can show that there exists a sequence of $u = (u(0), u(\tau), u(2\tau), \ldots, u(k\tau))$ and an integer $T = k\tau$ such that $x(t; x_0, u) \in \widetilde{S}^*$, $\forall t \geq T$. In fact, for any $T \in N$, if $x(T; x_0, u) \notin \widetilde{S}^*$ holds, let $x(T; x_0, u) = \delta_{2^n}^{i_T}$, $[x(T+1; x_0, u)] = Lu(t_l)\delta_{2^n}^{i_T} = \delta_{2^n}^{i_{T+1}}$. Repeating the procedure, according to the second condition of Definition 4.6, we get $\Psi(\delta_{2^n}^{i_0}) < \Psi(\delta_{2^n}^{i_1}) < \cdots < \Psi(\delta_{2^n}^{i_\alpha}) < \cdots < \Psi(\delta_{2^n}^{i_\beta}) < \cdots$ where $\forall \alpha, \beta \in N$, $i_\alpha \neq i_\beta$, $\alpha \neq \beta$. Obviously, it is a contraction to the fact that system (4.14) only have 2^n states. Above all, for sample point $x_0 \notin \widetilde{S}^*$, there exists an integer $T = k\tau$ and a control sequence u such that $x(T; x_0, u) \in \widetilde{S}^*$. Combining the first part, we can obtain that for any $x_0 \notin \widetilde{S}^*$, there exists a homologous sequence control u, such that $x(t; x_0, u) \in S$, $\forall t \geq T$. Thus, system (4.20) is stabilized to S by the control of SDSFC (4.18).

(Necessity) Now suppose PBCN (4.20) is S-stabilization under SDSFC (4.18). Then, for any $x(0) = \delta_{2^n}^i \in \Delta_{2^n}$, we define c_i as follows:

$$
\begin{cases} 0, & x_0 \in \widetilde{S}^* \\ \min\{k\tau : ((LHW_{[2^n]})^\tau \Phi_n{}^\tau)^k x_0 \in \widetilde{S}^*\}, & x_0 \notin \widetilde{S}^* \end{cases} \tag{4.24}
$$

It can be proved that $\Psi(x) = [\lambda_1\ \lambda_2\ \cdots\ \lambda_{2^n}]x$ is a control Lyapunov function of PBCN (4.20), where $\lambda_i = -c_i,\ i = 1, 2, \ldots, 2^n$. On the one hand, the PBCN is S-stabilization, for all the states $x_0 \in \widetilde{S}^*$ and $x_0 \in [x(t)]$, there exist a sequence of states$\{x_i, i = 1, 2, \ldots, l\} \in \widetilde{S}^*$, a sequence of corresponding states $\{x_{k_j}, k = 1, 2, \ldots, \tau-1, j = 1, 2, \ldots, k\} \in S$, and a control u^* such that $(Lu^*)^\tau x_0 = [x^0(t)]$ and $(Lu^*)^k x_0 = [x_k^0]$ for $\forall k \in [1, \tau)$. Then $\Psi((Lu^*)^\tau x_0) - \Psi([x^0(t)]) = 0$ and $\Psi((Lu^*)^k x_0) - \Psi([x_k^0]) = 0$ are obvious, the condition (i) of Definition 4.6 holds. On the other hand, for $x^i = \delta_{2^n}^i \notin \widetilde{S}^*$, there exist an integer m, a corresponding u and a sequence of states $\{x_j = \delta_{2^n}^{l_j} \in \widetilde{S}^*\}$ such that $[x_j] = (Lu)^{m\tau} x_0 = \{\delta_{2^n}^{l_j}\}$. Naturally, we have $c_i > c_{l_j}$ by the definition of c_i, consequently, $\lambda_i < \lambda_{l_j}$, which means $\Psi([x_j]) - \Psi(x_i) = \Psi([(Lu)^\tau x_i]) - \Psi(x_i) = \lambda_{l_j} - \lambda_i > 0$. Then, condition (ii) of Definition 4.6 can be obtained easily. In all, $\Psi(x) = [\lambda_1\ \lambda_2\ \cdots\ \lambda_{2^n}]x$ is a control Lyapunov function of PBCN (4.20). □

Remark 4.6 Different from conventional state feedback control, the states considered in system (4.20) should act as a sample point. If not, the controlled system may jump out of the set S, though it enters the largest sampled point control invariant set $\widetilde{S}^*$.

Now, we have showed the sample-data state feedback stabilization of PBCNs through the classical control Lyapunov function approach. Next, we will provide two different methods to design sample-data state feedback controller.

Sample-Data State Feedback Controller Design: The First Method
For simplicity, we define the following notations.

- $\{\lambda_\alpha = \lambda_\beta\}|U_{\alpha,\beta}$ means the equality $\lambda_\alpha = \lambda_\beta$ holds only when $\delta_{2^n}^{\lambda_\alpha}, \delta_{2^n}^{\lambda_\beta} \in \widetilde{S}^*$ and $u \in U_{\alpha,\beta}$, where $U_{\alpha,\beta} = \{u : \delta_{2^n}^{\lambda_\alpha} = (Lu)^\tau \delta_{2^n}^{\lambda_\beta}, u \in \Delta_{2^m}\}$.
- $\{\lambda_\alpha > \lambda_\beta\}|U_{\alpha,\beta}$ means the inequality $\lambda_\alpha > \lambda_\beta$ holds only when $u \in U_{\alpha,\beta}$, where $U_{\alpha,\beta} = \{u : \delta_{2^n}^{\lambda_\alpha} = (Lu)^\tau \delta_{2^n}^{\lambda_\beta}, u \in \Delta_{2^m}\}$. Obviously, $\delta_{2^n}^{\lambda_\beta} \notin \widetilde{S}^*$.
- $\Omega_0(\widetilde{S}^*) = \widetilde{S}^*$.
- $\Omega_1(\widetilde{S}^*) = \{\delta_{2^n}^{\lambda_\beta} : \forall \delta_{2^n}^{\lambda_\alpha} \in \widetilde{S}^*\ s.t.\ \{\lambda_\alpha > \lambda_\beta\}|U_{\alpha,\beta}\}$, which means $\forall \delta_{2^n}^{\lambda_\beta} \in \Omega_1(\widetilde{S}^*)$ can enter into $\widetilde{S}^*$ in τ steps.
- $\Omega_2(\widetilde{S}^*) = \{\delta_{2^n}^{\lambda_\beta} : \forall \delta_{2^n}^{\lambda_\alpha} \in \Omega_1(\widetilde{S}^*)\ s.t.\ \{\lambda_\alpha > \lambda_\beta\}|U_{\alpha,\beta}\}$.
- ...
- $\Omega_{k+1}(\widetilde{S}^*) = \{\delta_{2^n}^{\lambda_\beta} : \forall \delta_{2^n}^{\lambda_\alpha} \in \Omega_k(\widetilde{S}^*)\ s.t.\ \{\lambda_\alpha > \lambda_\beta\}|U_{\alpha,\beta}\}$.

Due to the equivalence of the SDSFC of PBCN (4.20) and the control Lyapunov function, we can obtain a necessary and sufficient condition of the existence of a control Lyapunov function.

Theorem 4.4 *System (4.20) has a control Lyapunov function, if and only if the following two conditions are satisfied simultaneously:*

(I) $\widetilde{S}^* \neq \varnothing$;
(II) there exists a positive integer T *such that* $\Omega_T(\widetilde{S}^*) = \Delta_{2^n}$.

Proof (Sufficiency) Assume $\widetilde{S}^*$ is not an empty set and there exists a positive integer T satisfying $\Omega_T(\widetilde{S}^*) = \Delta_{2^n}$, for $\forall x(0) \in \Delta_{2^n}$, there exists a control sequence u such that $x(T\tau) \in \widetilde{S}^*$. Furthermore, if $x(T\tau) \in \widetilde{S}^*$, by Definition 4.4 we can obtain that $x(T\tau + t) \in \widetilde{S}^*$ for $\forall t$. Thus, PBCN (4.20) is S-stabilized under SDSFC. Considering the above-mentioned equivalence, PBCN (4.20) holds a control Lyapunov function.

(Necessity) If PBCN (4.20) holds a control Lyapunov function, obviously, there exists an SDSFC (4.18) to ensure PBCN (4.20) is S-stabilized. Then, for $\forall x(0) \in \Delta_{2^n}$, there exists a positive integer T such that for all $t \geq T$, $x(t) \in S$. Obviously, $\widetilde{S}^* \neq \varnothing$. Next, assume condition (ii) is not satisfied, then $\Omega_T(\widetilde{S}^*) \neq \Delta_{2^n}$. In view of $S-$stabilization, there exists a state $x(t) \notin \widetilde{S}^*$, by Algorithm 18, we cannot find a sequence of control u to make sure x_i can keep in S after x_i entered S, which contradicts with PBCN (4.20) is S-stabilized. Thus, Theorem 4.4 is satisfied. □

Definition 4.8 If there exists a positive integer T, such that $\Omega_T(\widetilde{S}^*) = \Delta_{2^n}$, then $\cup_{i=1}^{T}\Omega_i(\widetilde{S}^*)$ is called an admissible set of control Lyapunov function.

From Theorem 4.4, we can easily know that only an admissible set of probability control Lyapunov function can define the SDSFC. Otherwise, the structural matrix of the SDSFC will lose some values.

Considering $\{\lambda_a = \lambda_b\}|U_{a,b}$, when $\delta_{2^n}^{\lambda_a}, \delta_{2^n}^{\lambda_b} \in \widetilde{S}^*$ and $\{\lambda_a > \lambda_b\}|U_{a,b}$, when $\delta_{2^n}^{\lambda_b} \notin \widetilde{S}^*$. We can obtain $U_j = \{H = \delta_{2^m}[p_1\ p_2\ \cdots\ p_{2^n}], \delta_{2^m}^{p_i} \in U_{i',i}\}$, where $\delta_{2^n}^{i} \in \Omega_l(\widetilde{S}^*)$, $\delta_{2^n}^{i'} \in \Omega_{l-1}(\widetilde{S}^*)$ is one of the sample-data state feedback controls, with its structural matrix of mapping $\Psi(x) : G = [\lambda_1\ \lambda_2\ \cdots\ \lambda_{2^n}]$.

Theorem 4.5 *The set consisting of all the SDSFCs of system (4.20) is* $U = \bigcup_{j=1}^{l} U_j$ *with its structure matrices of mapping* $\Psi(x) : G = [\lambda_1\ \lambda_2\ \cdots\ \lambda_{2^n}]$.

Proof Consider system $[x(t+1)] = Lu(t)[x(t)]$, where $u(t) = Hx(t)$, now, we proof such matrix H belongs to U. As the system is S-stabilized, for any $x(t) = \delta_{2^n}^{i} \notin \widetilde{S}^*$, there exists an integer l_i such that $\delta_{2^n}^{i} \in \Omega_{l_i}(\widetilde{S}^*)$ and at least a state $\delta_{2^n}^{i'} \in \Omega_{l_i-1}(\widetilde{S}^*)$ such that $\{\lambda_{i'} > \lambda_i\}|U_{i',i}$, which means there exists a control $u \in U_{i',i}$ such that $\delta_{2^n}^{i'} = (Lu)^{\tau}\delta_{2^n}^{i}$, i.e., $\delta_{2^n}^{i}$ enters $\widetilde{S}^*$ in $l_i\tau$ steps. Assume $H = \delta_{2^m}[p_1\ p_2\ \cdots\ p_{2^n}]$, then $u = Hx = \delta_{2^m}^{p_i}$. Obviously, $H \in U$. □

Sample-Data State Feedback Controller Design: The Second Method
In addition to the above method, we propose another method to define and solve the control Lyapunov function.

Theorem 4.6 *System (4.20) is globally stabilized to set S via SDSFC (4.18) if and only if there exists a positive integer t such that*

$$Col_i[((LHW_{[2^n]})^{\tau}\Phi_n^{\tau})^t] \in \widetilde{S}^*, \quad i = 1, 2, \ldots, 2^n. \tag{4.25}$$

Proof (Sufficiency) On the basis of the assumed condition above, for any initial state $x(0) \in \Delta_{2^n}$, there exists a positive integer k, such that $x(k\tau) = ((LHW_{[2^n]})^{\tau}$

$\Phi_n^\tau)^k x(0) \in \widetilde{S}^*$. Then for such $x(k\tau) = \delta_{2^n}^{i_k}$, by Definition 4.4, $x(k\tau+1) \in S, x(k\tau+2) \in S, \ldots, x((k+1)\tau) \in \widetilde{S}^*$. Thus, for $\forall t \geq k\tau$, we can obtain that $x(t) \in S$.

(Necessity) Assume that system (4.20) is globally stabilized to S via SDSFC (4.18). If there exist an initial state $x(0)$ and a positive integer k such that $x(k\tau) = ((LHW_{[2^n]})^\tau \Phi_n^\tau)^k\, x(0) \notin \widetilde{S}^*$, a constructive method can show that there may exists a positive integer t such that $x(k\tau)$ may jump out S in t steps though it has entered S before. □

Considering PBCN (4.20) and SDSFC (4.18) with its structural matrix $H = \delta_{2^m}[p_1\ p_2\ \cdots\ p_{2^n}]$.

First, we give a lemma.

Lemma 4.4 $LHW_{[2^n]}(\delta_{2^n}^i)^2 = \delta_{2^n}^{\alpha(p_i-1)2^n+i}$.

Proof Next, we give a brief proof process

$$
\begin{aligned}
LHW_{[2^n]}(\delta_{2^n}^i)^2 &= L(H\delta_{2^n}^i)\delta_{2^n}^i \\
&= L\delta_{2^m}^{p_i}\delta_{2^n}^i = L(\delta_{2^m}^{p_i}\delta_{2^n}^i) \\
&= L\delta_{2^{m+n}}^{(p_i-1)2^n+i} = \delta_{2^n}^{\alpha(p_i-1)2^n+i}.
\end{aligned}
\tag{4.26}
$$

□

Assume that there exists a positive integer k, such that $x(k\tau) = \delta_{2^n}^i$ and $x((k+1)\tau) = (LHW_{[2^n]})^\tau(\Phi_n)^\tau x(k\tau) = \delta_{2^n}^j$. Then, we have

$$
\begin{aligned}
\delta_{2^n}^j &= (LHW_{[2^n]})^\tau \Phi_n^\tau \delta_{2^n}^i \\
&= (LHW_{[2^n]})^\tau (\delta_{2^n}^i)^{\tau+1} \\
&= (LHW_{[2^n]})^{\tau-1}(LHW_{[2^n]}(\delta_{2^n}^i)^2)(\delta_{2^n}^i)^{\tau-1} \\
&= (LHW_{[2^n]})^{\tau-1}\delta_{2^n}^{\alpha(p_i-1)2^n+i}(\delta_{2^n}^i)^{\tau-1} \\
&= (LHW_{[2^n]})^{\tau-1}\delta_{2^n}^{\beta_i^1}(\delta_{2^n}^i)^{\tau-1}.
\end{aligned}
\tag{4.27}
$$

For $1 \leq k \leq \tau - 1$, we have

$$
\begin{aligned}
\delta_{2^n}^j &= (LHW_{[2^n]})^{\tau-k}\delta_{2^n}^{\beta_i^k}(\delta_{2^n}^i)^{\tau-k} \\
&= (LHW_{[2^n]})^{\tau-(k+1)}(LHW_{[2^n]}\delta_{2^n}^{\beta_i^k}\delta_{2^n}^i)(\delta_{2^n}^i)^{\tau-(k+1)} \\
&= (LHW_{[2^n]})^{\tau-(k+1)}\delta_{2^n}^{\alpha(p_i-1)2^n+\beta_i^k}(\delta_{2^n}^i)^{\tau-(k+1)} \\
&= (LHW_{[2^n]})^{\tau-(k+1)}\delta_{2^n}^{\beta_i^{k+1}}(\delta_{2^n}^i)^{\tau-(k+1)}.
\end{aligned}
\tag{4.28}
$$

By induction, we can obtain when $k = \tau - 1$, $\delta_{2^n}^{\beta_i^\tau} = \delta_{2^n}^j$.

Remark 4.7 Because the expression of L does not affect the results of $\delta_{2^n}^j$, the matrix L of PBCNs is expressed as $\delta_{2^n}[\alpha_1\ \alpha_2\ \cdots\ \alpha_{2^{m+n}}]$ here to simplify the derivation process.

Based on the above analysis, we can redefine the control Lyapunov function of PBCNs under SDSFC as follows.

Definition 4.9 (Control Lyapunov Function of PBCNs with Probability $\rho = 1$**)** For the known set S and corresponding set $\widetilde{S}^*$, if there exists a mapping $\Psi(x)$: $\Delta_{2^n} \to \sum_i R_i$ satisfying:

- For any $\delta_{2^n}^i \in \widetilde{S}^*$ and $\delta_{2^n}^i \wedge [x(t)] = \delta_{2^n}^i$, there exists a sequence of states $\{\delta_{2^n}^j, j \in [1, l]\} \in \widetilde{S}^*$ and a sequence of corresponding states $\{\delta_{2^n}^{k_m}, k \in [1, \tau - 1], m \in [1, n]\} \in S$ such that $\Psi(\delta_{2^n}^{\beta_i^\tau}) - \Psi([\delta_{2^n}^j]) = 0$ and $\Psi(\delta_{2^n}^{\beta_i^k}) - \Psi([\delta_{2^n}^{k_m}]) = 0$ for $\forall k \in [1, \tau)$. Here $[\delta_{2^n}^j] \wedge [x(t+1)] = [\delta_{2^n}^j]$ means the Boolean addition of $\delta_{2^n}^j$, $[\delta_{2^n}^{k_m}] \wedge [\delta_{2^n}^k] = [\delta_{2^n}^{k_m}]$ means the Boolean addition of $\delta_{2^n}^{k_m}$ and $(Lu)^\tau[x(t)] = [x(t+1)]$, $(Lu)^k[x(t)] = [\delta_{2^n}^k]$;
- For any state $\delta_{2^n}^i \notin \widetilde{S}^*$, we have $\Psi(\delta_{2^n}^{\beta_i^\tau}) - \Psi(\delta_{2^n}^i) > 0$.

Then, the mapping $\Psi(x) : \Delta_{2^n} \to \sum_i R_i$ is called a control Lyapunov function of PBCN (4.20).

Denote $\Omega_k(\widetilde{S}^*)$ as a set of states which can be driven to $\widetilde{S}^*$ in $k\tau$ steps with a series control sequence, in other words, for any $x(0) \in \Delta_{2^n}$,

$$\begin{cases} \Omega_1(\widetilde{S}^*) = \{x(0) : (LEW_{[2^n]})^\tau (\Phi_n)^\tau x(0) \in \widetilde{S}^*\}, \\ \Omega_{k+1}(\widetilde{S}^*) = \{x(0) : (LEW_{[2^n]})^\tau (\Phi_n)^\tau x(0) \in \Omega_k(\widetilde{S}^*)\}. \end{cases} \tag{4.29}$$

Denote $\widetilde{S}^* = \{\delta_{2^n}^{i_k}, k = 1, 2, \ldots, l\}$ and $\bar{S} = \{i_k, k = 1, 2, \ldots, l\}$. Afterwards, the following conclusion can be obtained directly:

$$\begin{cases} \Omega_1(\widetilde{S}^*) = \{\delta_{2^n}^i : \beta_i^\tau \in \bar{S}, 1 \le i \le 2^n\}, \ with\ corresponding \\ p_i \in [1, 2^m],\ \beta_i^1 = \alpha_{(p_i-1)2^n+i},\ \beta_i^{k+1} = \alpha_{(p_i-1)2^n+\beta_i^k}; \\ \Omega_k(\widetilde{S}^*) = \bigcup\{\delta_{2^n}^i : \delta_{2^n}^{\beta_i^\tau} \in \Omega_{(k-1)}(\widetilde{S}^*), 1 \le i \le 2^n\}. \end{cases} \tag{4.30}$$

Theorem 4.7 *PBCN (4.20) is globally stabilized to set* $S = \{\delta_{2^n}^{i_k}, k = 1, 2, \ldots, l\}$ *via SDSFC (4.18), then we have the following:*

(I) $\Omega_1(\widetilde{S}^*) \backslash \widetilde{S}^* \neq \varnothing$;
(II) There exists an integer $1 \le T \le 2^n$ *such that* $\Delta_{2^n} = \bigcup_{1 \le l \le T} \Omega_l(\widetilde{S}^*) \backslash \Omega_{l-1}(\widetilde{S}^*)$.

Proof

(I) Obviously, $\widetilde{S^*} \subseteq \Omega_1(\widetilde{S^*})$. If $\Omega_1(\widetilde{S^*})\backslash\widetilde{S^*} = \varnothing$, for any $x_i \notin \widetilde{S^*}$, then $x_i \notin \Omega_1(\widetilde{S^*})$, it cannot enter $\widetilde{S^*}$, which contradicts with S-stabilization.
(II) If $\Delta_{2^n} \neq \bigcup_{1\le l\le N} \Omega_l(\widetilde{S^*})\backslash\, \Omega_{l-1}(\widetilde{S^*})$, there exists at least a state

$$x_i \notin \Omega_l(\widetilde{S^*})\backslash\Omega_{l-1}(\widetilde{S^*}),$$

for any $l \in [1, N]$. Evidently, no such control u can steer x_i into $\widetilde{S^*}$. It contradicts with S-stabilization.

□

Based on the second method above, we can obtain the corresponding method to design SDSFC.

Theorem 4.8 *Assume that system (4.20) is globally stabilized to set S via SDSFC (4.18) with its structural matrix* $H = \delta_{2^m}[p_1\ p_2\ \cdots\ p_{2^n}]$, *then for any* $1 \le i \le 2^n$, *there exists a maximum integer* $1 \le l_i \le N$ *such that* $\delta_{2^n}^i \in \Omega_{l_i}(\widetilde{S^*})\backslash\Omega_{l_i-1}(\widetilde{S^*})$ *with* $\Omega_0(\widetilde{S^*}) = \widetilde{S^*}$. *We can obtain*

$$\begin{cases} \beta_i^\tau \in \bar{S}, & \text{for } l_i = 1, i \notin \bar{S}, \\ \delta_{2^n}^{\beta_i^\tau} \in \Omega_{l_i-1}(\widetilde{S^*})\backslash\Omega_{l_i-2}(\widetilde{S^*}), & \text{for } l_i \ge 2, i \notin \bar{S}. \end{cases} \tag{4.31}$$

It is a necessary and sufficient condition to make system (4.20) globally stabilized to set S through SDSFC (4.18).

Proof $\Delta_{2^n} = \bigcup_{1\le l\le N} \Omega_l(\widetilde{S^*})\backslash\Omega_{l-1}(\widetilde{S^*})$ from Theorem 4.7. Considering the probability, for $\forall i \le i \le 2^n$, it is easy to see that there may exist a sequence of integers $1 \le l_{i_j} \le N,\ j = 1, 2, \ldots, q$ such that $\delta_{2^n}^i \in \Omega_{l_i}(\widetilde{S^*})\backslash\Omega_{l_i-1}(\widetilde{S^*})$. It is not hard to choose the maximum l_i to ensure $\delta_{2^n}^i$ can enter into $\widetilde{S^*}$ entirely in $l_i\tau$ steps. When $l_i = 1$, for any $\delta_{2^n}^i \in S$, there exists a possible p_i such that $\delta_{2^n}^i \in \Omega_1(\widetilde{S^*})$. Through the definition of $\beta_i^k, k \ge 1$ in (4.28), it is known that $\alpha_{(p_i-1)2^n+\beta_i^{\tau-1}} \in \{i_k\}$ i.e. $\beta_i^\tau \in \{i_k\}$. Now, we define $G(i) = \{j : \alpha_j = i, 1 \le j \le 2^{m+n}\}, 1 \le i \le 2^n$, which actually means the (r, r)-th element of the matrix L.

Step 1. For $j_1 \in G(\bar{S})$ and $1 \le \beta_i^{\tau-1} \le 2^n$, there exists a $p_i \in [1, 2^n]$, such that $(p_i - 1)2^n + \beta_i^{\tau-1} = j_1$.

Step 2. We have $\beta_i^{\tau-1} = j_1 - (p_i - 1)2^n \in [1, 2^n]$. Considering $\beta_i^{\tau-1} = \alpha_{(p_i-1)2^n+\beta_i^{\tau-2}}$, and assuming $(p_i - 1)2^n + \beta_i^{\tau-1} = j_2$, one can get $j_2 \in G(j_1 - (p_i - 1)2^n)$. Then, $\beta_i^{\tau-2} = j_2 - (p_i - 1)2^n$, where $\beta_i^{\tau-2} \in [1, 2^n]$.

⋮

Step $\tau - 1$. Since $\Omega_1(\widetilde{S^*})\backslash\widetilde{S^*} \neq \varnothing$, there exists at least one $j_{\tau-1} \in G(j_{\tau-1} - (p_i-1)2^n)$. Besides, $j_{\tau-1} = (p_i-1)2^n+\beta_i^1$. One can get $\beta_i^1 = j_{\tau-1}-(p_i-1)2^n$.

Step τ. Since $\beta_i^1 = \alpha_{(p_i-1)2^n+i}$. Solving the equation $\alpha_{(p_i-1)2^n+i} = j_{\tau-1} - (p_i - 1)2^n$, one can get all possible p_i satisying $\beta_i^\tau \in \{i_k\}$.

When $l_i \geq 2$, for any i, considering $\beta_i^\tau \in \Omega_{(l_i-1)}(S) \setminus \Omega_{(l_i-2)}(S)$, and repeating the procedure above, p_i can be obtained in the same way. In all, the control matrices H are obtained. □

Remark 4.8 Here, we choose the maximum l_i not only to ensure that $\delta_{2^n}^i$ can enter into S entirely in $l_i\tau$ steps, but also to find all possible sample-data state feedback controllers as well as corresponding probabilistic control Lyapunov functions.

Next, we are going to solve the matrices H and obtain the matrices G simultaneously. For simplicity, we define the following notations.

- $\{\lambda_i = \lambda_{j_l}\}|_{\{\delta_{2^m}^{p_i}\}}, l \in [1, q]$ means the equality $\{\lambda_i = \lambda_{j_l}, l \in [1, q]\}$ holds only when $\delta_{2^n}^{\lambda_i}, \delta_{2^n}^{\lambda_{j_l}} \in \widetilde{S}^*$ and p_i is the solution to $(p_i - 1)2^n + i = j_l$;
- $\{\lambda_{j_l} \geq \lambda_i\}|_{\{\delta_{2^m}^{p_i}\}}, l \in [1, q]$ means the inequality $\{\lambda_{j_l} > \lambda_i, l = 1, 2, \ldots, q\}$ holds only for $l_i = 1$ and $j_l \in (\bar{S})$, when p_i is the solution to $\beta_i^\tau = j_l$;
- $\{\lambda_{j_l} > \lambda_i\}|_{\{\delta_{2^m}^{p_i}\}}, l = 1, 2, \ldots, q$ which means the equality $\{\lambda_{j_l} > \lambda_i, l \in [1, q]\}$ holds only for $l_i \geq 2$ and $j_l \in \Omega_{l_i-1}(\widetilde{S}^*) \setminus \Omega_{l_i-2}(\widetilde{S}^*)$, when p_i is the solution to $\delta_{2^n}^{\beta_i^\tau} = j_l$.

Considering

$$\{\lambda_i = \lambda_{j_l}\} \mid \{\alpha_{(p_i-1)2^n+i} = j_l\}, l \in [1, q],$$

when $\delta_{2^n}^{\lambda_i}, \delta_{2^n}^{\lambda_{j_l}} \in \widetilde{S}^*$;

$$\{\lambda_{j_l} > \lambda_i\} \mid \{\beta_i^\tau = j_l, l \in [1, q]\},$$

when $l_i = 1$ and $j_l \in \bar{S}$;

$$\{\lambda_{j_l} > \lambda_i\}|\{\delta_{2^n}^{\beta_i^\tau} = j_l, l \in [1, q]\},$$

when $l_i \geq 2$ and $j_l \in \Omega_{l_i-1}(\widetilde{S}^*) \setminus \Omega_{l_i-2}(\widetilde{S}^*)$, from the above procedure, we obtain one of the sample-data state feedback controllers $U_j = \{H = \delta_{2^m}[p_1\ p_2\ \cdots\ p_{2^n}]\}$ with its structure matrix of mapping $\Psi(x) : G = [\lambda_1\ \lambda_2\ \cdots\ \lambda_{2^n}]$.

Theorem 4.9 *The set consisting of all the SDSFC (4.18) of system (4.20) is* $U = \cup_{j=1}^l U_j$.

Proof Consider system $[x(t + 1)] = Lu(t)[x(t)]$, where $u(t) = Hx(t)$, now, we prove H belongs to U. As the system is S-stabilized, for any $x(t) = \delta_{2^n}^i \notin \widetilde{S}^*$, there exists an integer l_i such that $\delta_{2^n}^i \in \Omega_{l_i}(\widetilde{S}^*)$ and at least a state $\delta_{2^n}^{i'} \in \Omega_{l_i-1}(\widetilde{S}^*)$ such that $\{\lambda_{i'} > \lambda_i\}|\{\delta_{2^n}^{\beta_i^\tau} = i'\}$, which means there exists a control $u \in \delta_{2^m}^{p_i}$ such that $\delta_{2^n}^{i'} = (Lu)^\tau \delta_{2^n}^i$, i.e., $\delta_{2^n}^i$ enters $\widetilde{S}^*$ in $l_i\tau$ steps.

Let $H = \delta_{2^m}[p_1\ p_2\ \cdots\ p_{2^n}]$, we can find $u = Hx = \delta_{2^m}^{p_i}$. Obviously, $H \in U$. □

4.2.3 *Example and Simulations*

From the above analysis, the second method is found to be much more complex when the number of control inputs and logical states is large. But when they are small, the second method will be better. For explaining it, we give the following examples.

Consider an apoptosis network [6], which consists of there states (IAP, C3a,C8a) and two inputs. The dynamics of the considered apoptosis network can be described as

$$\begin{cases} x_1(t+1) = f_1(x_1(t), x_2(t), x_3(t), u_1(t), u_2(t)), \\ x_2(t+1) = f_2(x_1(t), x_2(t), x_3(t), u_1(t), u_2(t)), \\ x_3(t+1) = f_3(x_1(t), x_2(t), x_3(t), u_1(t), u_2(t)), \end{cases}$$

where

$$\begin{array}{ll} f_1^1 = \neg x_2(t) \wedge u_1(t), & \mathbf{P}(f_1^1) = 0.3 \\ f_1^2 = x_1(t), & \mathbf{P}(f_1^2) = 0.3 \\ f_2^1 = \neg x_1(t) \wedge (\neg x_3(t)), & \mathbf{P}(f_2^1) = 0.9 \\ f_2^2 = x_2(t), & \mathbf{P}(f_2^2) = 0.1 \\ f_3 = x_2(t) \vee u_1(t) \wedge u_2(t), & \mathbf{P}(f_3) = 1. \end{array} \tag{4.32}$$

Let $S = \{\delta_8^2, \delta_8^4, \delta_8^6, \delta_8^8\}$, and the Boolean models $x(t) = \ltimes_{i=1}^3 x_i(t)$ and $u(t) = \ltimes_{i=1}^2 u_i(t)$, using the semi-tensor product, we have

$$[x(t+1)] = Lu(t)[x(t)], \tag{4.33}$$

where L is

$$\begin{bmatrix} 1\,1\,1\,0\,1\,1\,1\,1\,1\,0\,1\,0\,1\,0\,1\,0\,0\,0\,0\,0\,0\,0\,0\,0\,0\,0\,0\,0\,0\,0\,0\,0 \\ 0\,0\,0\,1\,0\,0\,0\,0\,0\,1\,0\,1\,0\,1\,0\,1\,1\,1\,1\,1\,1\,1\,1\,1\,1\,1\,1\,1\,1\,1\,1\,0 \\ 1\,1\,1\,0\,1\,0\,1\,0\,1\,0\,1\,0\,1\,0\,1\,0\,0\,0\,0\,0\,0\,0\,0\,0\,0\,0\,0\,0\,0\,0\,0\,0 \\ 0\,0\,0\,1\,0\,0\,0\,0\,0\,1\,0\,1\,0\,0\,0\,0\,1\,1\,1\,1\,1\,0\,1\,0\,1\,1\,1\,1\,1\,0\,1\,0 \\ 1\,0\,1\,0\,1\,0\,1\,1\,1\,0\,1\,0\,1\,0\,1\,0\,0\,0\,0\,0\,0\,0\,0\,0\,0\,0\,0\,0\,0\,0\,0\,0 \\ 0\,0\,0\,0\,0\,0\,0\,0\,0\,1\,0\,1\,0\,1\,0\,1\,1\,1\,1\,0\,1\,0\,1\,0\,1\,1\,1\,1\,1\,1\,1\,1 \\ 1\,0\,1\,0\,1\,0\,1\,0\,1\,0\,1\,0\,1\,0\,1\,0\,0\,0\,0\,0\,0\,0\,0\,0\,0\,0\,0\,0\,0\,0\,0\,0 \\ 0\,0\,0\,0\,0\,0\,0\,0\,0\,1\,0\,1\,0\,0\,0\,0\,1\,1\,1\,0\,1\,0\,1\,0\,1\,1\,1\,1\,1\,0\,1\,1 \end{bmatrix}$$

Considering feedback law as follows:

$$u(t) = Hx(t_l), \quad t_l \leq t \leq t_{l+1}. \tag{4.34}$$

Let the feedback matrix H be

$$H = \delta_4[p_1\ p_2\ p_3\ p_4\ p_5\ p_6\ p_7\ p_8].$$

Besides, $\tau = 2$. Next, we construct a sample-data state feedback controller to make system (4.33) stabilized to a set $S = \{\delta_8^2, \delta_8^4, \delta_8^6, \delta_8^8\}$ globally. Then, by Algorithm 18, one can find the largest sampled point set $\widetilde{S}^* = S = \{\delta_8^2, \delta_8^4, \delta_8^6, \delta_8^8\}$. Next, we take δ_8^1, δ_8^3, δ_8^5, δ_8^7 into consideration. Through the same calculation, we can obtain that

$$\begin{aligned}
&\{\lambda_2 = \lambda_2, \lambda_2 = \lambda_4, \lambda_2 = \lambda_6, \lambda_2 = \lambda_8\}|\{\delta_4^2, \delta_4^4\}, \{\lambda_2 = \lambda_2, \lambda_2 = \lambda_4\}|\{\delta_4^3\};\\
&\{\lambda_4 = \lambda_2, \lambda_4 = \lambda_4, \lambda_4 = \lambda_6, \lambda_4 = \lambda_8\}|\{\delta_4^2, \delta_4^4\}, \{\lambda_4 = \lambda_2, \lambda_4 = \lambda_4\}|\{\delta_4^3\};\\
&\{\lambda_6 = \lambda_2, \lambda_6 = \lambda_4, \lambda_6 = \lambda_6, \lambda_6 = \lambda_8\}|\{\delta_4^2, \delta_4^4\}, \{\lambda_6 = \lambda_2, \lambda_6 = \lambda_4\}|\{\delta_4^3\};\\
&\{\lambda_8 = \lambda_2, \lambda_8 = \lambda_4, \lambda_8 = \lambda_6, \lambda_8 = \lambda_8\}|\{\delta_4^2\}, \{\lambda_8 = \lambda_2, \lambda_8 = \lambda_4\}|\{\delta_4^3\},\\
&\{\lambda_8 = \lambda_2, \lambda_8 = \lambda_6, \lambda_8 = \lambda_6\}|\{\delta_4^4\};\\
&\{\lambda_2 > \lambda_1, \lambda_4 > \lambda_1\}|\{\delta_4^3\}, \{\lambda_2 > \lambda_1, \lambda_4 > \lambda_1, \lambda_6 > \lambda_1, \lambda_8 > \lambda_1\}|\{\delta_4^4\};\\
&\{\lambda_2 > \lambda_3, \lambda_4 > \lambda_3, \lambda_6 > \lambda_3, \lambda_8 > \lambda_3\}|\{\delta_4^3, \delta_4^4\};\\
&\{\lambda_2 > \lambda_5, \lambda_4 > \lambda_5, \lambda_6 > \lambda_5, \lambda_8 > \lambda_5\}|\{\delta_4^3, \delta_4^4\};\\
&\{\lambda_2 > \lambda_7, \lambda_4 > \lambda_7, \lambda_6 > \lambda_7, \lambda_8 > \lambda_7\}|\{\delta_4^3, \delta_4^4\}.
\end{aligned} \tag{4.35}$$

According to the definition before, it is found that

$$\Omega_0(\widetilde{S}^*) = \widetilde{S}^* = \{\delta_8^2, \delta_8^4, \delta_8^6, \delta_8^8\},$$

and

$$\Omega_1(\widetilde{S}^*)\backslash\Omega_0(\widetilde{S}^*) = \{\delta_8^1, \delta_8^3, \delta_8^5, \delta_8^7\}.$$

Obviously, $\widetilde{S}^* \neq \varnothing$ and $\Omega_1(\widetilde{S}^*) = \Delta_{2^n}$, system (4.33) can be globally stabilized to the set S by SDSFC with Theorem 4.7.

In the following, it is easy to obtain that the number of all the controllers is 24, we do not list them all, but take one of them for example to solve

$$G = [\lambda_1\ \lambda_2\ \lambda_3\ \lambda_4\ \lambda_5\ \lambda_6\ \lambda_7\ \lambda_8].$$

$$\begin{aligned}
U_1 =\{&\{\lambda_2 = \lambda_2, \lambda_2 = \lambda_4, \lambda_2 = \lambda_6, \lambda_2 = \lambda_8\}|\{\delta_4^2, \delta_4^4\};\\
&\{\lambda_4 = \lambda_2, \lambda_4 = \lambda_4, \lambda_4 = \lambda_6, \lambda_4 = \lambda_8\}|\{\delta_4^2, \delta_4^4\};\\
&\{\lambda_6 = \lambda_2, \lambda_6 = \lambda_4, \lambda_6 = \lambda_6, \lambda_6 = \lambda_8\}|\{\delta_4^2, \delta_4^4\};\\
&\{\lambda_8 = \lambda_2, \lambda_8 = \lambda_4, \lambda_8 = \lambda_6, \lambda_8 = \lambda_8\}|\{\delta_4^2\};\\
&\{\lambda_2 > \lambda_1, \lambda_4 > \lambda_1\}|\{\delta_4^3\};\\
&\{\lambda_2 > \lambda_3, \lambda_4 > \lambda_3, \lambda_6 > \lambda_3, \lambda_8 > \lambda_3\}|\{\delta_4^3, \delta_4^4\};\\
&\{\lambda_2 > \lambda_5, \lambda_4 > \lambda_5, \lambda_6 > \lambda_5, \lambda_8 > \lambda_5\}|\{\delta_4^3, \delta_4^4\};\\
&\{\lambda_2 > \lambda_7, \lambda_4 > \lambda_7, \lambda_6 > \lambda_7, \lambda_8 > \lambda_7\}|\{\delta_4^3, \delta_4^4\}\}.
\end{aligned} \tag{4.36}$$

We can obtain

$$\begin{cases}
\lambda_2 = \lambda_2, \lambda_2 = \lambda_4, \lambda_2 = \lambda_6, \lambda_2 = \lambda_8,\\
\lambda_4 = \lambda_2, \lambda_4 = \lambda_4, \lambda_4 = \lambda_6, \lambda_4 = \lambda_8,\\
\lambda_6 = \lambda_2, \lambda_6 = \lambda_4, \lambda_6 = \lambda_6, \lambda_6 = \lambda_8,\\
\lambda_8 = \lambda_2, \lambda_8 = \lambda_4, \lambda_8 = \lambda_6, \lambda_8 = \lambda_8,\\
\lambda_2 > \lambda_1, \lambda_4 > \lambda_1,\\
\lambda_2 > \lambda_3, \lambda_4 > \lambda_3, \lambda_6 > \lambda_3, \lambda_8 > \lambda_3,\\
\lambda_2 > \lambda_5, \lambda_4 > \lambda_5, \lambda_6 > \lambda_5, \lambda_8 > \lambda_5,\\
\lambda_2 > \lambda_7, \lambda_4 > \lambda_7, \lambda_6 > \lambda_7, \lambda_8 > \lambda_7.
\end{cases} \tag{4.37}$$

Solving inequalities and equalities in (4.37), one solution of G is $\lambda_1 = \lambda_3 = \lambda_5 = \lambda_7 = 1$, $\lambda_2 = \lambda_4 = \lambda_6 = \lambda_8 = 2$, thus, the corresponding control Lyapunov function is

$$\Psi(x) = [1\ 2\ 1\ 2\ 1\ 2\ 1\ 2]x.$$

Finally, all the feasible feedback laws U are given by

$$U = \bigcup_{j=1}^{24} U_j = \{E = \delta_2[p_1\ p_2\ p_3\ p_4\ p_5\ p_6\ p_7\ p_8] : \\ p_1 \in \{3,4\}, p_2 \in \{2,3,4\}, p_3 \in \{3,4\}, p_4 \in \{2,3,4\}, \\ p_5 \in \{3,4\}, p_6 \in \{2,3,4\}, p_7 \in \{3,4\}, p_8 \in \{2,3,4\}\}. \tag{4.38}$$

Considering the Boolean model $x(t) = \ltimes_{i=1}^{2} x_i(t)$ and $u(t) = u_1(t)$, using the semi-tensor product, one have

$$[x(t+1)] = Lu(t)[x(t)], \tag{4.39}$$

where

$$L = \begin{bmatrix} 1\,0\,1\,0\,0\,1\,0\,1 \\ 1\,1\,0\,0\,0\,0\,1\,0 \\ 0\,0\,0\,1\,1\,0\,0\,0 \\ 0\,0\,0\,0\,1\,0\,0\,1 \end{bmatrix}.$$

Considering the feedback law as follows:

$$u(t) = Ex(t_l), \quad t_l \le t \le t_{l+1}. \tag{4.40}$$

Let the feedback matrix E be

$$E = \delta_2[p_1\ p_2\ p_3\ p_4].$$

Besides, $\tau = 2$. Then one can construct the SDSFC such that system (4.39) will be stabilized to the set $S = \{\delta_4^1, \delta_4^2\}$ globally.

Step 1. Finding the largest sampled point set $\widetilde{S}^*$ by Algorithm 18.
For $x(t_l) = \delta_4^1$, there exists a controller $u = \delta_2^1$ such that $x(t_l + 1) \in \{\delta_4^1, \delta_4^2\}$ and $x(t_l + 2) = \delta_4^1$ *or* δ_4^2.
For $x(t_l) = \delta_4^2$, there exists a controller $u = \delta_2^1$ such that $x(t_l + 1) = \delta_4^2$ and $x(t_l + 2) = \delta_4^2$.
Due to $\delta_4^1, \delta_4^2 \in S$, the sampled point set $S^* = S$.
Obviously, $\widetilde{S}^* = S^* = S = \{\delta_4^1, \delta_4^2\}$ and $\bar{S} = \{1, 2\}$.

Step 2. Since $\alpha_{j_1} \in \bar{S}$, then $j_1 \in D(\bar{S}) = \{1, 2, 3, 5, 6\}$. Solving $\beta_i^2 = \alpha_{(p_i-1)2^2+i} = j_1 - (p_i - 1)2^2$;
If $j_1 = 2,\ p_3 = 2$, that is, when $p_3 = 2$, δ_4^3 can reach δ_4^1 or δ_4^2 in two steps with probability $\rho = 1$; If $j_1 = 3,\ p_4 = 1$, that is, when $p_4 = 1$, δ_4^4 can reach δ_4^1 in two steps with probability $\rho < 1$; If $j_1 = 5,\ p_4 = 2$, that is, when $p_4 = 2$, δ_4^4

can reach δ_4^1 in two steps with probability $\rho < 1$, which can also reach δ_4^4 in a certain probability; If $j_1 = 6$, that is, when $p_3 = 3$, δ_4^3 can reach δ_4^1 in two steps with probability $\rho = 1$.

Step 3. Since we want to choose the maximum l_i, δ_4^4 cannot enter $\widetilde{S}^*$ in two steps with probability $\rho = 1$, which means it may have a bigger l_i. Regard δ_4^4, $\alpha_{j_1} = \{\delta_4^1, \delta_4^2, \delta_4^3\}$, then $j_2 \in D(1, 2, 3) = \{1, 2, 3, 4, 5, 6, 7, 8\}$. Solving $\alpha_{(p_i-1)2^2+i} = j_2 - (p_i - 1)2^2$. If $j_2 = 4$, $p_4 = 1$, which was considered before. If $j_2 = 8$, $p_4 = 2$, that is, when $p_4 = 2$, δ_4^4 can reach δ_4^1, δ_4^4 in no more than four steps with total probability $\rho = 1$. Thus, when $p_4 = 2$, δ_4^4 cannot enter S entirely for the existence of a circle from δ_4^4 to δ_4^4. Obviously, there exists a maximum integer 2 such that $\Omega_2(\widetilde{S}^*) = \Delta_4$.

Through the same calculation, we can obtain that when $l_i = 0$

$$\{\lambda_1 = \lambda_1, \lambda_1 = \lambda_2\}|\{\delta_2^1\}; \{\lambda_2 = \lambda_2\}|\{\delta_2^1\}; \tag{4.41}$$

When $l_i = 1$

$$\{\lambda_1 > \lambda_3, \lambda_2 > \lambda_3\}|\{\delta_2^1\}; \{\lambda_1 > \lambda_3\}|\{\delta_2^2\}; \{\lambda_1 > \lambda_4\}|\{\delta_2^1\}. \tag{4.42}$$

In the following, all possible SDSFCs are designed. First, all

$$U_j = \{H = \delta_2[p_1\ p_2\ p_3\ p_4]\}, \quad j = 1, 2$$

are given

$$\begin{aligned} U_1 =& \{\{\lambda_1 = \lambda_1, \lambda_1 = \lambda_2\}|\{\delta_2^1\}, \{\lambda_2 = \lambda_2\}|\{\delta_2^1\}, \\ & \{\lambda_1 > \lambda_3, \lambda_2 > \lambda_3\}|\{\delta_2^1\}, \{\lambda_1 > \lambda_4\}|\{\delta_2^1\}\}; \\ U_2 =& \{\{\lambda_1 = \lambda_1, \lambda_1 = \lambda_2\}|\{\delta_2^1\}, \{\lambda_2 = \lambda_2\}|\{\delta_2^1\}, \\ & \{\lambda_1 > \lambda_3\}|\{\delta_2^2\}, \{\lambda_1 > \lambda_4\}|\{\delta_2^1\}\}. \end{aligned} \tag{4.43}$$

Next, we take U_1 for example to solve $G = [\lambda_1\ \lambda_2\ \lambda_3\ \lambda_4]$

$$\begin{cases} \lambda_1 = \lambda_1,\ \lambda_1 = \lambda_2,\ \lambda_2 = \lambda_2, \\ \lambda_1 > \lambda_3,\ \lambda_2 > \lambda_3,\ \lambda_1 > \lambda_4. \end{cases} \tag{4.44}$$

Solving inequalities and equalities in (4.44), one solution of G is $\lambda_1 = \lambda_2 = 2$, $\lambda_3 = 1$, $\lambda_4 = 0$, thus, the corresponding control Lyapunov function is $\Psi(x) = [2\ 2\ 1\ 0]x$.

Finally, all the feasible feedback laws U are given by

$$U = \bigcup_{j=1}^{2} U_j = \{E = \delta_2[p_1\ p_2\ p_3\ p_4] : p_1 = 1, p_2 = 1, p_3 \in \{1, 2\}, p_4 = 1\}. \tag{4.45}$$

From above examples, it is obtained that both methods get the correct results. When the dimension of L is small, the second method has a clearer procedure, but if the size of L is relatively big, the second method is a little bit complex, at this time, the first method is a better choice.

4.3 Set Stabilization of Probabilistic Boolean Control Networks

In this section, we investigate the set stabilization of PBCNs under SDSFC within finite and infinite time, respectively. First, the algorithms are, respectively, proposed to find the sampled point set and the largest sampled point control invariant set of PBCNs by SDSFC. Based on this, a necessary and sufficient criterion is proposed for the global set stabilization of PBCNs by SDSFC within finite time. Moreover, the time-optimal sampled-data state feedback controller is designed. It is interesting that if the sampling period is changed, the time of global set stabilization of PBCNs may also change or even the PBCNs cannot achieve set stabilization. Second, a criterion for the global set stabilization of PBCNs by SDSFC within infinite time is obtained. Furthermore, all possible sampled-data state feedback controllers are obtained by using all the complete families of reachable sets. Finally, three examples are presented to illustrate the effectiveness of the obtained results.

4.3.1 Problem Formulation

A PBCN with n nodes is described as

$$x_i(t+1) = f_i^{\lambda_i}(x_1(t), \ldots, x_n(t), u_1(t), \ldots, u_m(t)),\ i = 1, 2, \ldots, n.$$

For $t \geq 0$, where $x(t) := (x_1(t), x_2(t), \ldots, x_n(t))^{\mathrm{T}} \in \mathcal{D}^n$ is a logical state and $u(t) := (u_1(t), u_2(t), \ldots, u_m(t))^{\mathrm{T}} \in \mathcal{D}^m$ is a control input, respectively. Logical functions $f_i^{\lambda_i} : \mathcal{D}^{m+n} \to \mathcal{D},\ i = 1, 2, \ldots, n,\ \lambda_i = 1, 2, \ldots, l_i$ is chosen randomly from a known finite set of Boolean functions $F_i = \{f_i^1, f_i^2, \ldots, f_i^{l_i}\}$. There are a total of $\prod_{i=1}^n l_i$ models. The λth model is denoted by $\Omega_\lambda = \{f_1^{\lambda_1}, f_2^{\lambda_2}, \ldots, f_n^{\lambda_n}\}$, and the probability of f_i being $f_i^{\lambda_i}$ is $\mathbf{P}(f_i = f_i^{\lambda_i}) = p_i^{\lambda_i},\ \lambda_i = 1, 2, \ldots, l_i$, where $\sum_{\lambda_i=1}^{l_i} p_i^{\lambda_i} = 1$. Then, the probability of Ω_λ is $\mathbf{P}(\Omega_\lambda \text{ is chosen}) = \prod_{i=1}^n p_i^{\lambda_i}$.

Under the framework of semi-tensor product, system can be alternatively expressed as the algebraic form $x_i(t+1) = M_i^{\lambda_i} u(t) x(t)$, where $x(t) = \ltimes_{i=1}^n x_i(t)$, $u(t) = \ltimes_{j=1}^m u_j(t)$ and $M_i^{\lambda_i}$ are structure matrices corresponding to logical functions $f_i^{\lambda_i}$, which can be chosen from a corresponding matrix set $M_i = \{M_i^1, M_i^2, \ldots, M_i^{l_i}\},\ i = 1, 2, \ldots, n$.

Let $\mathbb{E}x$ represent the overall expectation of x, then we have

$$\mathbb{E}x_i(t+1) = \widetilde{M}_i u(t)\mathbb{E}x(t)$$

and

$$\widetilde{M}_i = \sum_{j=1}^{l_1} p_i^j M_i^j.$$

Using the semi-tensor product, system finally leads to the following equation:

$$\mathbb{E}x(t+1) = Lu(t)\mathbb{E}x(t), \quad t \geq 0, \tag{4.46}$$

where $L = \widetilde{M}_1 \ltimes_{i=2}^{n} [(I_{2^{n+m}} \otimes \widetilde{M}_i)\Phi_{m+n}]$. Split L into 2^m equal blocks as: $L = [L_1\ L_2\ \cdots\ L_{2^m}]$, where $L_i \in \mathcal{L}_{2^n \times 2^n}$, $i = 1, 2, \ldots, 2^m$.

A sampled-data state feedback set stabilization problem for PBCN will be considered. In addition, the feedback control law to be designed for system is in the form of: $u_j(t) = h_j(x_1(t_l), \ldots, x_n(t_l))$, $t_l \leq t < t_{l+1}$, where h_j, $j = 1, 2, \ldots, m$ are Boolean functions mapping $\mathcal{D}^n$ to $\mathcal{D}$, and $\tau := t_{l+1} - t_l \in \mathbb{Z}_+$ are sampling instants, $t_l = l\tau \geq 0$, $l = 0, 1, \ldots$. Using the semi-tensor product, the dynamics of control system can be converted to the following form:

$$u(t) = Hx(t_l), \quad t_l \leq t < t_{l+1}, \tag{4.47}$$

where $H = \delta_{2^m}[p_1\ p_2\ \cdots\ p_{2^n}]$, $p_i \in \{1, 2, \ldots, 2^m\}$, $i = 1, 2, \ldots, 2^n$. When $\tau = 1$, the SDSFC has been studied in [1, 7].

Consider system (4.46) with control (4.47), when $0 \leq t \leq \tau$, the following equations hold,

$$\begin{aligned}
\mathbb{E}x(1) &= LHx(0)\mathbb{E}x(0), \\
&\ \ \vdots \\
\mathbb{E}x(t) &= LHx(0)\mathbb{E}x(t-1) = (LHx(0))^t\mathbb{E}x(0).
\end{aligned}$$

Similarly, when $\tau < t \leq 2\tau$, it holds that

$$\mathbb{E}x(t) = LHx(\tau)\mathbb{E}x(t-1) = (LHx(\tau))^{t-\tau}\mathbb{E}x(\tau).$$

Therefore, for any positive integer $t \in (l\tau, (l+1)\tau]$, we have

$$\mathbb{E}x(t) = (LHx(l\tau))^{t-l\tau}\mathbb{E}x(l\tau).$$

Definition 4.10 Given a set $S \subseteq \Delta_{2^n}$, system (4.46) is said to be finite-time stabilized to S globally with probability one via SDSFC (4.47) if, for any $x(0) \in \Delta_{2^n}$, there exist an SDSFC sequence $\mathbf{u}$ and an integer $T \geq 0$ such that $\mathbf{P}(x(t) \in S | x(0), \mathbf{u}) = 1$, for any $t \geq T$.

Definition 4.11 Given a set $S \subseteq \Delta_{2^n}$, system (4.46) is said to be infinite-time stabilized to S globally with probability one via SDSFC (4.47) if, for any $x(0) \in \Delta_{2^n}$, there exists an SDSFC sequence $\mathbf{u}$ such that

$$\lim_{t \to \infty} \mathbf{P}(x(t) \in S | x(0), \mathbf{u}) = 1.$$

4.3.2   Finite-Time Global S-Stabilization

First, some preiliminary definitions and concepts are presented before deriving S-stabilization criteria.

Definition 4.12 A subset $S^* \subseteq S$ is called a sampled point set of S of system (4.46) with sampling period τ, if for any $x(0) \in S^*$, there exists a SDSFC sequence $\mathbf{u}$ such that $\mathbf{P}(x(t) \in S | x(0), \mathbf{u}) = 1$ for any $t \in [1, \tau]$.

Definition 4.13 A subset $\widetilde{S} \subseteq S$ is called a sampled point control invariant set of S for system (4.46) via SDSFC (4.47), if for any initial state $x(t_l) \in \widetilde{S}$, there exists a SDSFC sequence $\mathbf{u}$ such that $\mathbf{P}(x(t) \in S | x(t_l), \mathbf{u}) = 1$ for any $t \in [1, \tau - 1]$ and $\mathbf{P}(x(\tau) \in \widetilde{S} | x(t_l), \mathbf{u}) = 1$.

Considering system (4.46) with control (4.47), Algorithm 19 is proposed to search the sampled point set S^* according to Definition 4.12 when S is given.

Algorithm 19 Determine the sampled point set S^* of set S

1: Input:$S^* := \varnothing$.
2: **for** each state $\delta_{2^n}^i \in S$ **do**
3:     **for** α from 1 to 2^m **do**
4:         $Flag := 1$
5:         **for** $j := 1$ to τ **do**
6:             **if** $\forall s_0 \in [1, 2^n]$, such that $[(L_\alpha)^j \delta_{2^n}^i]_{s_0} = 0$
7:                 or $\delta_{2^n}^{s_0} \notin S$ **then**
8:                 $Flag := 0$
9:             **end if**
10:         **end for**
11:         **if** Flag =1 **then**
12:             **return** $S^* := S^* \cup \{\delta_{2^n}^i\}$.
13:         **end if**
14:     **end for**
15: **end for**

If $S^* = \varnothing$, it is obvious that system (4.46) cannot achieve S-stabilization via SDSFC (4.47). According to the above analysis, the control $\mathbf{u}$ at time t is a state feedback control at sampled time t_l. We assume $\widetilde{S}^* = S^* = S = \{\delta_8^1, \delta_8^4, \delta_8^8\}$ and $x(0) = \delta_8^6 \notin S^*$. Then, we can find there exists $\delta_{2^m}^j = \delta_4^1$ such that $x(1) = L\delta_4^1 \delta_8^6 =$

$\delta_8^8 \in \widetilde{S}^*$, but $x(2) = L\delta_4^1 L\delta_4^1 \delta_8^6 = \delta_8^3 \notin S$. Therefore, comparing with conventional state feedback controllers, if a state does not act at a sampled point, it may still run out of S although it has ever entered $\widetilde{S}^*$, hence we should study sampled point set first.

From the definition of $\widetilde{S}$, it is obvious $\widetilde{S} \subseteq S^*$, and the union of any two invariant sets is still an invariant set, so here we denote $\widetilde{S}^*$ as the largest sampled point control invariant set of S. Instead of searching for $\widetilde{S}^*$ by traversing method, it is much easier to delete elements in S^* directly as follows.

Step 1 Find $\delta_{2^n}^{i_1} \in S^*$, for any input $\delta_{2^m}^{\alpha} \in \Delta_{2^m}$, there exist a positive integer $j \in [1, \tau - 1]$, $s_1, s_2 \in [1, 2^n]$, such that $[(L_\alpha)^j \delta_{2^n}^{i_1}]_{s_1} > 0$ and $\delta_{2^n}^{s_1} \notin S$ or $[(L_\alpha)^\tau \delta_{2^n}^{i_1}]_{s_2} > 0$ and $\delta_{2^n}^{s_2} \notin S^*$. Let S_1' be the set of all such $\delta_{2^n}^{i_1}$, if $S_1' = \varnothing$, then $\widetilde{S}^* = S^*$; otherwise let $S_1^* = S^* \backslash S_1'$, going to Step 2.

Step 2 Repeat similar processes in Step 1 until we find the set S_T^* satisfying for any $\delta_{2^n}^{i} \in S_T^*$, there exists $\delta_{2^m}^{\alpha} \in \Delta_{2^m}$ such that for any $j \in [1, \tau - 1]$, if $[(L_\alpha)^j \delta_{2^n}^{i}]_{s_1} > 0$ then $\delta_{2^n}^{s_1} \in S$, and if $[(L_\alpha)^\tau \delta_{2^n}^{i}]_{s_2} > 0$ then $\delta_{2^n}^{s_2} \in S^*$.

Next, Algorithm 20 is proposed to find the largest sampled point control invariant set $\widetilde{S}^*$.

Algorithm 20 Search the largest sampled point control invariant set $\widetilde{S}^*$ of set S

Input S^*
$\widetilde{S}^* = S^*$
for each state $\delta_{2^n}^{i} \in S^*$ **do**
  $Flag1 := 1$
  for $\alpha := 1$ to 2^m **do**
    $Flag2 := 0$
    for $j := 1$ to $\tau - 1$ **do**
      if $\exists\, s_1 \in [1, 2^n]$, such that $[(L_\alpha)^j \delta_{2^n}^{i}]_{s_1} > 0$
       and $\delta_{2^n}^{s_1} \notin S$ **then**
        $Flag2 := 1$
      end if
    end for
    if $\exists\, s_2 \in [1, 2^n]$, such that $[(L_\alpha)^j \delta_{2^n}^{\tau}]_{s_2} > 0$
      and $\delta_{2^n}^{s_2} \notin S^*$ **then**
      $Flag2 := 1$
    end if
    if Flag2 = 0 **then**
      $Flag1 := 0$
    end if
  end for
  if Flag1 = 1 **then**
    $\widetilde{S}^* := \widetilde{S}^* \backslash \{\delta_{2^n}^{i}\}$
  end if
end for
return $\widetilde{S}^*$

If $\widetilde{S^*} = \varnothing$, system (4.46) cannot achieve global S-stabilization under SDSFC with probability one. According to Algorithm 20, we have the largest sampled point control invariant set now.

Remark 4.9 The computational complexity of Algorithm 19 is $O(2^m \times |S|\tau)$, and the computational complexity of Algorithm 20 is $O(2^m \times |S^*|\tau)$. The time complexity of the control design methods increase exponentially since it is required to calculate matrix L with dimension $2^n \times 2^{m+n}$. Due to the high computational complexity, these methods can handle networks up to $10 - 15$ nodes.

Now, we study the set sequences in Definition 4.14, which are prepared to design the controller.

Definition 4.14 For every $\Omega_i(\widetilde{S^*}) \subseteq \Delta_{2^n}$, we denote $\{\Omega_i(\widetilde{S^*})\}$ as the sequence of sets, in which the states can be steered to $\widetilde{S^*}$ in $i\tau$ steps with probability one under some control sequences: $\Omega_0(\widetilde{S^*}) = \widetilde{S^*}$,

$$\begin{aligned}\Omega_{k+1}(\widetilde{S^*}) = \{a \in \Delta_{2^n} | &\text{there exists a } \mathbf{u} \text{ such that} \\ &\mathbf{P}(x((t+1)\tau) \in \Omega_k(\widetilde{S^*}) | x(t\tau) = a, \mathbf{u}) = 1\}.\end{aligned} \tag{4.48}$$

According to Definition 4.14, for every $\Omega_i(\widetilde{S^*}) \subseteq \Delta_{2^n}$, we have

$$\begin{cases} \Omega_1(\widetilde{S^*}) = \{\delta_{2^n}^i \in \Delta_{2^n} | [(L_{p_i})^\tau \delta_{2^n}^i] \circ (\mathbf{1}_{2^n} - \sum_{a \in \widetilde{S^*}} a) = \mathbf{0}_{2^n}\}, \\ \Omega_{k+1}(\widetilde{S^*}) = \{\delta_{2^n}^i \in \Delta_{2^n} | [(L_{p_i})^\tau \delta_{2^n}^i] \circ (\mathbf{1}_{2^n} - \sum_{a \in \Omega_k(\widetilde{S^*})} a) = \mathbf{0}_{2^n}\}. \end{cases}$$

Then, some fundamental properties for sets $\Omega_i(\widetilde{S^*}) \subseteq \Delta_{2^n}$ are obtained as follows.

Lemma 4.5 *The following properties hold:*

(I) $\widetilde{S^*} \subseteq \Omega_1(\widetilde{S^*})$ *and* $\Omega_l(\widetilde{S^*}) \subseteq \Omega_{l+1}(\widetilde{S^*})$ *for all* $l \geq 1$.
(II) If $\Omega_l(\widetilde{S^*}) = \Omega_{l+1}(\widetilde{S^*})$, *then* $\Omega_k(\widetilde{S^*}) = \Omega_l(\widetilde{S^*})$, *for all* $k \geq l$.
(III) If $\Omega_1(\widetilde{S^*}) = \widetilde{S^*}$, *then* $\Omega_l(\widetilde{S^*}) = \widetilde{S^*}$, *for all* $l \geq 1$.

Theorem 4.10 *System (4.46) is said to be finite-time stabilized to S globally with probability one under SDSFC (4.47), if and only if the following conditions are satisfied:*

(I) $\widetilde{S^*} \neq \varnothing$;
(II) there is a positive integer T *satisfying* $\Omega_T(\widetilde{S^*}) = \Delta_{2^n}$.

Proof (Sufficiency) Assume that conditions (I) and (II) are satisfied, which implies that any $x(0) \in \Delta_{2^n}$ can be steered to $\widetilde{S^*}$ within $T\tau$ steps with probability one. If $x(T\tau) \in \widetilde{S^*}$ with probability one, then $\mathbf{P}(x(T\tau + N) \in \widetilde{S^*} | x(0) \in \Delta_{2^n}, \mathbf{u}) = 1$ for any $N \in \mathbb{Z}_+$. Thus, system (4.46) achieves global S-stabilization with probability one.

(Necessity) If $\widetilde{S^*} = \varnothing$, it is obvious that condition (II) is not satisfied. Assume that system (4.46) is S-stabilization with probability one via sampled-data state feedback controller (4.47), thus for any initial state $x(0) = \delta_{2^n}^j \in \Delta_{2^n}$, there exists a positive integer T_j such that $\mathbf{P}(x((T_j\tau) \in \widetilde{S}^*|x(0) = \delta_{2^n}^j, \mathbf{u}) = 1$. That is to say, state $\delta_{2^n}^j$ can be steered to $\widetilde{S}^*$ within $T_j\tau$ steps, then $\delta_{2^n}^j \in \Omega_{T_j}(\widetilde{S}^*)$. Let $T = \max\{T_j | j = 1, 2, \ldots, 2^n\}$, according to Lemma 4.5 and the arbitrariness of j, we have $\Omega_T(\widetilde{S}^*) = \Delta_{2^n}$. Thus, conditions (I) and (II) are satisfied. □

Hence, for every initial state $x(0) = \delta_{2^n}^j \in \Delta_{2^n}$, there exists a unique positive integer T_j such that $\delta_{2^n}^j \in \Omega_{T_j}(\widetilde{S}^*)\backslash\Omega_{T_j-1}(\widetilde{S}^*)$ with $\Omega_0(\widetilde{S}^*) = \widetilde{S}^*$. Then, set Δ_{2^n} can be represented as the union of disjoint sets: $\Delta_{2^n} = \widetilde{S}^* \cup (\Omega_1(\widetilde{S}^*)\backslash\widetilde{S}^*) \cup \cdots \cup (\Omega_T(\widetilde{S}^*)\backslash\Omega_{T-1}(\widetilde{S}^*))$.

Theorem 4.11 *Consider system (4.46) and a given S, if there exists an SDSFC law (4.47), which can stabilize system (4.46) to the set S with probability one, then the element p_i in the sampled-data state feedback matrix $H = \delta_{2^m}[p_1\ p_2\ \cdots\ p_{2^n}]$ satisfies:*

$$\begin{cases} \text{for any } i \in [1, \tau-1], ((L_{p_j})^i \delta_{2^n}^j) \circ (\mathbf{1}_{2^n} - \sum_{a \in S} a) = \mathbf{0}_{2^n} \\ \text{and } ((L_{p_j})^\tau \delta_{2^n}^j) \circ (\mathbf{1}_{2^n} - \sum_{a \in \widetilde{S}^*} a) = \mathbf{0}_{2^n}, \quad \text{when } T_j = 0; \\ ((L_{p_j})^\tau \delta_{2^n}^j) \circ (\mathbf{1}_{2^n} - \sum_{a \in \Omega_{T_j-1}(\widetilde{S}^*)} a) = \mathbf{0}_{2^n}, \quad \text{when } T_j \geq 1. \end{cases}$$

Proof According to Theorem 4.10, let $x(0) = \delta_{2^n}^j \in \Delta_{2^n}$, the corresponding unique integer is T_j. If $T_j = 1$, then $((L_{p_j})^i \delta_{2^n}^j) \circ (\mathbf{1}_{2^n} - \sum_{a \in S} a) = \mathbf{0}_{2^n}$ for any $i \in [1, \tau-1]$. It implies $\mathbf{P}(x(i) \in S|\delta_{2^n}^j \in \Delta_{2^n}, \delta_{2^m}^{p_j}) = 1$ for any $i \in [1, \tau-1]$. Besides, $((L_{p_j})^\tau \delta_{2^n}^j) \circ (\mathbf{1}_{2^n} - \sum_{a \in \widetilde{S}^*} a) = \mathbf{0}_{2^n}$ implies $\mathbf{P}(x(\tau) \in \widetilde{S}^*|\delta_{2^n}^j \in \Delta_{2^n}, \delta_{2^m}^{p_j}) = 1$. If $T_j \geq 1$, then $((L_{p_j})^\tau \delta_{2^n}^j) \circ (\mathbf{1}_{2^n} - \sum_{a \in \Omega_{T_j-1}(\widetilde{S}^*)} a) = \mathbf{0}_{2^n}$. It implies $\mathbf{P}(x(\tau) \in \Omega_{T_j-1}(\widetilde{S}^*)|\delta_{2^n}^j, \delta_{2^m}^{p_j}) = 1$.

Hence, the sampled-data state feedback controller (4.47) stabilizing system (4.46) to the set S globally with probability one is designed by the law. □

Consider the probabilistic Boolean control model given as follows:

$$\begin{cases} x_1(t+1) = f_1^{\lambda_1}(x_1(t), x_2(t), x_3(t), u_1(t), u_2(t)), \\ x_2(t+1) = f_2^{\lambda_2}(x_1(t), x_2(t), x_3(t), u_1(t), u_2(t)), \\ x_3(t+1) = f_3^{\lambda_3}(x_1(t), x_2(t), x_3(t), u_1(t), u_2(t)), \end{cases}$$

where $x_1,\ x_2,\ x_3$ are state variables and $u_1,\ u_2$ are control inputs. Here, the sampling period is 2, $l_1 = 3$, $l_2 = l_3 = 4$ and $p_1^1 = p_2^1 = p_2^4 = \frac{1}{3}$, $p_1^2 = p_2^2 = p_2^3 = p_3^3 = \frac{1}{6}$, $p_1^3 = p_3^4 = \frac{1}{2}$, $p_3^1 = \frac{1}{4}$, $p_3^2 = \frac{1}{12}$. Let $x(t) = \ltimes_{i=1}^3 x_i(t)$ and $u(t) = \ltimes_{i=1}^2 u_i(t)$, using the semi-tensor product, we have

$$\mathbb{E}x(t+1) = Lu(t)\mathbb{E}x(t), \tag{4.49}$$

where

$$
\begin{aligned}
L =[&\delta_8^1, \frac{1}{6}(\delta_8^1+\delta_8^3)+\frac{1}{3}(\delta_8^2+\delta_8^4), \frac{1}{6}(\delta_8^1+\delta_8^5)+\frac{1}{3}(\delta_8^2+\delta_8^6),\\
&\frac{2}{3}\delta_8^2+\frac{1}{3}\delta_8^3, \frac{2}{3}\delta_8^5+\frac{1}{3}\delta_8^7, \delta_8^8, \frac{1}{6}(\delta_8^2+\delta_8^4)+\frac{1}{3}(\delta_8^6+\delta_8^8), \delta_8^3,\\
&\delta_8^1, \frac{1}{3}\delta_8^5+\frac{2}{3}\delta_8^7, \delta_8^4, \frac{1}{2}\delta_8^1+\frac{1}{2}\delta_8^5, \delta_8^7, \frac{1}{6}(\delta_8^1+\delta_8^5)+\frac{1}{3}(\delta_8^2+\delta_8^6),\\
&\delta_8^1, \frac{1}{3}\delta_8^6+\frac{2}{3}\delta_8^8, \frac{1}{2}\delta_8^4+\frac{1}{2}\delta_8^8, \frac{1}{6}(\delta_8^1+\delta_8^5)+\frac{1}{3}(\delta_8^2+\delta_8^6), \delta_8^8, \delta_8^1,\\
&\frac{2}{3}\delta_8^5+\frac{1}{3}\delta_8^7, \frac{1}{6}(\delta_8^5+\delta_8^7)+\frac{1}{3}(\delta_8^6+\delta_8^8), \frac{1}{3}\delta_8^1+\frac{2}{3}\delta_8^3, \delta_8^1, \delta_8^3,\\
&\frac{1}{6}(\delta_8^1+\delta_8^3)+\frac{1}{3}(\delta_8^2+\delta_8^4), \delta_8^1, \frac{1}{3}\delta_8^7+\frac{2}{3}\delta_8^8, \frac{1}{2}\delta_8^1+\frac{1}{2}\delta_8^5,\\
&\frac{1}{2}\delta_8^7+\frac{1}{2}\delta_8^8, \frac{1}{3}\delta_8^3+\frac{2}{3}\delta_8^4, \frac{1}{4}\delta_8^5+\frac{3}{4}\delta_8^6].
\end{aligned}
$$

According to system (4.49), the following sampled-data state feedback controller will be designed in the form of:

$$
u(t) = Hx(t_l), \quad t_l < t \leq t_{l+1}. \tag{4.50}
$$

Let H be presented by $H = \delta_4[p_1\ p_2\ \cdots\ p_8]$, where $p_i \in \{1, 2, 3, 4\}$. Now, we design a sampled-data state feedback controller to make system (4.49) global stabilization to the set $\{\delta_8^1, \delta_8^4, \delta_8^8\}$ with probability one.

Step 1 Find the sampled point set by Algorithm 19. For $x(t_l) = \delta_8^1$, there exists $p_1 = 1$ such that $\mathbf{P}(x(t_l+1) = \delta_8^1 \in S) = 1$ and $\mathbf{P}(x(t_l+2) = \delta_8^1 \in S) = 1$. For $x(t_l) = \delta_8^4$, there exists $p_4 = 3$ such that $\mathbf{P}(x(t_l+1) = \delta_8^1 \in S) = 1$ and $\mathbf{P}(x(t_l+2) = \delta_8^4) = \mathbf{P}(x(t_l+2) = \delta_8^8) = \frac{1}{2}$, so $\mathbf{P}(x(t_l+2) \in S) = 1$. For $x(t_l) = \delta_8^8$, there exists $p_8 = 3$ such that $\mathbf{P}(x(t_l+1) = \delta_8^1 \in S) = 1$ and $\mathbf{P}(x(t_l+2) = \delta_8^4) = \mathbf{P}(x(t_l+2) = \delta_8^8) = \frac{1}{2}$, so $\mathbf{P}(x(t_l+2) \in S) = 1$. Based on Algorithm 20, it is easy to find $\widetilde{S}^* = S^* = S = \{\delta_8^1, \delta_8^4, \delta_8^8\}$.

Step 2 According to Theorem 4.11, it is obtained that $\Omega_1(\widetilde{S}^*)\backslash\widetilde{S}^* = \{\delta_8^3, \delta_8^5, \delta_8^7\}$ and $\Omega_2(\widetilde{S}^*)\backslash\Omega_1(\widetilde{S}^*) = \{\delta_8^2, \delta_8^6\}$. Since $\widetilde{S}^* \neq \varnothing$ and $\Omega_2(\widetilde{S}^*) = \Delta_{2^n}$, system (4.49) can stabilize to set S globally via SDSFC (4.50) with probability one.

Step 3 According to Theorem 4.11, determine p_i, $i = 1, 2, \ldots, 8$. For $i \in \{1, 4, 8\}$, p_i satisfies for any $t \in [1, \tau - 1]$, $((L_{p_i})^t \delta_{2^n}^i) \circ (\mathbf{1}_{2^n} - \sum_{a \in S} a) = \mathbf{0}_{2^n}$ and $((L_{p_i})^\tau \delta_{2^n}^i) \circ (\mathbf{1}_{2^n} - \sum_{a \in \widetilde{S}^*} a) = \mathbf{0}_{2^n}$. It is found that $p_1 = 1$ or 2, $p_4 = 3$, and $p_8 = 3$. For $i \in \{3, 5, 7\}$, p_i satisfies $((L_{p_i})^\tau \delta_{2^n}^i) \circ (\mathbf{1}_{2^n} - \sum_{a \in \widetilde{S}^*} a) = \mathbf{0}_{2^n}$. It is found that $p_3 = 3$, $p_5 = 2$ and $p_7 = 2$. For $i \in \{2, 6\}$, p_i satisfies $((L_{p_i})^\tau \delta_{2^n}^i) \circ (\mathbf{1}_{2^n} - \sum_{a \in \Omega_1(\widetilde{S}^*)} a) = \mathbf{0}_{2^n}$. It is found that $p_2 = 2$ and $p_6 = 1$.

Step 4 Finally, the time-optimal SD state feedback laws H is designed by

$$H = \delta_4[1\ 2\ 3\ 3\ 2\ 1\ 2\ 3].$$

Hence, it is obvious that for any initial states $\delta_{2^n}^i \in \Delta_{2^n}$, we have $P[x(t) \in S|\delta_{2^n}^i, \mathbf{u}] = 1$, for any $t \geq 4$.

Remark 4.10 If we change the sampling period, the sampled point set and the largest sampled point control invariant set of (4.46) may change, and the time of global S-stabilization with probability one via SDSFC (4.47) may change as well. For example, if the sampling period is three, then S^* and $\widetilde{S}^*$ are consistent with sampling period is two. However, $\Omega_1(\widetilde{S}^*)\backslash\widetilde{S}^* = \{\delta_8^2, \delta_8^3, \delta_8^5, \delta_8^7\}$ and $\Omega_2(\widetilde{S}^*)\backslash\Omega_1(\widetilde{S}^*) = \{\delta_8^2\}$, which means that the state δ_8^2 stabilizes to S is faster than sampling period is two but the δ_8^6 is slower. Hence, if we change sampling period, the time of system (4.46) for achieving global S-stabilization with probability one via SDSFC (4.47) may change, and system (4.46) may even cannot stabilize to S globally with probability one within finite time.

4.3.3 Infinite-Time S-Stabilization

Different from finite-time S-stabilization, we can find all possible sampled-data state feedback controllers of system (4.46) in infinite time by making some modification in the definition of the sequence of sets $\{\Omega_i(\widetilde{S}^*)\}$.

Definition 4.15 For every $\Theta_i(\widetilde{S}^*) \subseteq \Delta_{2^n}$, we denote $\{\Theta_i(\widetilde{S}^*)\}$ as the sequence of sets, in which the states have the probability to be steered to $\widetilde{S}^*$ within $i\tau$ steps under some control sequences

$$\Theta_0(\widetilde{S}^*) = \widetilde{S}^*,$$

$$\begin{aligned}&\Theta_{k+1}(\widetilde{S}^*)\\ =&\{a \in \Delta_{2^n}|\text{there exists a } \mathbf{u} \text{ such that } \mathbf{P}(x((t+1)\tau) \in \Theta_k(\widetilde{S}^*)|x(t\tau) = a, \mathbf{u}) > 0\}.\end{aligned}$$

Remark 4.11 Notice that the sequence $\{\Theta_i(\widetilde{S}^*)\}$ has the same conclusion as in Lemma 4.5, such as $\widetilde{S}^* \subseteq \Theta_1(\widetilde{S}^*)$ and $\Theta_l(\widetilde{S}^*) \subseteq \Theta_{l+1}(\widetilde{S}^*)$ for all $l \geq 1$. The proof process is also similar to Lemma 4.5, thus the proof is omitted here.

Theorem 4.12 *System (4.46) is said to be infinite-time stabilized to S globally with probability one via SDSFC (4.47), if and only if the following conditions are satisfied:*

(I) $\widetilde{S}^* \neq \varnothing$;
(II) there exists an integer N such that $\Theta_N(\widetilde{S}^*) = \Delta_{2^n}$.

Proof (Sufficiency) Without loss of generality, we assume that $\widetilde{S^*} = \{\delta_{2^n}^{2^n-1}, \delta_{2^n}^{2^n}\}$, $|\widetilde{S^*}| = 2$. If $|\widetilde{S^*}| > 2$, we assume $|\widetilde{S^*}| = k$. If $k = 3$, we assume that $\widetilde{S^*} = \{\delta_{2^n}^{2^n-2}, \delta_{2^n}^{2^n-1}, \delta_{2^n}^{2^n}\}$. If conditions (i) and (ii) hold, for any $x(t_l) = \delta_{2^n}^i \in \Delta_{2^n}$, there exists $u(t) = \delta_{2^m}^j$, such that $Col_i(L) = Col_i((L_j)^\tau)$, so we can get a matrix L, which can be represented as

$$L = \begin{bmatrix} \Gamma^{\mathrm{T}} & 0_{2^n-3} & 0_{2^n-3} & 0_{2^n-3} \\ \alpha_1^{\mathrm{T}} & \beta_1 & \lambda_1 & \gamma_1 \\ \alpha_2^{\mathrm{T}} & \beta_2 & \lambda_2 & \gamma_2 \\ \alpha_3^{\mathrm{T}} & \beta_3 & \lambda_3 & \gamma_3 \end{bmatrix}.$$

Since each column of L sums to $\mathbf{1}$, we assume $\alpha_1'^{\mathrm{T}} = \alpha_1^{\mathrm{T}} + \alpha_2^{\mathrm{T}}$, $\alpha_2'^{\mathrm{T}} = \alpha_3^{\mathrm{T}}$, $\lambda = \frac{1}{2}(\beta_1 + \beta_2) + \frac{1}{2}(\lambda_1 + \lambda_2)$, $\gamma = \gamma_1 + \gamma_2$. Then, $1 - \lambda = \frac{1}{2}\beta_3 + \frac{1}{2}\lambda_3$, $1 - \gamma = \gamma_3$. We can obtain a matrix $L' \in \mathcal{L}_{(2^n-1)\times(2^n-1)}$,

$$L' = \begin{bmatrix} \Gamma^{\mathrm{T}} & 0_{2^n-3} & 0_{2^n-3} \\ \alpha_1'^{\mathrm{T}} & \lambda & \gamma \\ \alpha_2'^{\mathrm{T}} & 1-\lambda & 1-\gamma \end{bmatrix},$$

and a new invariant set $\widetilde{S^*}' = \{\delta_{2^n-1}^{2^n-2}, \delta_{2^n-1}^{2^n-1}\}$ with $|\widetilde{S^*}'| = 2$. Hence, if $|\widetilde{S^*}| > 2$, by viewing one state of $\widetilde{S^*}$ as $\delta_{2^n}^{2^n}$ and the others as $\delta_{2^n}^{2^n-1}$, we obtain a new invariant set $\widetilde{S^*}'$ with $|\widetilde{S^*}'| = 2$. Let

$$L = \begin{bmatrix} \Gamma^{\mathrm{T}} & 0_{2^n-2} & 0_{2^n-2} \\ \alpha_1^{\mathrm{T}} & \lambda & \gamma \\ \alpha_2^{\mathrm{T}} & 1-\lambda & 1-\gamma \end{bmatrix},$$

then we have

$$L^t = \begin{bmatrix} (\Gamma^{\mathrm{T}})^t & 0_{2^n-2} & 0_{2^n-2} \\ \alpha_{1t} & \lambda_t & \gamma_t \\ \alpha_{2t} & 1-\lambda_t & 1-\gamma_t \end{bmatrix},$$

where $\alpha_{1t} = (\alpha_1^{\mathrm{T}})(\Gamma^T)^{t-1} + \sum_{i=0}^{t-2}(\lambda\alpha_1^{\mathrm{T}} + \gamma\alpha_2^{\mathrm{T}})(\Gamma^{\mathrm{T}})^i$ and $\alpha_{2t} = (\alpha_2^{\mathrm{T}})(\Gamma^{\mathrm{T}})^{t-1} + \sum_{i=0}^{t-2}((1-\lambda)\alpha_1^{\mathrm{T}} + (1-\gamma)\alpha_2^{\mathrm{T}})(\Gamma^{\mathrm{T}})^i$. Since each column of L^t sums to $\mathbf{1}$, it holds that $\Gamma^{2^n}\mathbf{1}_{2^n-2} + \sum_{i=0}^{2^n-1}\Gamma^i(\alpha_1 + \alpha_2) = \mathbf{1}_{2^n-2}$.

In order to prove that system (4.46) can achieve infinite-time global S-stabilization with probability one via SDSFC (4.47), it is equivalent to prove that

$$\lim_{t\to\infty}\sum_{i=0}^{t-1}\Gamma^i(\alpha_1 + \alpha_2) = \mathbf{1}_{2^n-2}.$$

Since $\sum_{i=0}^{t-1}\Gamma^i(\alpha_1+\alpha_2)$ is nondecreasing with the k increases and $\sum_{i=0}^{t-1}\Gamma^i(\alpha_1+\alpha_2) < \mathbf{1}_{2^n-2}$, hence $\sum_{i=0}^{t-1}\Gamma^i(\alpha_1+\alpha_2)$ is convergent. Since $L^{(t+1)2^n}=L^{t2^n}L^{2^n}$, we have $\sum_{i=0}^{(t+1)2^n-1}\Gamma^i(\alpha_1+\alpha_2)=\sum_{i=0}^{t2^n-1}\Gamma^{2^n}\Gamma^i(\alpha_1+\alpha_2)+\sum_{i=0}^{2^n-1}\Gamma^i(\alpha_1+\alpha_2)$. Define $\eta(t)=\sum_{i=0}^{t2^n-1}\Gamma^i(\alpha_1+\alpha_2)-\mathbf{1}_{2^n-2}$, then we have

$$\begin{aligned}\eta(t+1)=&\sum_{i=0}^{(t+1)2^n-1}\Gamma^i(\alpha_1+\alpha_2)-\mathbf{1}_{2^n-2}\\=&\Gamma^{2^n}\sum_{i=0}^{t2^n-1}\Gamma^i(\alpha_1+\alpha_2)+\sum_{i=0}^{2^n-1}\Gamma^i(\alpha_1+\alpha_2)-\mathbf{1}_{2^n-2}\\=&\Gamma^{2^n}\eta(t).\end{aligned}$$

According to the condition (ii), we know that for all $t\geq N$ have $(Row_{2^n}(L^t)+Row_{2^n-1}(L^t))>0$, which implies the sum of each column of Γ^{2^n} is strictly less than $\mathbf{1}$. Then, Γ^{2^n} is strictly Schur stable. Then, we have $\lim_{t\to\infty}\eta(t)=\mathbf{0}_{2^n-2}$, hence, we have $\lim_{t\to\infty}\sum_{i=0}^{t-1}\Gamma^i(\alpha_1+\alpha_2)=\mathbf{1}_{2^n-2}$.

(Necessity) If $\widetilde{S}^*=\emptyset$, condition (ii) is not satisfied. Assume that system (4.46) is said to be infinite-time stabilized to S globally with probability one via SDSFC (4.47). Thus, for any initial state $x(0)=\delta_{2^n}^j\in\Delta_{2^n}$, there exists a positive integer T_j such that $\mathbf{P}(x((T_j\tau)\in\widetilde{S}^*|x(0)=\delta_{2^n}^j,\mathbf{u})>0$. That is to say, state $\delta_{2^n}^j$ has a positive probability to be steered to $\widetilde{S}^*$ in $T_j\tau$ steps, then $\delta_{2^n}^j\in\Theta_{T_j}(\widetilde{S}^*)$. According to the arbitrariness of j, we can find a unique positive integer $N=\max\{T_j|j=1,2,\ldots,2^n\}$ satisfying $\Theta_N(\widetilde{S}^*)=\Delta_{2^n}$. Thus, conditions (i) and (ii) are satisfied.

□

Remark 4.12 According to Definition 4.15 of $\{\Theta_i(\widetilde{S}^*)\}$, for each state $\delta_{2^n}^i\in\Delta_{2^n}$, there is a unique positive integer $0\leq k\leq N$ such that $\delta_{2^n}^i\in\Theta_k(\widetilde{S}^*)\backslash\Theta_{k-1}(\widetilde{S}^*)$, thus here we have $N\leq 2^n-|\Theta_0(\widetilde{S}^*)|=2^n-|\widetilde{S}^*|$.

Remark 4.13 The infinite-time global S-stabilization with probability one for system (4.46) under SDSFC (4.47) is discussed here, hence the impact within τ steps can be ignored. Therefore, in the proof of Theorem 4.12, we obtain each column of the matrix in units of τ steps. If $\tau=1$, this problem can be regarded as system (4.46) stabilizing to S globally with probability one within infinite time via state feedback control.

Next we will consider all sampled-data state feedback controllers for system (4.46) if it can stabilize to the set S with probability one in infinite time. Given a set $S\subseteq\Delta_{2^n}$, according to Theorem 4.12, we can find a unique integer N^* such that $\Theta_{N^*}(\widetilde{S}^*)=\Delta_{2^n}$, and $\Theta_N(\widetilde{S}^*)=\Delta_{2^n}$ for all $N\geq N^*$.

Definition 4.16 For a finite-time sequence of reachable sets $\{\Gamma_i | i = 1, 2, \ldots, N\}$, we define a sequence of sets

$$\begin{cases} R(\Gamma_1) = \{x_0 \in \Delta_{2^n} | \exists \mathbf{u}, s.t. \mathbf{P}(x(\tau) \in \Gamma_1 | x_0, \mathbf{u}) > 0\}, \\ R(\Gamma_k) = \{x_0 \in \Delta_{2^n} | \exists \mathbf{u}, s.t. \mathbf{P}(x(\tau) \in \Gamma_k | x_0, \mathbf{u}) > 0\}. \end{cases}$$

Then we can get all sequences of reachable sets as follows.

Step 1 Let $\Gamma_0 = \widetilde{S}^*$, we can choose any nonempty set $\Gamma_1 \subseteq \Theta_1(\widetilde{S}^*) \backslash \Gamma_0$. For all $k \geq 2$, we can choose any nonempty set $\Gamma_k \subseteq R(\Gamma_{k-1}) \cap [\Theta_k(\widetilde{S}^*) \backslash (\bigcup_{i=0}^{k-1} \Gamma_i)]$. Continue this process until we find a positive integer N satisfying $N^* \leq N \leq 2^n - |S^*|$, such that

$$R(\Gamma_{N-1}) \cap [\Theta_N(\widetilde{S}^*) \backslash (\bigcup_{i=0}^{N-1} \Gamma_i)] \neq \varnothing,$$

and

$$R(\Gamma_N) \cap [\Theta_{N+1}(\widetilde{S}^*) \backslash (\bigcup_{i=0}^{N} \Gamma_i)] = \varnothing.$$

Thus, we can obtain a finite-time sequence of reachable sets $\{\Gamma_i | i = 1, \ldots, N\}$, which has $\Gamma_i \cap \Gamma_j = \varnothing$, $\Gamma_i \neq \varnothing$ for any $i \neq j$. It is obvious that all states in Γ_k can reach $\widetilde{S}^*$ in $k\tau$ steps.

Step 2 Repeat the process of Step 1 until we cannot find a new finite-time sequence of reachable sets $\{\Gamma_i | i = 1, 2, \ldots, N\}$.

Definition 4.17 The finite-time sequence $\{\Gamma_i | i = 0, 1, 2, \ldots, N\}$ is said to be a complete family of reachable sets for system (4.46), if $\bigcup_{i=0}^{N} \Gamma_i = \Delta_{2^n}$.

We define the complete families of reachable sets as

$$C_i := \{\Gamma_i | i = 0, 1, 2, \cdots, N_i\}, \quad i = 1, 2, \ldots, \sigma.$$

Theorem 4.13 *If system (4.46) has complete families of reachable sets, then the element p_i in the sampled-data state feedback matrix $H = \delta_{2^m}[p_1\ p_2\ \cdots\ p_{2^n}]$ of system (4.46) satisfies:*

$$\begin{cases} ((L_{p_i})^\tau \delta_{2^n}^i) \circ (\mathbf{1}_{2^n} - \sum_{a \in \Gamma_0} a) = \mathbf{0}_{2^n}, \quad \delta_{2^n}^i \in \Gamma_0; \\ ((L_{p_i})^\tau \delta_{2^n}^i) \circ (\mathbf{1}_{2^n} - \sum_{a \in \bigcup_{j=0}^{k-1} \Gamma_j} a) = \mathbf{0}_{2^n} \text{ and} \\ ((L_{p_i})^\tau \delta_{2^n}^i) \circ (\mathbf{1}_{2^n} - \sum_{a \in \bigcup_{j=0}^{k-2} \Gamma_j} a) \neq \mathbf{0}_{2^n}, \quad \delta_{2^n}^i \in \Gamma_k. \end{cases}$$

Proof According to Theorem 4.12 and the above steps, we can get families of reachable sets $C_i := \{\Gamma_i | i = 0, 1, \ldots, N_i\}$, $i = 1, 2, \ldots, \sigma$. For each $C_i =$

$\{\Gamma_0, \Gamma_1, \ldots, \Gamma_{N_i}\}$, we have $\bigcup_{j=0}^{N_i} \Gamma_j = \Delta_{2^n}$. In the following proof, we assume $(L_{p_i})^\tau \delta_{2^n}^i = \sum_{s=1}^q m_s \delta_{2^n}^s$, $\sum_{s=1}^q m_s = 1$ and $m_s > 0$ for any $s \in [1, q]$ first.

If $\delta_{2^n}^i \in \Gamma_0$, then we can find all possible integers $1 \leq p_i \leq 2^m$, such that $((L_{p_i})^\tau \delta_{2^n}^i) \circ (\mathbf{1}_{2^n} - \sum_{a \in \Gamma_0} a) = \mathbf{0}_{2^n}$, which implies for each $s \in [1, q]$, $\delta_{2^n}^s \in \Gamma_0$. If $\delta_{2^n}^i \in \Gamma_k$, we can find all possible integers $1 \leq p_i \leq 2^m$, such that $((L_{p_i})^\tau \delta_{2^n}^i) \circ (\mathbf{1}_{2^n} - \sum_{a \in \bigcup_{j=0}^{k-1} \Gamma_j} a) = \mathbf{0}_{2^n}$ and $((L_{p_i})^\tau \delta_{2^n}^i) \circ (\mathbf{1}_{2^n} - \sum_{a \in \bigcup_{j=0}^{k-2} \Gamma_j} a) \neq \mathbf{0}_{2^n}$. It implies $\delta_{2^n}^s \in \bigcup_{j=0}^{k-1} \Gamma_j$ for each $s \in [1, q]$ and there exists a $s \in [1, q]$ such that $\delta_{2^n}^s \in \Gamma_{k-1}$.

Hence, we can find the corresponding sampled-data state feedback controllers for the reachable sets $C_i = \{\Gamma_0, \Gamma_1, \ldots, \Gamma_{N_i}\}$ of system (4.46), defined as

$$\Phi_i = \{H_{ij}\} = \{\delta_{2^m}[p_1 \; p_2 \; \cdots \; p_{2^n}]\}.$$

Then we can obtain all possible sampled-data state feedback controllers for system (4.46). □

If $\bigcup_{i=0}^N \Gamma_i \neq \Delta_{2^n}$, the sampled-data state feedback matrix cannot be obtained from the above steps. Hence, only the complete families of reachable sets can be used to obtain all possible sampled-data state feedback controllers for system (4.46).

Theorem 4.14 *If there exist complete families of reachable sets for system (4.46) stabilizing to the set S globally, then all possible sampled-data state feedback controllers are in the set:* $\Phi = \bigcup_{i=1}^\sigma \Phi_i$.

Proof According to Theorem 4.13, every element in Φ is an eligible sampled-data state feedback controller for system (4.46), thus we only need to prove that every eligible sampled-data state feedback controller for system (4.46) belongs to Φ.

Now we assume that there exists an eligible sampled-data state feedback controller, $H = \delta_2[p_1 \; p_2 \; \cdots \; p_{2^n}]$ but $H \notin \Phi$. Since system (4.46) achieves global S-stabilization with probability one within infinite time via the above controller, thus for any $x(0) = \delta_{2^n}^i \in \Delta_{2^n}$, there exists a control sequence $\mathbf{u}$ such that $\lim_{t\to\infty} \mathbf{P}(x(t) \in S | x(0), \mathbf{u}) = 1$. It means for any $x(0) = \delta_{2^n}^i \in \Delta_{2^n}$, for any $\varepsilon > 0$, there exists a $t_{i(\varepsilon)}$, such that for any $t \geq t_{i(\varepsilon)}$, we have $\| \mathbf{P}(x(t) \in S | x(0), \mathbf{u}) - 1 \| < \varepsilon$. Let $T = \max\{t_{i(\varepsilon)} | i = 1, 2, \ldots, 2^n\}$, then for any $\delta_{2^n}^i \in \Delta_{2^n}$, there exists a unique $T_{i(\varepsilon)}$, $0 \leq T_{i(\varepsilon)} \leq \frac{T}{\tau} + 1$, such that $\delta_{2^n}^i \in \Gamma_{T_{i(\varepsilon)}}$. Thus, for every $\delta_{2^n}^i \in \Gamma_{T_{i(\varepsilon)}}$, we have $\| \mathbf{P}(x(t) \in S | \delta_{2^n}^i, \mathbf{u}) - 1 \| < \varepsilon$ for any $t \geq T_{i(\varepsilon)}\tau$.

It is obvious that for any $i \neq j$, $\Gamma_i \cap \Gamma_j = \varnothing$, $\Gamma_i \neq \varnothing$. Due to the arbitrariness of i, there is a positive integer m such that $\bigcup_{t=0}^m \Gamma_t = \Delta_{2^n}$. Then, we get a finite-time sequence of reachable sets $\{\Gamma_1, \ldots, \Gamma_m\}$. Moreover, it is easy to see that $\Gamma_1 \subseteq \Theta_1(\widetilde{S}^*) \backslash \Gamma_0$, $\Gamma_k \subseteq R(\Gamma_{k-1}) \cap [\Theta_k(\widetilde{S}^*) \backslash (\bigcup_{i=0}^{k-1} \Gamma_i)]$, $2 \leq k \leq m$. Hence, $\{\Gamma_1, \Gamma_2, \ldots, \Gamma_m\}$ is a complete family of reachable sets for system (4.46), which implies that there exists a positive integer i such that $H \in \Phi_i \subset \Phi$. This is a contradiction to the $H \notin \Phi$. Therefore, all possible sampled-data state feedback controllers for system (4.46) belong to $\Phi = \bigcup_{i=1}^{\sigma} \Phi_i$. □

Consider system (4.46) with $S = \{\delta_4^3, \delta_4^4\}$, the model is given as follows:

$$L = [\tfrac{1}{2}(\delta_4^1 + \delta_4^2)\ \tfrac{1}{6}(\delta_4^1 + \delta_4^3) + \tfrac{1}{3}(\delta_4^2 + \delta_4^4)\ \delta_4^1\ \delta_4^2\ \tfrac{1}{2}(\delta_4^3 + \delta_4^4)\ \tfrac{1}{2}(\delta_4^1 + \delta_4^2)\ \delta_4^3\ \delta_4^4]. \tag{4.51}$$

According to system (4.51), the following sampled-data state feedback controller will be designed in the form of:

$$u(t) = Hx(t_l), \quad t_l \le t \le t_{l+1}. \tag{4.52}$$

Denote $H = \delta_2[p_1\ p_2\ p_3\ p_4]$, where $p_i \in \{1, 2\}$, and sampling period is 2. Now we are going to obtain all possible SD state feedback controllers for system (4.51) infinite-time stabilizing to the set $\{\delta_4^3, \delta_4^4\}$ globally with probability one.

Step 1 Find the sampled point set by Algorithm 19.
For $x(t_l) = \delta_4^3$, there exists $p_3 = 2$ such that $\mathbf{P}(x(t_l + 1) = \delta_4^3 \in S | x(t_l) = \delta_4^3, u(t) = \delta_2^2) = 1$ and $\mathbf{P}(x(t_l + 2) \in S | x(t_l) = \delta_4^3, u(t) = \delta_2^2) = 1$. For $x(t_l) = \delta_4^4$, there exists $p_4 = 2$ such that $\mathbf{P}(x(t_l+1) = \delta_4^4 \in S | x(t_l) = \delta_4^4, u(t) = \delta_2^2) = 1$ and $\mathbf{P}(x(t_l + 2) \in S | x(t_l) = \delta_4^4, u(t) = \delta_2^2) = 1$. Based on Algorithm 20, it is easy to find $\widetilde{S}^* = S^* = S = \{\delta_4^3, \delta_4^4\}$.

Step 2 According to Definition 4.15, we can obtain that $\Theta_1(\widetilde{S}^*)\backslash\widetilde{S}^* = \{\delta_4^1, \delta_4^2\}$ and $\Gamma_0 = \{\delta_4^3, \delta_4^4\}$. Thus, we can choose $\Gamma_1 \subseteq \Theta_1(\widetilde{S}^*)\backslash\Gamma_0$ as $\Gamma_1 = \{\delta_4^1, \delta_4^2\}$, $\Gamma_1 = \{\delta_4^1\}$, or $\Gamma_1 = \{\delta_4^2\}$.
Then, we have

$$R(\Gamma_1) \cap [\Theta_2(\widetilde{S}^*)\backslash(\Gamma_0 \cup \Gamma_1)] = \begin{cases} \varnothing, & \Gamma_1 = \{\delta_4^1, \delta_4^2\}, \\ \{\delta_4^2\}, & \Gamma_1 = \{\delta_4^1\}, \\ \{\delta_4^1\}, & \Gamma_1 = \{\delta_4^2\}. \end{cases}$$

For $\Gamma_1 = \{\delta_4^1\}$, we have $\Gamma_2 = \{\delta_4^2\}$ and $R(\Gamma_2) \cap [\Theta_3(\widetilde{S}^*)\backslash(\Gamma_0 \cup \Gamma_1 \cup \Gamma_2)] = \varnothing$.
For $\Gamma_1 = \{\delta_4^2\}$, we have $\Gamma_2 = \{\delta_4^1\}$ and $R(\Gamma_2) \cap [\Theta_3(\widetilde{S}^*)\backslash(\Gamma_0 \cup \Gamma_1 \cup \Gamma_2)] = \varnothing$.
Hence, we have total three families of reachable sets as

$$C_1 = \left\{\Gamma_0 = \left\{\delta_4^3, \delta_4^4\right\}, \Gamma_1 = \left\{\delta_4^1, \delta_4^2\right\}\right\},$$

$$C_2 = \left\{\Gamma_0 = \left\{\delta_4^3, \delta_4^4\right\}, \Gamma_1 = \left\{\delta_4^1\right\}, \Gamma_2 = \left\{\delta_4^2\right\}\right\},$$

$$C_3 = \left\{\Gamma_0 = \left\{\delta_4^3, \delta_4^4\right\}, \Gamma_1 = \left\{\delta_4^2\right\}, \Gamma_2 = \left\{\delta_4^1\right\}\right\}.$$

Step 3 According to Theorem 4.13, we can obtain the corresponding sampled-data state feedback controllers

$$\Phi_1 = \{\delta_2[1\ 1\ 2\ 2], \delta_2[1\ 2\ 2\ 2], \delta_2[2\ 1\ 2\ 2], \delta_2[2\ 2\ 2\ 2]\},$$

$$\Phi_2 = \{\delta_2[1\ 1\ 2\ 2], \delta_2[1\ 2\ 2\ 2], \delta_2[2\ 1\ 2\ 2], \delta_2[2\ 2\ 2\ 2]\},$$

$$\Phi_3 = \{\delta_2[1\ 1\ 2\ 2], \delta_2[1\ 2\ 2\ 2]\}.$$

Hence, we can obtain all possible sampled-data state feedback controllers for system (4.51), which are $\Phi = \cup_{i=1}^{3} \Phi_i$.

Next, we consider system (4.51) stabilizing to S globally with probability one within finite time.

Step 1 We get set $\widetilde{S^*} = S^* = S = \{\delta_4^3, \delta_4^4\}$.
Step 2 According to Theorem 4.11, it is obtained that

$$\Omega_1(\widetilde{S^*}) \backslash \widetilde{S^*} = \{\delta_4^1\}$$

and

$$\Omega_2(\widetilde{S^*}) \backslash \Omega_1(\widetilde{S^*}) = \varnothing.$$

Since $\Omega_1(\widetilde{S^*}) \neq \Delta_{2^n}$, system (4.51) cannot stabilize to S via SDSFC (4.52) globally with probability one within finite time. Hence, we cannot obtain the time-optimal sampled-data state feedback controller for system (4.51).

Remark 4.14 According to the above example, we can find four sampled-data state feedback controllers such that system (4.51) can stabilize to S globally with probability one within infinite time. However, there does not exist a sampled-data state feedback controller such that the system stabilizes to S globally with probability one within finite time. Hence, the existence of sampled-data state feedback controllers for infinite-time and finite-time S-stabilization are not equivalent.

Consider an apoptosis network [8], the dynamic of the considered apoptosis network can be represented as follows:

$$\begin{cases} x_1(t+1) = f_1^{\lambda_1}(x_1(t), x_2(t), x_3(t), u_1(t), u_2(t)), \\ x_2(t+1) = f_2^{\lambda_2}(x_1(t), x_2(t), x_3(t), u_1(t), u_2(t)), \\ x_3(t+1) = f_3^{\lambda_3}(x_1(t), x_2(t), x_3(t), u_1(t), u_2(t)). \end{cases}$$

Here, x_1 is state variable presents the concentration level of apoptosis proteins (IAP), x_2 is state variable presents the active level of caspase 3 (C3a), x_3 is state variable presents the active level of caspase 8 (C8a), and u_1, u_2 are the control inputs. Here

$$
\begin{array}{ll}
f_1^1 = \neg x_2(t) \wedge u_1(t) & \mathbf{P}(f_1^1) = 0.3, \\
f_1^2 = x_1(t) & \mathbf{P}(f_1^2) = 0.7, \\
f_2^1 = \neg x_1(t) \wedge (\neg x_3(t)) & \mathbf{P}(f_2^1) = 0.9, \\
f_2^2 = x_2(t) & \mathbf{P}(f_2^2) = 0.1, \\
f_3 = x_2(t) \vee u_1(t) \wedge u_2(t) & \mathbf{P}(f_3) = 1.
\end{array}
$$

Using the semi-tensor product, we have

$$
\mathbb{E}x(t+1) = Lu(t)\mathbb{E}x(t), \tag{4.53}
$$

where

$$
\begin{aligned}
L = [&0.07\delta_8^1 + 0.63\delta_8^3 + 0.03\delta_8^5 + 0.27\delta_8^7 \; 0.1\delta_8^1 + 0.9\delta_8^3 \; \cdots \\
&0.07\delta_8^2 + 0.63\delta_8^4 + 0.03\delta_8^6 + 0.27\delta_8^8 \; 0.9\delta_8^6 + 0.1\delta_8^8].
\end{aligned}
$$

Here, only part information of matrix L is given due to page limitation. According to system (4.53), the following sampled-data state feedback controller will be designed in the form of:

$$
u(t) = Hx(t_l), \quad t_l < t \le t_{l+1}. \tag{4.54}
$$

Let H be presented by $H = \delta_4[p_1 \; p_2 \; \cdots \; p_8]$, where $p_i \in \{1, 2, 3, 4\}$ and sampling period is 2. Now, we design an SDSFC such that system (4.53) will achieve global stabilization to the set $S = \{\delta_8^2, \delta_8^4, \delta_8^6, \delta_8^8\}$ with probability one.

First, we are going to consider system (4.53) stabilizing to S globally with probability one within finite time. It is easy to find $\Omega_1(\widetilde{S}^*) = \Delta_{2^n}$. Hence, we can obtain the time-optimal sampled-data state feedback controller $H = \delta_4[3\;3\;3\;3\;3\;2\;3\;4]$ for system (4.53).

Second, we are going to obtain all possible sampled-data state feedback controllers for system (4.53) infinite-time stabilizing to the set S globally with probability one.

Step 1 Based on the above analysis, we have found $\widetilde{S}^* = S^* = S = \{\delta_8^2, \delta_8^4, \delta_8^6, \delta_8^8\}$.

Step 2 According to Definition 4.15, we can obtain that

$$
\Theta_1(\widetilde{S}^*)\backslash\widetilde{S}^* = \{\delta_8^1, \delta_8^3, \delta_8^5, \delta_8^7\}
$$

and $\Gamma_0 = \{\delta_8^2, \delta_8^4, \delta_8^6, \delta_8^8\}$. Thus we can choose $\Gamma_1 \subseteq \Theta_1(\widetilde{S}^*)\backslash\Gamma_0$ as $\Gamma_1 = \{\delta_8^1\}$, $\Gamma_1 = \{\delta_8^3\}$, $\Gamma_1 = \{\delta_8^5\}$, $\Gamma_1 = \{\delta_8^7\}$, $\Gamma_1 = \{\delta_8^1, \delta_8^3\}$, $\Gamma_1 = \{\delta_8^1, \delta_8^5\}$, $\Gamma_1 = \{\delta_8^1, \delta_8^7\}$, $\Gamma_1 = \{\delta_8^3, \delta_8^5\}$, $\Gamma_1 = \{\delta_8^3, \delta_8^7\}$, $\Gamma_1 = \{\delta_8^5, \delta_8^7\}$, $\Gamma_1 = \{\delta_8^1, \delta_8^3, \delta_8^5\}$, $\Gamma_1 = \{\delta_8^1, \delta_8^5, \delta_8^7\}$, $\Gamma_1 = \{\delta_8^1, \delta_8^3, \delta_8^7\}$ or $\Gamma_1 = \{\delta_8^3, \delta_8^5, \delta_8^7\}$. Taking every Γ_1 into account, we have total 74 families of reachable sets by discussing all kinds of Γ_1.

Step 3 According to Theorem 4.13, we can obtain all possible sampled-data state feedback controllers $\Phi : \Phi = \bigcup_{i=1}^{74} \Phi_i = \{H_{ij} = \delta_4[p_1\ p_2\ \cdots\ p_8]\}$, which satisfies $p_1 \in \{3, 4\}$, $p_2 \in \{2, 3, 4\}$, $p_3 \in \{3, 4\}$, $p_4 \in \{1, 2, 3, 4\}$, $p_5 \in \{3, 4\}$, $p_6 \in \{2, 3, 4\}$, $p_7 \in \{3, 4\}$, $p_8 \in \{2, 3, 4\}$. Hence, we have obtained 1728 sampled-data state feedback controllers globally stabilizing system (4.53) to the set S within infinite-time with probability one.

4.4 Summary

In this chapter, we first used semi-tensor product method to study the SDSFC problem of PBCNs, which is the investigation of gene regulatory networks from a different point of view. Based on the algebraic representation of logical functions, a necessary and sufficient condition has been given to guarantee the existence of an SDSFC for the stabilization of PBCNs, and the controller has been designed under the presented conditions. Since the considered system is a PBCN, which is more general and complicated than a BCN, the analysis and controller design process is more difficult and challenging than that in [2]. Moreover, the method used in this section is different and simpler than that given by [2] in designing controllers. At the same time, the work of this section can be applied to some Boolean models of biological networks. For example, the models of the lac operon in Escherichia coli [9] and the apoptosis networks [8]. Moreover, we use the SDSFC for the stabilization problem of a 7-gene network containing the genes WNT5A, pirin, S100P, RET1, MART1, HADHB and STC2 [4].

Then, we have studied the partial stabilization problem of PBCNs under SDSFC and raised a control Lyapunov function approach for the problem. First, the partial stabilization problem has been converted into global set stabilization one. Next, control Lyapunov function under SDSFC and its structural matrix G have been defined. It is found that the existence of control Lyapunov function is equivalent to that of SDSFC. A necessary and sufficient condition for the existence of control Lyapunov function under SDSFC is obtained, and all possible SDSFC and their corresponding structural matrices of a control Lyapunov function have been designed by two different methods. In the first method, admissible control Lyapunov function has been defined. In the second one, we have redefined the control Lyapunov function under SDSFC of PBCNs.

Besides, the computational complexity of Algorithm 18 is $O(|\widetilde{S}^*|\tau \times 2^m)$ and the computational complexity of the process of searching for all controllers as well as control Lyapunov functions is $O(|S|\tau \times 2^m)$ in this section. Computational complexity is one of the current challenges in the study of control theory for BCNs or PBCNs. It is obvious that computational complexity grows exponentially. Thus, the proposed method in this section can be applied to the networks of no more than 15 nodes at present. In fact, it is difficult to effective reduce the high computational complexity, for the control design methods are based on the state transition matrix L. In future work, we will pay attention to the reduction of the computational complexity to expand the scope of application.

Finally, we have investigated the finite-time and infinitetime set stabilization of PBCNs under SDSFC, respectively. Algorithms have been respectively proposed to find the sampled point set and the largest sampled point control invariant set. A necessary and sufficient condition has been proposed for finite-time global set stabilization of PBCNs by SDSFC and the time-optimal sampled-data state-feedback controller has been designed. Further, a necessary and sufficient condition for infinite-time global set stabilization of PBCNs has been obtained and all possible sampled-data state-feedback controllers have been designed by using all the complete families of reachable sets. Finally, three examples have been presented to illustrate the effectiveness of the obtained results. In the future, the application of the proposed control method to some biological systems with more nodes will be considered. Moreover, we will continue to study other control design methods to reduce the computational complexity meanwhile saving the control cost. Besides, the sampling period considered in this article is fixed at all the time, thus one interesting research topic includes that the sampling period is not periodic.

References

1. Li, R., Yang, M., Chu, T.: State feedback stabilization for probabilistic Boolean networks. Automatica **50**(4), 1272–1278 (2014)
2. Liu, Y., Cao, J., Sun, L., et al.: Sampled-data state feedback stabilization of Boolean control networks. Neural Comput. **28**(4), 778–799 (2016)
3. Li, H., Wang, Y.: Output feedback stabilization control design for Boolean control networks. Automatica **49**(12), 3641–3645 (2013)
4. Pal, R., Datta, A., Bittner, M.L., et al.: Intervention in context-sensitive probabilistic Boolean networks. Bioinformatics **21**(7), 1211–1218 (2005)
5. Li, H., Ding, X.: A control Lyapunov function approach to feedback stabilization of logical control networks. SIAM J. Control Optim. **57**(2), 810–831 (2019)
6. Tong, L., Liu, Y., Li, Y., et al.: Robust control invariance of probabilistic Boolean control networks via event-triggered control. IEEE Access **6**, 37767–37774 (2018)
7. Li, R., Yang, M., Chu, T.: State feedback stabilization for Boolean control networks. IEEE Trans. Autom. Control **58**(7), 1853–1857 (2013)
8. Chaves, M.: Methods for qualitative analysis of genetic networks. In: 2009 European Control Conference (ECC), pp. 671–676 (2009)
9. Veliz-Cuba, A., Stigler, B.: Boolean models can explain bistability in the *lac* operon. J. Comput. Biol. **18**(6), 783–794 (2011)

Part III
Aperiodic Sampled-Data Control

Chapter 5
Stabilization of Aperiodic Sampled-Data Boolean Control Networks

Abstract In this chapter, we study the stabilization of Boolean control networks (BCNs) under aperiodic sampled-data control (ASDC).

5.1 Stabilization of Boolean Control Networks Under Aperiodic Sampled-Data Control

In this section, a novel technique is provided for the global stability analysis of Boolean control networks (BCNs) under aperiodic sampled-data control (ASDC). The sampling period is allowed to be taken from a limited number of values. Using semi-tensor product, a BCN under ASDC can be converted into a switched Boolean network (BN). Here, the switched BN can only switch at sampling instants, which does not mean that the switches occur at each sampling instant. For the switched BN, we consider two cases: (i) switched BN with all stable subsystems; (ii) switched BN containing both stable subsystems and unstable subsystems. For these two cases, the techniques of switching-based Lyapunov function and the average dwell time method are used to derive sufficient conditions for global stability of BCNs under ASDC, respectively. The upper bounds of the cost function, which is defined to guarantee that the considered BCN is not only globally stabilized but also can assurance the performance at an appropriate level, are determined, respectively. Moreover, an algorithm is presented to construct the ASDCs. Finally, the obtained results are demonstrated by several examples.

5.1.1 *Problem Formulation*

A BCN under ASDC can be described as follows:

$$\begin{aligned} X(t+1) &= f(X(t), U(t)), \\ U(t) &= e(X(t_k)), \quad t_k \le t < t_{k+1}, \end{aligned} \tag{5.1}$$

Y. Liu et al., *Sampled-data Control of Logical Networks*,
https://doi.org/10.1007/978-981-19-8261-3_5

where $X(t)$ and $U(t)$ represent the n-dimensional state variable and m-dimensional ASDC input variable at discrete time t, taking values in $\mathcal{D}^n$ and $\mathcal{D}^m$, respectively. t_k for $k = 0, 1, \ldots$ are sampling instants. The maps $f : \mathcal{D}^{n+m} \to \mathcal{D}^n$ and $e : \mathcal{D}^n \to \mathcal{D}^m$ are logical functions.

Upon representing the logical variables $X(t)$ and $U(t)$ by means of their vector forms $x(t) \in \Delta_{2^n}$ and $u(t) \in \Delta_{2^m}$, respectively, the BCN (5.1) can be given as below,

$$x(t+1) = Lu(t)x(t), \tag{5.2}$$

where $L \in \mathcal{L}_{2^n \times 2^{n+m}}$. Similar to the above analysis, the algebraic form of ASDC can be described as

$$u(t) = Kx(t_k), \quad t_k \le t < t_{k+1}, \tag{5.3}$$

where $K \in \mathcal{L}_{2^m \times 2^n}$.

In this section, the sampling period is defined as follows.

Let $h_k \triangleq t_{k+1} - t_k$, where $h_k \in Z_h \triangleq \{i_1, \ldots, i_l\}$, $i_1 < i_2 < \cdots < i_l$ and $i_j,\ j = 1, \ldots, l$, are positive integers. Then h_k can take one value from these l values.

Consider system (5.2) and the controller (5.3), we have

$$\begin{aligned}
x(t_k+1) &= Lu(t_k)x(t_k) = LW_{[2^n,2^m]}x(t_k)u(t_k),\\
x(t_k+2) &= Lu(t_k+1)x(t_k+1) = (LW_{[2^n,2^m]})^2 x(t_k)u(t_k)u(t_k+1),\\
&\ \vdots\\
x(t_{k+1}) &= Lu(t_{k+1}-1)x(t_{k+1}-1) = (LW_{[2^n,2^m]})^{t_{k+1}-t_k} x(t_k)u(t_k)\cdots u(t_{k+1}-1),
\end{aligned}$$

where $k = 0, 1, \ldots$. Since $t_{k+1} - t_k = h_k$ and $u(t_k) = u(t_k+1) = \cdots = u(t_{k+1}-1)$, we can discretize system (5.2) with sampling period h_k as follows:

$$\begin{aligned}
x(t_{k+1}) &= (LW_{[2^n,2^m]})^{h_k} x(t_k)\Phi_m^{h_k-1}u(t_k),\\
&\triangleq \tilde{L}^{h_k}x(t_k)\Phi_m^{h_k-1}u(t_k),
\end{aligned} \tag{5.4}$$

where $k = 0, 1, \ldots$. It is worth noting that $\tilde{L}^{h_k}$ and $\Phi_m^{h_k-1}$ critically depend on h_k. Here, h_k takes l values, then $\tilde{L}^{h_k}$ and $\Phi_m^{h_k-1}$ take l distinct values, respectively. Thus, we can regard system (5.4) as a switched BN with l subsystems.

Define $\sigma(t_k) \in Z_\sigma \triangleq \{1, 2, \ldots, l\}$ as a piecewise constant switching signal. Let $A_{\sigma(t_k)} = \tilde{L}^{i_{\sigma(t_k)}}$ and $B_{\sigma(t_k)} = \Phi_m^{i_{\sigma(t_k)}-1}$, then system (5.4) can be rewritten as below,

$$x(t_{k+1}) = A_{\sigma(t_k)}x(t_k)B_{\sigma(t_k)}u(t_k), \tag{5.5}$$

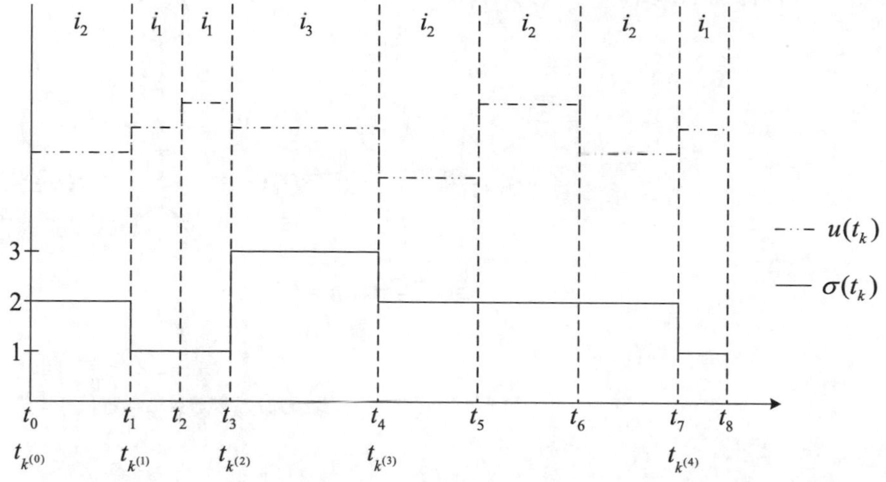

Fig. 5.1 An example of the switching law

where $k = 0, 1, \ldots$. During the sampling interval $[t_k, t_{k+1})$, h_k takes value $i_j,\ j \in Z_\sigma$, then $\sigma(t_k) = j$ and system (5.5) resides in the jth-subsystem. So only when h_k varies, system (5.5) switches.

It is worthwhile to note that switches may not occur at every sampling instant. The following example is given to explain this point.

Assume that the sampling period h_k takes three values $i_1 = 2,\ i_2 = 4,\ i_3 = 6$, then system (5.5) has three subsystems and $\sigma(t_k) \in Z_\sigma = \{1, 2, 3\}$.

In Fig. 5.1, $t_k,\ k = 1, \ldots, 8$ are the sampling instants. But the switches only occur at $t_{k(j)},\ j = 1, \ldots, 4$, where $t_{k(1)} = t_1,\ t_{k(2)} = t_3,\ t_{k(3)} = t_4$ and $t_{k(4)} = t_7$, which means that switches may not occur at every sampling instant.

For any $t_k > t_0 = 0$, a switching sequence $t_0 = t_{k(0)} < t_{k(1)} < \cdots < t_{k(l)} < t_k$ during interval $[t_0, t_k)$ is assumed. When $t \in [t_{k(J)}, t_{k(j+1)}),\ j = 1, 2, \ldots, i$, the $\sigma(t_{k(j)})$th subsystem is activated; that is, switching sequence corresponding to the switching signal $\sigma(t_k)$ is given as follows:

$$\{(\sigma(t_{k(0)}), t_{k(0)}), \ldots, (\sigma(t_{k(i)}), t_{k(i)})|\sigma(t_{k(j)}) \in Z_\sigma,\ j = 0, 1, \ldots, i,\ i \geq 1\},$$

where $t_{k(0)} = t_0$ and $t_{k(j)} \in \{t_1, \ldots, t_{k-1}\},\ j = 1, \ldots, i$.

Consider system (5.5) and controller (5.3), we have

$$\begin{aligned} x(t_{k+1}) &= A_{\sigma(t_k)} x(t_k) B_{\sigma(t_k)} u(t_k) \\ &= A_{\sigma(t_k)} x(t_k) B_{\sigma(t_k)} K x(t_k) \\ &= A_{\sigma(t_k)} (I_{2^n} \otimes B_{\sigma(t_k)} K) \Phi_n x(t_k) \\ &\triangleq F_{\sigma(t_k)} x(t_k), \end{aligned} \tag{5.6}$$

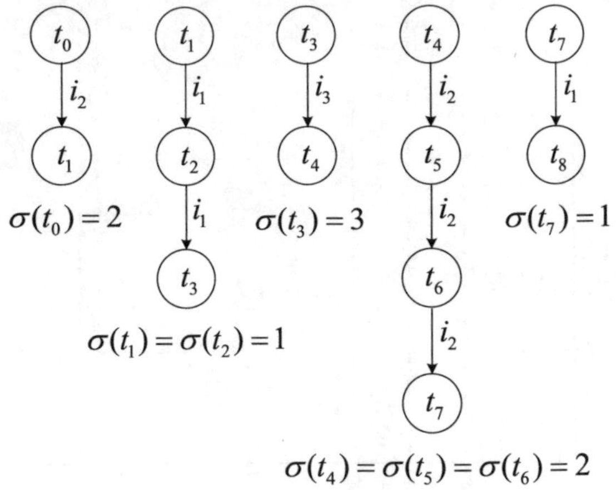

Fig. 5.2 Constructed forest of Fig. 5.1

where $k = 0, 1, \ldots$ and $F_{\sigma(t_k)} = A_{\sigma(t_k)}(I_{2^n} \otimes B_{\sigma(t_k)}K)\Phi_n$.

Remark 5.1 For the switching law in Fig. 5.1, we can construct a forest according to the following procedure:

1. draw the roots t_0, t_1, t_3, t_4, t_7 (which are the switching instants);
2. draw the edge from t_0 to t_1;
3. draw the edge from t_1 to t_2 and the edge from t_2 to t_3;
4. $\cdots$

In Fig. 5.2, each directed tree can be used to express a subsystem of switched system (5.5). In a directed tree, the distance between the two adjacent nodes is the same, and it means that each subsystem of system (5.5) has a constant sampling period in Z_h.

5.1.2 Global Stability

By means of the average-dwell time method, the global stability of system (5.2) is analyzed.

A system is said to be globally stable at $X_e = (x_1^e, x_2^e, \ldots, x_n^e)$, if for any initial state, the state of the system converges to X_e. Without loss of generality, here, we assume $X_e = (x_1^e, x_2^e, \ldots, x_n^e) = (0, 0, \ldots, 0)$ (a coordinate transformation can guarantee this [1]).

Definition 5.1 (Meng et al. [2] and Hespanda and Morse [3]) For a switching signal σ and $t > t_0$, let $N_\sigma(t_0, t)$ be the switching numbers of σ over interval $[t_0, t)$.

If there exist an $N_0 \geq 0$ and an $\tau_a > 0$ such that

$$N_\sigma(t_0, t) \leq N_0 + \frac{t - t_0}{\tau_a},$$

holds for any t, then the switching signal σ is said to have an average dwell time not less than τ_a and N_0 is called a chatter bound.

Remark 5.2 According to the previous analysis, system (5.6) can only switch at the sampling instants, that is, $N_\sigma(t_0, t_k) \leq k$ for all k.

Definition 5.2 The set $\{\beta_j | j \in Z_\sigma\}$ is called a set of Lyapunov coefficients of system (5.6) if the following conditions are satisfied

$$\beta_j^{\mathrm{T}} \delta_{2^n}^{2^n} = 0, \tag{5.7}$$

$$\beta_j^{\mathrm{T}} \delta_{2^n}^{r} > 0, \quad r = 1, 2, \ldots, 2^n - 1, \tag{5.8}$$

$$\beta_j^{\mathrm{T}} (F_j - \lambda_j^{2i_j} I_{2^n}) \delta_{2^n}^{2^n} = 0, \tag{5.9}$$

$$\beta_j^{\mathrm{T}} (F_j - \lambda_j^{2i_j} I_{2^n}) \delta_{2^n}^{r} < 0, \quad r = 1, 2, \ldots, 2^n - 1, \tag{5.10}$$

$$\beta_i \leq \mu \beta_j, \quad i, j \in Z_\sigma, \ \mu \geq 1, \tag{5.11}$$

where $0 < \lambda_j < 1$, if the jth subsystem is stable; $\lambda_j \geq 1$, if the jth subsystem is unstable, and $i_j \in Z_h$ is the sampling period.

In the following theorem, a sufficient condition is given for the global stability of system (5.2) and we consider system (5.6) with all stable subsystems, i.e., $0 < \lambda_j < 1, \ j \in Z_\sigma$. Throughout this section, we assume the average dwell time of (5.6) not less than τ_a.

Theorem 5.1 *Given $\alpha_j \geq 0$, $\sum_{j=1}^{l} \alpha_j = 1$. If there exists a set of Lyapunov coefficients $\{\beta_j | j \in Z_\sigma\}$ as defined in Definition 5.2, such that*

$$\tau_a > \frac{\ln \sqrt{\mu}}{\sum_{j=1}^{l} \alpha_j \ln \lambda_j^{-1}}, \tag{5.12}$$

then system (5.2) is globally stable under feedback (5.3) at X_e with a decay rate $\theta(\tau_a, \alpha_j) = \mu^{\frac{1}{2\tau_a}} \prod_{j=1}^{l} \lambda_j^{\alpha_j}$, where α_j are the activation frequencies of the sampling periods.

Proof We first prove that system (5.6) is globally stable at X_e. Define a Lyapunov function of system (5.6) as follows:

$$V_{\sigma(t_k)}(t_k) = \beta^{\mathrm{T}}_{\sigma(t_k)} x(t_k). \tag{5.13}$$

Now, we denote $t_{k^{(1)}}, t_{k^{(2)}}, \ldots, t_{k^{(i)}}$ by the switching instants.

For any $t \in [t_{k^{(j-1)}}, t_{k^{(j)}}),\ j = 1, \ldots, i,\ t_{k^{(0)}} = t_0 = 0$, we have two cases: (i) $t_{k^{(j-1)}+1} < t_{k^{(j)}}$; (ii) $t_{k^{(j-1)}+1} = t_{k^{(j)}}$. For the first case $t_{k^{(j-1)}+1} < t_{k^{(j)}}$, we have $\sigma(t_{k^{(j-1)}}) = \sigma(t_{k^{(j-1)}+1})$, and then

$$\begin{aligned}
& V_{\sigma(t_{k^{(j-1)}})}(t_{k^{(j-1)}+2}) \\
=& \beta^{\mathrm{T}}_{\sigma(t_{k^{(j-1)}})} x(t_{k^{(j-1)}+2}) \\
=& \beta^{\mathrm{T}}_{\sigma(t_{k^{(j-1)}})} F_{\sigma(t_{k^{(j-1)}+1})} x(t_{k^{(j-1)}+1}) \\
=& \beta^{\mathrm{T}}_{\sigma(t_{k^{(j-1)}+1})} F_{\sigma(t_{k^{(j-1)}+1})} x(t_{k^{(j-1)}+1}) \\
=& \lambda_{\sigma(t_{k^{(j-1)}+1})}^{2(t_{k^{(j-1)}+2} - t_{k^{(j-1)}+1})} V_{\sigma(t_{k^{(j-1)}+1})}(t_{k^{(j-1)}+1}) + E_{\sigma(t_{k^{(j-1)}+1})} x(t_{k^{(j-1)}+1}) \\
\leq& \lambda_{\sigma(t_{k^{(j-1)}})}^{2(t_{k^{(j-1)}+2} - t_{k^{(j-1)}+1})} V_{\sigma(t_{k^{(j-1)}+1})}(t_{k^{(j-1)}+1}) \\
=& \lambda_{\sigma(t_{k^{(j-1)}})}^{2(t_{k^{(j-1)}+2} - t_{k^{(j-1)}+1})} \beta^{\mathrm{T}}_{\sigma(t_{k^{(j-1)}})} x(t_{k^{(j-1)}+1}) \\
=& \lambda_{\sigma(t_{k^{(j-1)}})}^{2(t_{k^{(j-1)}+2} - t_{k^{(j-1)}+1})} (\lambda_{\sigma(t_{k^{(j-1)}})}^{2(t_{k^{(j-1)}+1} - t_{k^{(j-1)}})} V_{\sigma(t_{k^{(j-1)}})}(t_{k^{(j-1)}}) + E_{\sigma(t_{k^{(j-1)}})} x(t_{k^{(j-1)}})) \\
\leq& \lambda_{\sigma(t_{k^{(j-1)}})}^{2(t_{k^{(j-1)}+2} - t_{k^{(j-1)}})} V_{\sigma(t_{k^{(j-1)}})}(t_{k^{(j-1)}}),
\end{aligned} \tag{5.14}$$

where

$$\begin{aligned}
& E_{\sigma(t_{k^{(j-1)}+1})} x(t_{k^{(j-1)}+1}) \\
=& (\beta^{\mathrm{T}}_{\sigma(t_{k^{(j-1)}+1})} F_{\sigma(t_{k^{(j-1)}+1})} - \lambda_{\sigma(t_{k^{(j-1)}+1})}^{2(t_{k^{(j-1)}+2} - t_{k^{(j-1)}+1})} \beta^{\mathrm{T}}_{\sigma(t_{k^{(j-1)}+1})}) x(t_{k^{(j-1)}+1}) \\
=& \beta^{\mathrm{T}}_{\sigma(t_{k^{(j-1)}+1})} (F_{\sigma(t_{k^{(j-1)}+1})} - \lambda_{\sigma(t_{k^{(j-1)}+1})}^{2(t_{k^{(j-1)}+2} - t_{k^{(j-1)}+1})} I_{2^n}) x(t_{k^{(j-1)}+1}) \\
\leq& 0,
\end{aligned}$$

and

$$
\begin{aligned}
&E_{\upsilon(t_{k(j-1)})}x(t_{k(j-1)})\\
=&(\beta^{\mathrm{T}}_{\sigma(t_{k(j-1)})}F_{\sigma(t_{k(j-1)})}-\lambda_{\sigma(t_{k(j-1)})}^{2(t_{k(j-1)+1}-t_{k(j-1)})}\beta^{\mathrm{T}}_{\sigma(t_{k(j-1)})})x(t_{k(j-1)})\\
=&\beta^{\mathrm{T}}_{\sigma(t_{k(j-1)})}(F_{\sigma(t_{k(j-1)})}-\lambda_{\sigma(t_{k(j-1)})}^{2(t_{k(j-1)+1}-t_{k(j-1)})}I_{2^n})x(t_{k(j-1)})\\
\leq&0.
\end{aligned}
$$

Similarly, one can get that $V_{\sigma(t_{k(j-1)})}(t_{k(j)})\leq\lambda_{\sigma(t_{k(j-1)})}^{2(t_{k(j)}-t_{k(j-1)})}V_{\sigma(t_{k(j-1)})}(t_{k(j-1)})$.

For the second case $t_{k(j-1)+1}=t_{k(j)}$, we have

$$
\begin{aligned}
&V_{\sigma(t_{k(j-1)})}(t_{k(j)})\\
=&\beta^{\mathrm{T}}_{\sigma(t_{k(j-1)})}x(t_{k(j)})\\
=&\beta^{\mathrm{T}}_{\sigma(t_{k(j-1)})}F_{\sigma(t_{k(j-1)})}x(t_{k(j-1)})\\
=&\lambda_{\sigma(t_{k(j-1)})}^{2(t_{k(j)}-t_{k(j-1)})}V_{\sigma(t_{k(j-1)})}(t_{k(j-1)})+E_{\sigma(t_{k(j-1)})}x(t_{k(j-1)})\\
\leq&\lambda_{\sigma(t_{k(j-1)})}^{2(t_{k(j)}-t_{k(j-1)})}V_{\sigma(t_{k(j-1)})}(t_{k(j-1)}).
\end{aligned}
$$

In summary, we have

$$
V_{\sigma(t_{k(j-1)})}(t_{k(j)})\leq\lambda_{\sigma(t_{k(j-1)})}^{2(t_{k(j)}-t_{k(j-1)})}V_{\sigma(t_{k(j-1)})}(t_{k(j-1)}),\quad j=1,\ldots,i. \tag{5.15}
$$

Since $\sigma(t_{k(j-1)})=\sigma(t_{k(j)-1})$, one can get that

$$
V_{\sigma(t_{k(j)-1})}(t_{k(j)})\leq\lambda_{\sigma(t_{k(j-1)})}^{2(t_{k(j)}-t_{k(j-1)})}V_{\sigma(t_{k(j-1)})}(t_{k(j-1)}),\quad j=1,\ldots,i. \tag{5.16}
$$

By (5.11), one has

$$
V_{\sigma(t_{k(j)})}(t_{k(j)})\leq\mu V_{\sigma(t_{k(j)-1})}(t_{k(j)}),\quad j=1,\ldots,i. \tag{5.17}
$$

Here, we only consider switching instants during interval $[t_0,t_k)$, where $t_{k(i)}<t_k$. Thus, we do not need to determine $\sigma(t_k)$ and for convenience, we assume $\sigma(t_k)=\sigma(t_{k-1})=\cdots=\sigma(t_{k-(k-k^{(i)})})=\sigma(t_{k(i)})$. Based on (5.14), we can get the following equation by induction,

$$
V_{\sigma(t_k)}(t_k)\leq\lambda_{\sigma(t_{k(i)})}^{2(t_k-t_{k(i)})}V_{\sigma(t_{k(i)})}(t_{k(i)}). \tag{5.18}
$$

Then, combining with inequalities (5.16), (5.17) and (5.18) gives that

$$
\begin{aligned}
V_{\sigma(t_k)}(t_k) &\le \lambda_{\sigma(t_{k(i)})}^{2(t_k - t_{k(i)})} V_{\sigma(t_{k(i)})}(t_{k(i)}) \\
&\le \lambda_{\sigma(t_{k(i)})}^{2(t_k - t_{k(i)})} \mu V_{\sigma(t_{k(i)-1})}(t_{k(i)}) \\
&\le \mu \lambda_{\sigma(t_{k(i)})}^{2(t_k - t_{k(i)})} \lambda_{\sigma(t_{k(i-1)})}^{2(t_{k(i)} - t_{k(i-1)})} V_{\sigma(t_{k(i-1)})}(t_{k(i-1)}) \\
&\le \cdots \\
&\le \mu^{N_\sigma(t_0, t_k)} \lambda_{\sigma(t_{k(i)})}^{2(t_k - t_{k(i)})} \cdots \lambda_{\sigma(t_0)}^{2(t_{k(1)} - t_0)} V_{\sigma(t_0)}(t_0) \qquad (5.19) \\
&= \mu^{N_\sigma(t_0, t_k)} \prod_{j=1}^{l} \lambda_j^{2\alpha_j (t_k - t_0)} V_{\sigma(t_0)}(t_0) \\
&\le \mu^{\frac{t_k - t_0}{\tau_a}} \prod_{j=1}^{l} \lambda_j^{2\alpha_j (t_k - t_0)} V_{\sigma(t_0)}(t_0) \\
&= \theta^{2t_k}(\tau_a, \alpha_j) V_{\sigma(t_0)}(t_0),
\end{aligned}
$$

where $\theta(\tau_a, \alpha_j) = \mu^{\frac{1}{2\tau_a}} \prod_{j=1}^{l} \lambda_j^{\alpha_j}$.

If $\theta(\tau_a, \alpha_j) < 1$, then $\mu^{\frac{1}{2\tau_a}} \prod_{j=1}^{l} \lambda_j^{\alpha_j} < 1$, and $\tau_a > \frac{\ln \sqrt{\mu}}{\sum_{j=1}^{l} \alpha_j \ln \lambda_j^{-1}}$.

Therefore, it can be concluded that $V_{\sigma(t_k)}(t_k)$ converges to zero as $k \to \infty$, if the average dwell time τ_a satisfies (5.12). Invoking the construction of V in (5.13) and conditions (5.7), (5.8), one obtains that $x(t_k) \to \delta_{2^n}^{2^n}$ as $k \to \infty$, i.e., system (5.6) is globally stable with respect to the point $\delta_{2^n}^{2^n}$. It implies that $x(t) \to \delta_{2^n}^{2^n}$ as $t \to \infty$. The proof is thus completed. □

In Theorem 5.1, the global stability of system (5.6) with all stable subsystems is studied. Now we consider the case that system (5.6) contains both stable subsystems and unstable subsystems.

Suppose that the index set of stable subsystems is φ_s and the index set of unstable subsystems is φ_u. Let $|\varphi_s| = r$ and $|\varphi_u| = l - r$, which means that the number of stable subsystems in system (5.6) is r and the number of unstable subsystems in system (5.6) is $l - r$. Now, denote $f_s \triangleq \sum_{j \in \varphi_s} \alpha_j$ and $f_u \triangleq \sum_{j \in \varphi_u} \alpha_j$ as the activation frequency of the stable subsystems and the activation frequency of the unstable subsystems, respectively.

In the following theorem, another sufficient condition for global stability of system (5.2) is given.

Theorem 5.2 *Given* $\alpha_j \geq 0$, $\sum_{j=1}^{l} \alpha_j = 1$. *If there exists a set of Lyapunov coefficients* $\{\beta_j | j \in Z_\sigma\}$ *as defined in Definition 5.2, such that* $\prod_{j=1}^{l} \lambda_j^{\alpha_j} < 1$ *holds, and*

$$\tau_a > \frac{\ln \sqrt{\mu}}{\ln \lambda_s^{-1} + f_u \ln(\lambda_s \lambda_u^{-1})}, \tag{5.20}$$

$$f_u < \frac{\ln \lambda_s^{-1}}{\ln(\lambda_u \lambda_s^{-1})}, \tag{5.21}$$

where $f_s \triangleq \sum_{j \in \varphi_s} \alpha_j$ *and* $f_u \triangleq \sum_{j \in \varphi_u} \alpha_j$, *then system (5.2) is globally stable under feedback (5.3) at* X_e *with a decay rate* $\hat{\theta}(\tau_a, f_u) = \mu^{\frac{1}{2\tau_a}} \lambda_s (\lambda_u \lambda_s^{-1})^{f_u}$, *where* $\lambda_s = \max_{j \in \varphi_s}\{\lambda_j\}$ *and* $\lambda_u = \max_{j \in \varphi_u}\{\lambda_j\}$.

Proof From the proof of Theorem 5.1, we can obtain that

$$\begin{aligned} V_{\sigma(t_k)}(t_k) &\leq \mu^{\frac{t_k}{\tau_a}} \prod_{j=1}^{l} \lambda_j^{2\alpha_j t_k} V_{\sigma(t_0)}(t_0) \\ &= (\mu^{\frac{1}{2\tau_a}} \prod_{j=1}^{l} \lambda_j^{\alpha_j})^{2t_k} V_{\sigma(t_0)}(t_0). \end{aligned}$$

Now, we only need to guarantee $\mu^{\frac{1}{2\tau_a}} \prod_{j=1}^{l} \lambda_j^{\alpha_j} < 1$, such that $\lim_{k \to \infty} V_{\sigma(t_k)}(t_k) = 0$.

By $f_s = \sum_{j \in \varphi_s} \alpha_j$, $f_u = \sum_{j \in \varphi_u} \alpha_j$ and $f_s + f_u = 1$, one can obtain that

$$\begin{aligned} \mu^{\frac{1}{2\tau_a}} \prod_{j=1}^{l} \lambda_j^{\alpha_j} &= \mu^{\frac{1}{2\tau_a}} \lambda_j^{\sum_{j \in \varphi_s} \alpha_j} \lambda_j^{\sum_{j \in \varphi_u} \alpha_j} \\ &\leq \mu^{\frac{1}{2\tau_a}} \lambda_s^{\sum_{j \in \varphi_s} \alpha_j} \lambda_u^{\sum_{j \in \varphi_u} \alpha_j} \\ &= \mu^{\frac{1}{2\tau_a}} \lambda_s^{f_s} \lambda_u^{f_u} \\ &= \mu^{\frac{1}{2\tau_a}} \lambda_s^{1-f_u} \lambda_u^{f_u}. \end{aligned}$$

To guarantee $\mu^{\frac{1}{2\tau_a}} \prod_{j=1}^{l} \lambda_j^{\alpha_j} < 1$, we only need to ensure $\mu^{\frac{1}{2\tau_a}} \lambda_s^{1-f_u} \lambda_u^{f_u} < 1$.

Since $\lambda_u,\ \lambda_s > 0$ and $\mu \geq 1$, one has

$$\begin{aligned} \mu^{\frac{1}{2\tau_a}} &< \lambda_s^{-(1-f_u)}\lambda_u^{-f_u}, \\ \frac{1}{\tau_a}\ln\sqrt{\mu} &< (1-f_u)\ln\lambda_s^{-1} + f_u\ln\lambda_u^{-1}. \end{aligned} \tag{5.22}$$

By (5.21), it can be deduced that the right hand side of (5.22) is $(1-f_u)\ln\lambda_s^{-1} + f_u\ln\lambda_u^{-1} > 0$. Thus, the inequality (5.22) is valid. Combing (5.22) and $\ln\sqrt{\mu} > 0$ gives that $\tau_a > \frac{\ln\sqrt{\mu}}{\ln\lambda_s^{-1} + f_u\ln(\lambda_s\lambda_u^{-1})}$, which means that when (5.20) holds, then we have $\mu^{\frac{1}{2\tau_a}}\prod_{j=1}^{l}\lambda_j^{\alpha_j} < 1$ and $\hat{\theta}(\tau_a, f_u) \triangleq \mu^{\frac{1}{2\tau_a}}\lambda_s^{1-f_u}\lambda_u^{f_u} = \mu^{\frac{1}{2\tau_a}}\lambda_s(\lambda_u\lambda_s^{-1})^{f_u}$.

Thus this completes the proof. □

Remark 5.3 According to Theorem 5.2, if system (5.6) contains both stable subsystems and unstable subsystems, according to inequality (5.21), only when f_u is very small, the global stability of system (5.6) can be guaranteed. Moreover, with the decrease of f_u, the decay rate $\hat{\theta}(\tau_a, f_u) = \mu^{\frac{1}{2\tau_a}}\lambda_s(\lambda_u\lambda_s^{-1})^{f_u}$ is also reduced.

5.1.3 Guaranteed Cost Analysis

In this subsection, we derive some conditions to guarantee not only the global stability of system (5.2) but also the performance with an adequate level. Motivated by optimal control of BCNs [4], we design a cost function as follows:

$$J = \sum_{k=0}^{\infty}\xi^{\mathrm{T}}u(t_k)x(t_k), \tag{5.23}$$

where $\xi > 0$ and $\xi^{\mathrm{T}}K\Phi_n\delta_{2^n}^{2^n} = 0$. If controller (5.3) guarantees that system (5.6) is globally stable and that J has an upper bound, then (5.3) is called the guaranteed cost controller and that system (5.6) achieves the guaranteed cost performance level.

Definition 5.3 The set $\{\beta_j | j \in Z_\sigma\}$ is called a set of Lyapunov coefficients for the cost function $J = \sum_{k=0}^{\infty}\xi^{\mathrm{T}}u(t_k)x(t_k)$ of system (5.6) if it satisfies the conditions (5.7), (5.8), (5.9), (5.11) and

$$(\beta_j^{\mathrm{T}}(F_j - \lambda_j^{2i_j}I_{2^n}) + \xi^{\mathrm{T}}K\Phi_n)\delta_{2^n}^r < 0, \quad r = 1, 2, \ldots, 2^n - 1, \tag{5.24}$$

where $0 < \lambda_j < 1$, if the jth subsystem is stable; $\lambda_j \geq 1$, if the jth subsystem is unstable, and $i_j \in Z_h$ is the sampling period.

In the following theorem, we consider the case that system (5.6) just contains stable subsystems, and a sufficient condition is given to guarantee that the cost function has an upper bound.

Theorem 5.3 *Given $\alpha_j \geq 0$, $\sum_{j=1}^{l} \alpha_j = 1$. If there exists a set of Lyapunov coefficients $\{\beta_j | j \in Z_\sigma\}$ for the cost function $J = \sum_{k=0}^{\infty} \xi^{\mathrm{T}} u(t_k) x(t_k)$ as defined in Definition 5.3, such that (5.12) holds, then system (5.2) is globally stable under feedback (5.3) with decay rate $\theta(\tau_a, \alpha_j)$ and cost function satisfies $J \leq \frac{1-\lambda_0^2}{(1-\theta^2(\tau_a,\alpha_j))} V_{\sigma(t_0)}(t_0)$, where $\lambda_0 = \min_{j \in Z_\sigma} \{\lambda_j\} < 1$.*

Proof Since (5.24) guarantees that (5.10), one can conclude from Theorem 5.1 that system (5.2) is globally stable with decay rate $\theta(\tau_a, \alpha_j)$.

So we only need to prove $J \leq \frac{1-\lambda_0^2}{(1-\theta^2(\tau_a,\alpha_j))} V_{\sigma(t_0)}(t_0)$.

Consider Lyapunov function (5.13), by (5.24), one has

$$
\begin{aligned}
& V_{\sigma(t_{k(j-1)})}(t_{k(j-1)+1}) \\
=& \beta^{\mathrm{T}}_{\sigma(t_{k(j-1)})} x(t_{k(j-1)+1}) \\
=& \beta^{\mathrm{T}}_{\sigma(t_{k(j-1)})} F_{\sigma(t_{k(j-1)})} x(t_{k(j-1)}) \\
=& \tilde{E}_{\sigma(t_{k(j-1)})} x(t_{k(j-1)}) + \lambda_{\sigma(t_{k(j-1)})}^{2(t_{k(j-1)+1} - t_{k(j-1)})} V_{\sigma(t_{k(j-1)})}(t_{k(j-1)}) - J(t_{k(j-1)}),
\end{aligned}
$$

where

$$
\begin{aligned}
& \tilde{E}_{\sigma(t_{k(j-1)})} x(t_{k(j-1)}) \\
=& (\beta^{\mathrm{T}}_{\sigma(t_{k(j-1)})} (F_{\sigma(t_{k(j-1)})} - \lambda_{\sigma(t_{k(j-1)})}^{2(t_{k(j-1)+1} - t_{k(j-1)})} I_{2^n}) + \xi^{\mathrm{T}} K \Phi_n) x(t_{k(j-1)}) \\
\leq& 0,
\end{aligned}
$$

So

$$
V_{\sigma(t_{k(j-1)})}(t_{k(j-1)+1}) \begin{cases} < \lambda_{\sigma(t_{k(j-1)})}^{2(t_{k(j-1)+1} - t_{k(j-1)})} V_{\sigma(t_{k(j-1)})}(t_{k(j-1)}) - J(t_{k(j-1)}), \\ = 0 \text{ if } x(t_{k(j-1)}) = \delta_{2^n}^{2^n}, \quad \text{otherwise.} \end{cases}
$$

For any $t \in [t_{k(j-1)}, t_{k(j)})$, $j = 1, \ldots, i$, $t_{k(0)} = t_0$, we have two cases: (i) $t_{k(j-1)+1} < t_{k(j)}$; (ii) $t_{k(j-1)+1} = t_{k(j)}$. For the first case $t_{k(j-1)+1} < t_{k(j)}$, we have $\sigma(t_{k(j-1)}) =$

$\sigma(t_{k^{(j-1)}+1})$, and then

$$\begin{aligned}
&V_{\sigma(t_{k^{(j-1)}})}(t_{k^{(j-1)}+2})\\
=&\beta^{\mathrm{T}}_{\sigma(t_{k^{(j-1)}})}x(t_{k^{(j-1)}+2})\\
=&\beta^{\mathrm{T}}_{\sigma(t_{k^{(j-1)}})}F_{\sigma(t_{k^{(j-1)}+1})}x(t_{k^{(j-1)}+1})\\
\leq&\lambda_{\sigma(t_{k^{(j-1)}})}^{2(t_{k^{(j-1)}+2}-t_{k^{(j-1)}+1})}V_{\sigma(t_{k^{(j-1)}})}(t_{k^{(j-1)}+1})-J(t_{k^{(j-1)}+1})\\
\leq&\lambda_{\sigma(t_{k^{(j-1)}})}^{2(t_{k^{(j-1)}+2}-t_{k^{(j-1)}})}V_{\sigma(t_{k^{(j-1)}})}(t_{k^{(j-1)}})-\lambda_{\sigma(t_{k^{(j-1)}})}^{2(t_{k^{(j-1)}+2}-t_{k^{(j-1)}+1})}J(t_{k^{(j-1)}})-J(t_{k^{(j-1)}+1}).
\end{aligned}\tag{5.25}$$

For the second case $t_{k^{(j-1)}+1}=t_{k^{(j)}}$, we have

$$V_{\sigma(t_{k^{(j-1)}})}(t_{k^{(j)}})\leq\lambda_{\sigma(t_{k^{(j-1)}})}^{2(t_{k^{(j)}}-t_{k^{(j-1)}})}V_{\sigma(t_{k^{(j-1)}})}(t_{k^{(j-1)}})-J(t_{k^{(j-1)}}),\tag{5.26}$$

by (5.25) and (5.26), one obtains that

$$\begin{aligned}
&V_{\sigma(t_{k^{(j-1)}})}(t_{k^{(j)}})\\
\leq&\lambda_{\sigma(t_{k^{(j-1)}})}^{2(t_{k^{(j)}}-t_{k^{(j-1)}})}V_{\sigma(t_{k^{(j-1)}})}(t_{k^{(j-1)}})-\sum_{s=0}^{k^{(j)}-k^{(j-1)}-1}\lambda_{\sigma(t_{k^{(j-1)}})}^{2(t_{k^{(j)}}-t_{k^{(j-1)}+s+1})}J(t_{k^{(j-1)}+s}).
\end{aligned}\tag{5.27}$$

Since $\sigma(t_{k^{(j-1)}})=\sigma(t_{k^{(j)}-1})$, it implies that

$$\begin{aligned}
&V_{\sigma(t_{k^{(j)}-1})}(t_{k^{(j)}})\\
\leq&\lambda_{\sigma(t_{k^{(j-1)}})}^{2(t_{k^{(j)}}-t_{k^{(j-1)}})}V_{\sigma(t_{k^{(j-1)}})}(t_{k^{(j-1)}})-\sum_{s=0}^{k^{(j)}-k^{(j-1)}-1}\lambda_{\sigma(t_{k^{(j-1)}})}^{2(t_{k^{(j)}}-t_{k^{(j-1)}+s+1})}J(t_{k^{(j-1)}+s}).
\end{aligned}\tag{5.28}$$

Then, we have

$$V_{\sigma(t_k)}(t_k)\leq\lambda_{\sigma(t_{k^{(i)}})}^{2(t_k-t_{k^{(i)}})}V_{\sigma(t_{k^{(i)}})}(t_{k^{(i)}})-\sum_{s=0}^{k-k^{(i)}-1}\lambda_{\sigma(t_{k^{(i)}})}^{2(t_k-t_{k^{(i)}+s+1})}J(t_{k^{(i)}+s}).\tag{5.29}$$

Combining with the inequalities (5.17), (5.28) and (5.29), one has that

$$
\begin{aligned}
& V_{\sigma(t_k)}(t_k) \\
& \le \lambda_{\sigma(t_k)}^{2(t_k - t_{k^{(i)}})} V_{\sigma(t_{k^{(i)}})}(t_{k^{(i)}}) - \sum_{s=0}^{k-k^{(i)}-1} \lambda_{\sigma(t_{k^{(i)}})}^{2(t_k - t_{k^{(i)}+s+1})} J(t_{k^{(i)}+s}) \\
& \le \lambda_{\sigma(t_{k^{(i)}})}^{2(t_k - t_{k^{(i)}})} \mu V_{\sigma(t_{k^{(i)}-1})}(t_{k^{(i)}}) - \sum_{s=0}^{k-k^{(i)}-1} \lambda_{\sigma(t_{k^{(i)}})}^{2(t_k - t_{k^{(i)}+s+1})} J(t_{k^{(i)}+s}) \\
& \le \cdots \\
& \le \mu^{N_\sigma(t_0,t_k)} \lambda_{\sigma(t_{k^{(i)}})}^{2(t_k - t_{k^{(i)}})} \lambda_{\sigma(t_{k^{(i-1)}})}^{2(t_{k^{(i)}} - t_{k^{(i-1)}})} \cdots \lambda_{\sigma(t_0)}^{2(t_{k^{(1)}} - t_0)} V_{\sigma(t_0)}(t_0) - \Omega \\
& \le \theta^{2t_k}(\tau_a, \alpha_j) V_{\sigma(t_0)}(t_0) - \Omega \\
& \le \theta^{2t_k}(\tau_a, \alpha_j) V_{\sigma(t_0)}(t_0) - \sum_{s=0}^{k-1} \mu^{N_\sigma(t_s,t_k)} \lambda_0^{2(t_k - t_{s+1})} J(t_s) \\
& \le \theta^{2t_k}(\tau_a, \alpha_j) V_{\sigma(t_0)}(t_0) - \sum_{s=0}^{k-1} \lambda_0^{2(t_k - t_{s+1})} J(t_s),
\end{aligned}
\tag{5.30}
$$

where

$$
\mu^{N_\sigma(t_0,t_k)} \lambda_{\sigma(t_{k^{(i)}})}^{2(t_k - t_{k^{(i)}})} \cdots \lambda_{\sigma(t_0)}^{2(t_{k^{(1)}} - t_0)} V_{\sigma(t_0)}(t_0) \le \theta^{2t_k}(\tau_a, \alpha_j) V_{\sigma(t_0)}(t_0),
$$

$\lambda_0 = \min_{j \in Z_\sigma}\{\lambda_j\} < 1$ and

$$
\begin{aligned}
\Omega = & \mu^{N_\sigma(t_0,t_k)} \lambda_{\sigma(t_{k^{(i)}})}^{2(t_k - t_{k^{(i)}})} \cdots \lambda_{\sigma(t_{k^{(1)}})}^{2(t_{k^{(2)}} - t_{k^{(1)}})} \sum_{s=0}^{k^{(1)}-1} \lambda_{\sigma(t_0)}^{2(t_{k^{(1)}} - t_{1+s})} J(t_s) \\
& + \mu^{N_\sigma(t_0,t_k)-1} \lambda_{\sigma(t_{k^{(i)}})}^{2(t_k - t_{k^{(i)}})} \cdots \lambda_{\sigma(t_{k^{(2)}})}^{2(t_{k^{(3)}} - t_{k^{(2)}})} \sum_{s=0}^{k^{(2)}-k^{(1)}-1} \lambda_{\sigma(t_{k^{(1)}})}^{2(t_{k^{(2)}} - t_{k^{(1)}+s+1})} J(t_{k^{(1)}+s}) \\
& + \cdots + \mu^0 \sum_{s=0}^{k-k^{(i)}-1} \lambda_{\sigma(t_{k^{(i)}})}^{2(t_k - t_{k^{(i)}+s+1})} J(t_{k^{(i)}+s}) \\
\ge & \sum_{s=0}^{k-1} \mu^{N_\sigma(t_s,t_k)} \lambda_0^{2(t_k - t_{s+1})} J(t_s).
\end{aligned}
$$

By $V_{\sigma(t_k)}(t_k) \geq 0$, one obtains that

$$\sum_{s=0}^{k-1} \lambda_0^{2(t_k - t_{s+1})} J(t_s) \leq \theta^{2t_k}(\tau_a, \alpha_j) V_{\sigma(t_0)}(t_0),\ t_0 = 0. \tag{5.31}$$

Summing both sides of (5.31) from $k = 1$ to $+\infty$, one obtains that

$$\begin{aligned}
&\sum_{k=1}^{+\infty}\sum_{s=0}^{k-1} \lambda_0^{2(t_k - t_{s+1})} J(t_s) \\
=&\sum_{s=0}^{+\infty} J(t_s)\Big(\sum_{k=s+1}^{+\infty} \lambda_0^{2(t_k - t_{s+1})}\Big) \\
=&(1-\lambda_0^2)^{-1}\sum_{s=0}^{+\infty} J(t_s) \\
\leq&\sum_{t_k=0}^{+\infty} \theta^{2t_k}(\tau_a, \alpha_j) V_{\sigma(t_0)}(t_0) \\
=&(1-\theta^2(\tau_a, \alpha_j))^{-1} V_{\sigma(t_0)}(t_0),
\end{aligned}$$

which implies that $J = \sum_{s=0}^{+\infty} J(t_s) \leq \frac{1-\lambda_0^2}{(1-\theta^2(\tau_a,\alpha_j))} V_{\sigma(t_0)}(t_0)$. The proof is completed. □

Subsequently, consideration of the case that system (5.6) contains r stable subsystems and $l-r$ unstable subsystems, a sufficient condition is given to guarantee that the J has an upper bound.

Theorem 5.4 *Given $\alpha_j \geq 0$, $\sum_{j=1}^{l} \alpha_j = 1$. If there exists a set of Lyapunov coefficients $\{\beta_j | j \in Z_\sigma\}$ for the $J = \sum_{k=0}^{\infty} \xi^{\mathrm{T}} u(t_k) x(t_k)$ as defined in Definition 5.3, such that the inequalities $\prod_{j=1}^{l} \lambda_j^{\alpha_j} < 1$ and (5.20) hold, then system (5.2) is globally stable under feedback (5.3) with decay rate $\hat{\theta}(\tau_a, f_u)$ and $J \leq \frac{1-\lambda_0^2}{(1-\hat{\theta}^2(\tau_a,\alpha_j))} V_{\sigma(t_0)}(t_0)$.*

Proof Follows from the proofs of Theorems 5.2 and 5.3. □

5.1.4 Controller Design

Now, we design ASDC (5.3) under which system (5.2) can be globally stabilized to $\delta_{2^n}^{2^n}$.

System (5.2) is globally stabilized to $\delta_{2^n}^{2^n}$ under ASDC (5.3), if for any initial state $x(0) \in \Delta_{2^n}$, there exists a sampling instant t_k, such that $x(t) = \delta_{2^n}^{2^n}$, $t \geq t_k$ and $x(t_{k-1}) \neq \delta_{2^n}^{2^n}$, where $t_k - t_{k-1} = h_{k-1}$, $h_{k-1} \in Z_h$, $k = 0, 1, 2, \ldots$. Here, we assume that the sampling periods are $t_k - t_{k-1} = j_1$, $t_{k-1} - t_{k-2} = j_2, \ldots, t_1 - t_0 = j_k$, where $j_1, j_2, \ldots, j_k$ only represent the selected positive integer from Z_h and for $\forall c, w \in \{1, 2, \ldots, k\}$, j_c can be larger than j_w, or less than j_w, or even equal to j_w.

Consider system (5.2) and controller (5.3), we get

$$\begin{aligned} x(t_k) &= (LW_{[2^n,2^m]})^{t_k - t_{k-1}} x(t_{k-1}) u(t_{k-1}) \cdots u(t_k - 1) \\ &= (LW_{[2^n,2^m]})^{j_1} (LW_{[2^n,2^m]})^{t_{k-1} - t_{k-2}} x(t_{k-2}) u(t_{k-2}) \cdots u(t_k - 1) \\ &= \cdots \\ &= (LW_{[2^n,2^m]})^{j_1} (LW_{[2^n,2^m]})^{j_2} \cdots (LW_{[2^n,2^m]})^{j_k} x(0) u(0) \cdots u(t_k - 1) \\ &\triangleq \bar{L}^{j_1} \bar{L}^{j_2} \cdots \bar{L}^{j_k} x(0) u(0) \cdots u(t_k - 1), \end{aligned} \tag{5.32}$$

where $x(t) = \delta_{2^n}^{2^n}$, $t \geq t_k$, $x(t_{k-1}) \neq \delta_{2^n}^{2^n}$ and $\bar{L} \triangleq LW_{[2^n,2^m]}$.

Split $\bar{L}^{j_c}$, $c = 1, 2, \ldots, k$ into 2^n equal blocks as

$$\bar{L}^{j_c} = [\bar{L}_1^{j_c} \; \bar{L}_2^{j_c} \; \cdots \; \bar{L}_{2^n}^{j_c}],$$

where $\bar{L}_i^{j_c} \in \mathcal{L}_{2^n \times 2^{mj_c}}$, $i = 1, 2, \ldots, 2^n$.

Assume $K = \delta_{2^m}[p_1 \; p_2 \; \cdots \; p_{2^n}]$, where controller (5.3) is in form of $u(t) = Kx(t_k)$, $t \in [t_k, t_{k+1})$, $k = 0, 1, \ldots$. Define sets by

$$\begin{aligned} R_{t_k}(\delta_{2^n}^{2^n}) = \{\delta_{2^n}^i :\ &\text{there exists } q_i \in \{(p_i - 1)\frac{1 - 2^{mj_1}}{1 - 2^m} + 1 \mid p_i = 1, 2, \ldots, 2^m\}, \\ &\text{such that } Col_{q_i}(\bar{L}_i^{j_1}) = \delta_{2^n}^{2^n}\} \setminus \{\delta_{2^n}^{2^n}\}, \end{aligned}$$

$$\begin{aligned} R_{t_{k-1}}(\delta_{2^n}^{2^n}) = \{\delta_{2^n}^i :\ &\text{there exists } q_i \in \{(p_i - 1)\frac{1 - 2^{mj_2}}{1 - 2^m} + 1 \mid p_i = 1, 2, \ldots, 2^m\}, \\ &\text{such that } Col_{q_i}(\bar{L}_i^{j_2}) \in R_{t_k}(\delta_{2^n}^{2^n})\} \setminus [R_{t_k}(\delta_{2^n}^{2^n}) \cup \{\delta_{2^n}^{2^n}\}], \end{aligned}$$

$$\vdots$$

$$\begin{aligned} R_{t_1}(\delta_{2^n}^{2^n}) = \{\delta_{2^n}^i :\ &\text{there exists } q_i \in \{(p_i - 1)\frac{1 - 2^{mj_k}}{1 - 2^m} + 1 \mid p_i = 1, 2, \ldots, 2^m\}, \\ &\text{such that } Col_{q_i}(\bar{L}_i^{j_k}) \in R_{t_2}(\delta_{2^n}^{2^n})\} \setminus [\cup_{2 \leq l \leq k} R_{t_l}(\delta_{2^n}^{2^n}) \cup \{\delta_{2^n}^{2^n}\}]. \end{aligned}$$

Theorem 5.5 *If system (5.2) can be globally stabilized to $\delta_{2^n}^{2^n}$ under ASDC (5.3), then*

(I) $\delta_{2^n}^{2^n} = LK\delta_{2^n}^{2^n}\delta_{2^n}^{2^n}$.
(II) There exists a k, such that $\Delta_{2^n} = [\cup_{1\leq l\leq k} R_{t_l}(\delta_{2^n}^{2^n})] \cup \{\delta_{2^n}^{2^n}\}$.

Proof If system (5.2) can be globally stabilized to $\delta_{2^n}^{2^n}$ by controller (5.3), then there exists $T > 0$ such that $\forall t \geq T,\ x(t) = \delta_{2^n}^{2^n}$. Suppose that $T = t_k$, we have $\delta_{2^n}^{2^n} = x(t_k + 1) = Lu(t_k)x(t_k) = LKx(t_k)x(t_k) = LK\delta_{2^n}^{2^n}\delta_{2^n}^{2^n}$. Therefore, (i) holds. If $\delta_{2^n}^{i} \notin [\cup_{1\leq l\leq k} R_{t_l}(\delta_{2^n}^{2^n})] \cup \{\delta_{2^n}^{2^n}\}$, then for initial state $x(0) = \delta_{2^n}^{i}$, system (5.2) can not converge to $\delta_{2^n}^{2^n}$ by controller (5.3). Indeed, system (5.2) can be globally stabilized to $\delta_{2^n}^{2^n}$, so the contradiction appears. Hence (II) holds as well. □

Base on Theorem 5.5, we present an algorithm to get the feedback matrix K as follows.

Algorithm 21 Get the feedback matrix K.

Step 1. Solving $\delta_{2^n}^{2^n} = LK\delta_{2^n}^{2^n}\delta_{2^n}^{2^n}$ to get p_{2^n}. If there is no solution, then K does not exist.
Step 2. For any initial state $x(0) = \delta_{2^n}^{i},\ i = 1, 2, \ldots, 2^n - 1$, if $\delta_{2^n}^{i} \in R_{t_c}(\delta_{2^n}^{2^n}),\ c \in \{1, 2, \ldots, k\}$, then there exists $q_i \in \{(p_i - 1)\frac{1-2^{mj_{k+1-c}}}{1-2^m} + 1 \mid p_i = 1, 2, \ldots, 2^m\}$, such that $Col_{q_i}(\bar{L}_i^{j_{k+1-c}}) \in R_{t_{c+1}}(\delta_{2^n}^{2^n})$, get p_i. If $\delta_{2^n}^{i} \notin \cup_{1\leq l\leq k} R_{t_l}(\delta_{2^n}^{2^n})$, then K does not exist.
Step 3. The feedback matrix $K = \delta_{2^m}[p_1\ p_2\ \cdots\ p_{2^n}]$ can be obtained.

Theorem 5.6 *The feedback law (5.3) with the state feedback matrix K obtained by Algorithm 21 can globally stabilize system (5.2) to* $X_e = \delta_{2^n}^{2^n}$.

Proof For $x(0) = \delta_{2^n}^{2^n}$, because there exists p_{2^n} such that $\delta_{2^n}^{2^n} = LK\delta_{2^n}^{2^n}\delta_{2^n}^{2^n}$, we have $x(t) \equiv \delta_{2^n}^{2^n},\ t \geq 0$. For $x(0) = \delta_{2^n}^{i} \in R_{t_k}(\delta_{2^n}^{2^n})$, there exists p_i such that $\delta_{2^n}^{2^n} = \bar{L}\delta_{2^n}^{i}\delta_{2^m}^{p_i}\cdots\delta_{2^m}^{p_i} = Col_{q_i}(\bar{L}_i^{j_1})$, hence we have $x(t) \equiv \delta_{2^n}^{2^n},\ t \geq j_1$. For $x(0) = \delta_{2^n}^{i} \in R_{t_{k-1}}(\delta_{2^n}^{2^n})$, there exists p_i such that $\bar{L}\delta_{2^n}^{i}\delta_{2^m}^{p_i}\cdots\delta_{2^m}^{p_i} = Col_{q_i}(\bar{L}_i^{j_2}) \in R_{t_k}(\delta_{2^n}^{2^n})$, so one has $x(t) \equiv \delta_{2^n}^{2^n},\ t \geq j_1 + j_2 \ldots$ By induction, we can conclude that for any initial state $x(0) = \delta_{2^n}^{i},\ i = 1, 2, \ldots, 2^n$, the state of system (5.2) converges to $X_e = \delta_{2^n}^{2^n}$. □

Remark 5.4 When the feedback matrix $K = \delta_{2^m}[p_1\ p_2\ \cdots\ p_{2^n}]$ is designed, it is necessary to make sure that $Col_i(L_{p_i})^{i_j} \neq \delta_{2^n}^{i}$, where i_j is the sampling period and the jth subsystem is stable. Otherwise, inequality (5.10) can not be satisfied.

5.1.5 Examples

This example is given to illustrate Theorem 5.2.

We consider a BCN model from [5].

$$\begin{cases} x_1(t+1) = \neg u_1(t) \wedge (x_2(t) \vee x_3(t)), \\ x_2(t+1) = \neg u_1(t) \wedge u_2(t) \wedge x_1(t), \\ x_3(t+1) = \neg u_1(t) \wedge (u_2(t) \vee (u_3(t) \wedge x_1(t))), \end{cases} \tag{5.33}$$

In this model, there are 6 variables. x_1: the lac mRNA; x_2 and x_3: high concentration level and medium concentration level of the lactose, respectively; u_1: the extracellular glucose; u_2 and u_3: high concentration level and medium concentration level of the extracellular lactose, respectively.

Here, let x_1, x_2 and x_3 be state variables; u_1, u_2 and u_3 be control inputs.

Set $x(t) = \ltimes_{i=1}^{3} x_i(t)$ and $u(t) = \ltimes_{i=1}^{3} u_i(t)$, we get

$$x(t+1) = Lu(t)x(t),$$

where

$$\begin{aligned} L = \delta_8[&8\ 8\ 8\ 8\ 8\ 8\ 8\ 8\ 8\ 8\ 8\ 8\ 8\ 8\ 8\ 8\ 8\ 8\ 8\ 8\ 8\ 8\ 8\ 8\ 8\ 8\ 8\ 8\ 8\ 8\ 8\ 8 \\ &1\ 1\ 1\ 5\ 3\ 3\ 3\ 7\ 1\ 1\ 1\ 5\ 3\ 3\ 3\ 7\ 3\ 3\ 3\ 7\ 4\ 4\ 4\ 8\ 4\ 4\ 4\ 8\ 4\ 4\ 4\ 8]. \end{aligned}$$

The ASDC for BCN (5.33) is given in the form (5.3) as

$$u(t) = Kx(t_k), \quad t_k \leq t < t_{k+1},$$

where $K = \delta_8[8\ 1\ 4\ 7\ 4\ 6\ 8\ 8]$.

Consider sampling period $h_k \in \{1, 2, 3\}$. Then we can get switched BN as below,

$$x(t_{k+1}) = F_{\sigma(t_k)}x(t_k),$$

where $F_1 = \delta_8[4\ 8\ 8\ 7\ 8\ 3\ 4\ 8]$ and $F_2 = \delta_8[8\ 8\ 8\ 4\ 8\ 1\ 8\ 8]$. We can observe that these two subsystems are unstable. While $F_3 = \delta_8[8\ 8\ 8\ 7\ 8\ 1\ 8\ 8]$, and it means that the third one is stable. According to Theorem 5.2, choosing $\lambda_1 = 1.5,\ \lambda_2 = 1.2,\ \lambda_3 = 0.9$, and then we have $\lambda_s = \lambda_3 = 0.9,\ \lambda_u = \lambda_1 = 1.5,\ f_s = \alpha_3,\ f_u = \alpha_1 + \alpha_2$.

By calculation, one can observe that the inequalities in (5.7), (5.8), (5.9), (5.10) and (5.11) in Theorem 5.2 hold for $\mu = 1.05$,

$$\beta_1 = (130, 132, 300, 290, 137, 501, 129, 0)^{\mathrm{T}},$$

$$\beta_2 = (131, 138, 286, 277, 142, 525, 130, 0)^{\mathrm{T}},$$

$$\beta_3 = (132, 137, 288, 290, 140, 501, 131, 0)^{\mathrm{T}}.$$

In addition, one can compute that $f_u < \frac{\ln \lambda_s^{-1}}{\ln(\lambda_u \lambda_s^{-1})} \approx 0.2063$.

Choose $f_u = 0.2,\ f_s = 1 - f_u = 0.8$, and then

$$\tau_a > \frac{\ln \sqrt{\mu}}{\ln \lambda_s^{-1} + f_u \ln(\lambda_s \lambda_u^{-1})} \approx 7.6345.$$

Choose $\tau_a = 8$. Hence, by Theorem 5.2, system (5.33) is globally stable with respect to δ_8^8 with a decay rate

$$\hat{\theta}(\tau_a, f_u) \approx 0.9999.$$

Assume that the first two subsystems are activated 2 units of time and the third subsystem is activated 8 units of time. When the initial state is $x(0) = \delta_8^4$, Fig. 5.3 shows the state trajectory of (5.33). Also, the corresponding switching signal $\sigma(t_k)$ and controller $u(t)$ are shown in Figs. 5.4 and 5.5, respectively. This example illustrates the validity of Theorem 5.2.

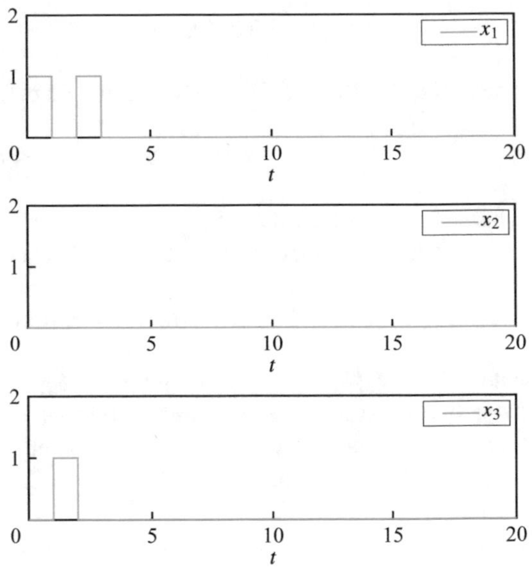

Fig. 5.3 The state trajectory of (5.33) with the initial state $x(0) = \delta_8^4$

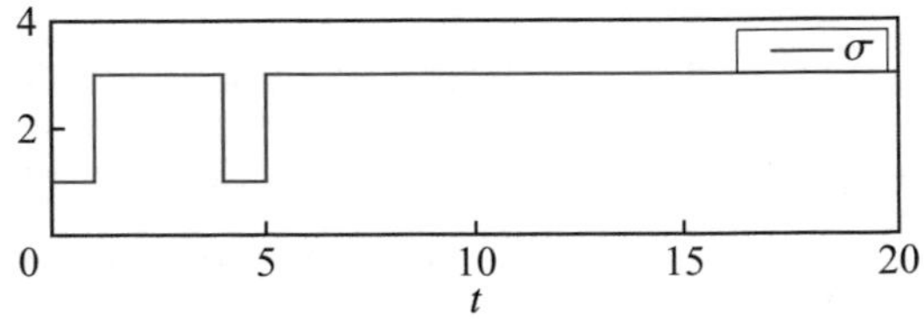

Fig. 5.4 For the initial state $x(0) = \delta_8^4$, the trajectory of switching signal $\sigma(t_k)$

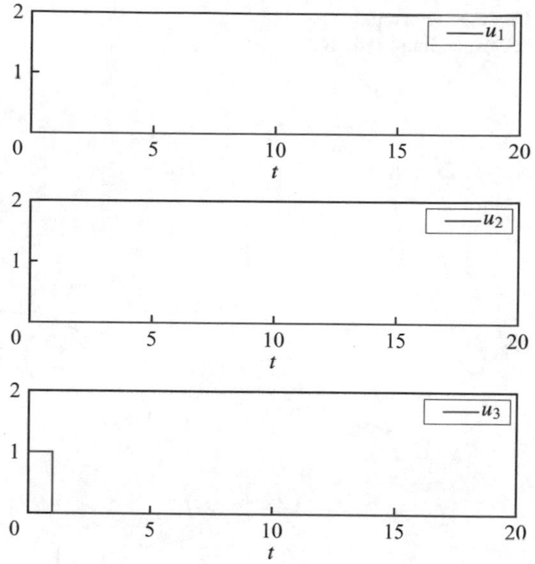

Fig. 5.5 For the initial state $x(0) = \delta_8^4$, the trajectory of controller $u(t)$

This example is given to illustrate Algorithm 21.

Consider system (5.2) with $2^n = 8,\ 2^m = 2$. Suppose that

$$L = \delta_8[4\ 7\ 2\ 8\ 7\ 6\ 4\ 4\ 3\ 3\ 1\ 1\ 6\ 5\ 2\ 8],$$

and $t_k - t_{k-1} = 2$, $t_{k-1} - t_{k-2} = 1$, $t_{k-2} - t_{k-3} = 3$, $t_{k-3} - t_{k-4} = 2$, $t_{k-4} - t_{k-5} = 1$, $t_{k-5} - t_{k-6} = 3, \ldots$. One has

$$\begin{aligned}\bar{L}^3 = \delta_8[&4\ 8\ 4\ 3\ 7\ 3\ 4\ 3\ 8\ 1\ 7\ 3\ 7\ 3\ 4\ 3\ 4\ 2\ 2\ 1\ 8\ 1\ 2\ 1\ 8\ 1\ 4\ 8\ 8\ 1\ 2\ 1\\ &8\ 1\ 7\ 3\ 6\ 5\ 7\ 6\ 6\ 5\ 7\ 6\ 4\ 2\ 6\ 5\ 4\ 8\ 4\ 3\ 4\ 2\ 2\ 1\ 4\ 8\ 4\ 3\ 8\ 1\ 4\ 8],\\ \bar{L}^2 = \delta_8[&8\ 1\ 2\ 1\ 4\ 2\ 2\ 1\ 7\ 3\ 4\ 3\ 4\ 8\ 4\ 3\ 4\ 2\ 6\ 5\ 6\ 5\ 7\ 6\ 8\ 1\ 7\ 3\ 8\ 1\ 4\ 8],\\ \bar{L}^1 = \delta_8[&4\ 3\ 7\ 3\ 2\ 1\ 8\ 1\ 7\ 6\ 6\ 5\ 4\ 2\ 4\ 8].\end{aligned}$$

Then $R_{t_3}(\delta_8^8) = \{\delta_8^1, \delta_8^7\}$, $R_{t_2}(\delta_8^8) = \{\delta_8^2, \delta_8^3, \delta_8^4, \delta_8^5\}$, $R_{t_1}(\delta_8^8) = \{\delta_8^6\}$, i.e., there exists an $k = 3$, such that $\Delta_8 = [\cup_{1 \le l \le 3} R_{t_l}(\delta_8^8)] \cup \{\delta_8^8\}$.

Next, the controller according to Algorithm 21 is designed.

Step 1. Solving $\delta_8^8 = LK\delta_8^8\delta_8^8$ to get $p_8 = 2$

Step 2. For initial state $x(0) = \delta_8^1$, $\delta_8^1 \in R_{t_3}(\delta_8^8)$, there exists $q_1 \in \{1, 4\}$, such that $Col_1(\bar{L}_1^2) = \delta_8^8$, get $p_1 = 1$. For initial state $x(0) = \delta_8^7$, $\delta_8^7 \in R_{t_3}(\delta_8^8)$, then there exists $q_7 \in \{1, 4\}$, such that $Col_1(\bar{L}_7^2) = \delta_8^8$, get $p_7 = 1$. For initial state $x(0) = \delta_8^2$, $\delta_8^2 \in R_{t_2}(\delta_8^8)$, there exists $q_2 \in \{1, 2\}$, such that $Col_1(\bar{L}_2^1) = \delta_8^7$, get $p_2 = 1$. Similarly, one can obtain $p_3 = p_4 = p_6 = 2$ and $p_5 = 1$.

Step 3. Calculate controller $u(t)$ with $K = \delta_2[1\ 1\ 2\ 2\ 1\ 2\ 1\ 2]$.

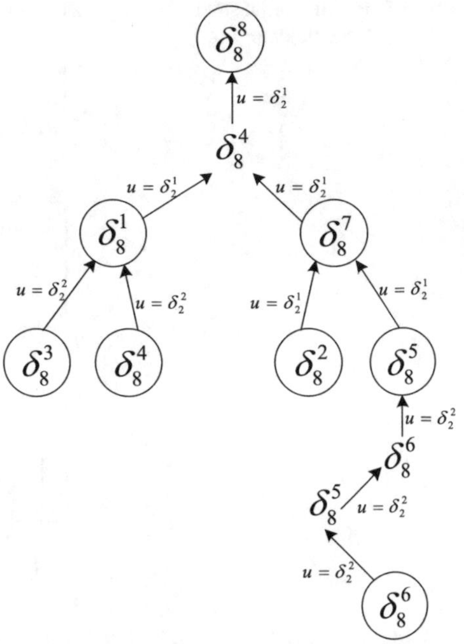

Fig. 5.6 Diagraph corresponding to the BCN

Finally, when $u(t) = Kx_1(t)x_2(t)x_3(t) = \delta_2[1\ 1\ 2\ 2\ 1\ 2\ 1\ 2]x(t)$, one can obtain a logical expression of u as $u(t) = (x_1(t) \wedge x_1(t)) \vee \{(\neg x_1(t)) \wedge [(x_2(t) \wedge x_3(t)) \vee (\neg x_2(t) \wedge x_3(t))]\}$.

Figure 5.6 shows that system (5.2) is globally stabilized to δ_8^8 with $2^n = 8,\ 2^m = 2$ under the designed controller $u(t)$. So this example illustrates the validity of Algorithm 21.

5.2 Stabilization of Aperiodic Sampled-Data Boolean Control Networks with All Modes Unstable

In this section, we aim to further study the global stability of Boolean control networks (BCNs) under aperiodic sampled-data control (ASDC). According to our previous work, it is known that a BCN under ASDC can be transformed into a switched Boolean network (BN), and further global stability of the BCN under ASDC can be obtained by studying the global stability of the transformed switched BN. Unfortunately, since the major idea of our previous work is to use stable subsystems to offset the state divergence caused by unstable subsystems, the switched BN considered has at least one stable subsystem. The central thought in this paper is that switching behavior also has good stabilization; i.e., the switched BN can also be stable with appropriate switching laws designed, even if all subsystems are unstable. This is completely different from that in our previous work.

Specifically, for this case, the dwell time should be limited within a pair of upper and lower bounds. By means of the discretized Lyapunov function and dwell time, a sufficient condition for global stability is obtained. Finally, the above results are demonstrated by a biological example.

5.2.1 Convert a Boolean Control Network Under Aperiodic Sampled-Data Control into a Switched Boolean Network

Consider the following BCN under ASDC:

$$\begin{aligned} X(t+1) &= f(X(t), U(t)), \\ U(t) &= e(X(t_k)), \quad t_k \le t < t_{k+1}, \end{aligned} \tag{5.34}$$

where $X(t) \in \mathcal{D}^n$ is the state variable, $U(t) \in \mathcal{D}^m$ is the ASDC input variable, and $t_k,\ k = 0, 1, \ldots$ are sampling instants. The mappings $f : \mathcal{D}^{n+m} \to \mathcal{D}^n$ and $e : \mathcal{D}^n \to \mathcal{D}^m$ are logical functions.

Then we can represent logical functions f and e by their unique structure matrices M and E, respectively. Here $X(t)$ and $U(t)$ are represented by their vector forms $x(t) \in \Delta_{2^n}$ and $u(t) \in \Delta_{2^m}$, respectively, System (5.34) is given as follows:

$$x(t+1) = Mu(t)x(t), \tag{5.35}$$

$$u(t) = Ex(t_k), \quad t_k \le t < t_{k+1}, \tag{5.36}$$

where $M \in \mathcal{L}_{2^n \times 2^{n+m}}$ and $E \in \mathcal{L}_{2^m \times 2^n}$.

Denote $h_k \triangleq t_{k+1} - t_k$ the kth sampling interval, where $h_k \in Z_h \triangleq \{i_1, i_2, \ldots, i_l\}$, $i_1 < i_2 < \cdots < i_l$ and $i_j,\ j = 1, 2, \ldots, l$ are positive integers. Then system (5.35) under ASDC (5.36) can be translated into a switched BN, which can be described as follows:

$$\begin{aligned} x(t_{k+1}) &= (MW_{[2^n,2^m]})^{h_k} x(t_k) \Phi_m^{h_k-1} u(t_k) \\ &= (MW_{[2^n,2^m]})^{h_k} (I_{2^n} \otimes \Phi_m^{h_k-1} E) \Phi_n x(t_k) \\ &\triangleq F_{\sigma(t_k)} x(t_k), \end{aligned} \tag{5.37}$$

where the switching signal $\sigma(t_k) \in Z_\sigma \triangleq \{1, 2, \ldots, l\}$. Note that the switched BN (5.37) switches only at the sampling instant, but the switch may not occur at every sampling instant. The corresponding details have been well discussed in [6]. Thus, the switching time sequence is given below:

$$0 = t_0 = t_{k^{(0)}} < t_{k^{(1)}} < t_{k^{(2)}} < \cdots < t_{k^{(i)}} < t_n,$$

where $t_{k^{(j)}} \in \{t_0, t_1, \ldots, t_n\}$, $j = 0, 1, \ldots, i$, and $t_0, t_1, \ldots, t_n$ are sampling instants.

5.2.2 Main Results

In this part, we analyze the global stability of system (5.35) under ASDC (5.36). We consider the case that all subsystems of the corresponding switched BN (5.37) are unstable.

Definition 5.4 System (5.35) is said to be globally stable at x_e, if for any initial state $x(0) \in \Delta_{2^n}$, the corresponding trajectory $x(t)$ converges to x_e.

Here, we assume $x_e = \delta_{2^n}^{2^n}$ (a coordinate transformation in [1] can ensure this).

Lemma 5.1 *System (5.35) is globally stable at x_e if and only if the corresponding switched BN (5.37) is globally stable at x_e and $\delta_{2^n}^{2^n} = ME\delta_{2^n}^{2^n}\delta_{2^n}^{2^n}$.*

Here, we assume that system (5.35) satisfies $\delta_{2^n}^{2^n} = ME\delta_{2^n}^{2^n}\delta_{2^n}^{2^n}$.

Let $Z \triangleq \{k^{(0)}, k^{(1)}, \ldots, k^{(i-1)}\}$. Define

$$\tau_{k^{(j)}} = k^{(j+1)} - k^{(j)},$$

as the dwell time, where $j = 0, 1, \ldots, i-1$ and $\tau_{k^{(j)}} \in [\tau_{\min}, \tau_{\max}]$ with $\tau_{\min} = \inf_{k^{(j)} \in Z} \tau_{k^{(j)}}$ and $\tau_{\max} = \sup_{k^{(j)} \in Z} \tau_{k^{(j)}}$.

Remark 5.5 Note that the dwell time should be limited within a pair of upper and lower bounds to ensure global stability, since too small or too large dwell time may lead the switched BN (5.37) containing all unstable subsystems to be unstable.

Definition 5.5 The set of vectors $\{\beta_{a,q} | a \in Z_\sigma, q = 0, 1, \ldots, L\}$ is defined as a set of Lyapunov coefficients of the switched BN (5.37) if for $\forall r = 1, 2, \ldots, 2^n - 1$ and $\forall a, b \in Z_\sigma$, the following equations/inequalities are satisfied:

$$\beta_{a,q}^{\mathrm{T}} \delta_{2^n}^{2^n} = 0, \quad q = 0, 1, \ldots, L, \tag{5.38}$$

$$\beta_{a,q}^{\mathrm{T}} \delta_{2^n}^{r} > 0, \quad q = 0, 1, \ldots, L, \tag{5.39}$$

$$[\beta_{a,q}^{\mathrm{T}}(F_a - I) + \frac{1}{h}(\beta_{a,q+1}^{\mathrm{T}} - \beta_{a,q}^{\mathrm{T}})F_a - \lambda_a \beta_{a,q}^{\mathrm{T}}]\delta_{2^n}^{r} < 0, \quad q = 0, 1, \ldots, L-1, \tag{5.40}$$

$$[\beta_{a,q+1}^{\mathrm{T}}(F_a - I) + \frac{1}{h}(\beta_{a,q+1}^{\mathrm{T}} - \beta_{a,q}^{\mathrm{T}})F_a - \lambda_a \beta_{a,q+1}^{\mathrm{T}}]\delta_{2^n}^{r} < 0, \quad q = 0, 1, \ldots, L-1, \tag{5.41}$$

$$F_a \delta_{2^n}^{2^n} = \delta_{2^n}^{2^n}, \tag{5.42}$$

$$[\beta_{a,L}^{\mathrm{T}}(F_a - I) - \lambda_a \beta_{a,L}^{\mathrm{T}}]\delta_{2^n}^r < 0, \tag{5.43}$$

$$\beta_{b,0}^{\mathrm{T}} \leq \mu_b \beta_{a,L}^{\mathrm{T}}, \quad 0 < \mu_b < 1, \ a \neq b, \tag{5.44}$$

where $\lambda_a > 0$ and $h = \lfloor \frac{\tau_{\min}}{L} \rfloor$.

A result on the global stability of system (5.35) is obtained below. Here, note that all subsystems in (5.37) can be unstable.

Theorem 5.7 *Consider system (5.35) under ASDC (5.36). If there exists a set of Lyapunov coefficients* $\{\beta_{a,q} > 0 | a \in Z_\sigma, q = 0, 1, \ldots, L\}$ *as defined in Definition 5.5 and a constant* $\tau_{\max} \geq \tau_{\min}$*, such that for any* $a, b \in Z_\sigma$, $a \neq b$, *the following inequality holds:*

$$\ln \mu_b + \tau_{\max} \ln(\lambda_a + 1) < 0, \tag{5.45}$$

then system (5.35) is globally stable at x_e.

Proof We first prove that the switched BN (5.37) is globally stable at x_e. For any $\sigma(t_k) \in Z_\sigma$, we construct following the Lyapunov function of the switched BN (5.37)

$$V_{\sigma(t_k)}(t_k) = \beta_{\sigma(t_k)}^{\mathrm{T}}(k) x(t_k). \tag{5.46}$$

Now, we denote $t_{k^{(1)}}, t_{k^{(2)}}, \ldots, t_{k^{(i)}}$ the switching instants.

For any $t \in [t_{k^{(j)}}, t_{k^{(j+1)}})$, $j = 0, 1, \ldots i-1$, the ath subsystem is activated, i.e., $\sigma(t_{k^{(j)}}) = a$.

The interval $[k^{(j)}, k^{(j)} + \tau^*)$ where $\tau^* = L\lfloor \frac{\tau_{\min}}{L} \rfloor$ is divided into L segments described as

$$N_{k^{(j)},q} = [k^{(j)} + qh, k^{(j)} + (q+1)h), \quad q = 0, 1, \ldots, L-1$$

of equal length $h = \lfloor \frac{\tau_{\min}}{L} \rfloor$. Thus $[k^{(j)}, k^{(j)} + \tau^*) = \bigcup_{q=0}^{L-1} N_{k^{(j)},q}$. The vector function $\beta_a(k)$ where $k \in [k^{(j)}, k^{(j)} + \tau^*)$ is chosen to be linear within each segment $N_{k^{(j)},q}$, $q = 0, 1, \ldots, L-1$. Let

$$\beta_{a,q} = \beta_a(k^{(j)} + qh) > 0, \quad q = 0, 1, \ldots, L-1.$$

Then

$$\begin{aligned}\beta_a(k) &= \beta_a(k^{(j)} + qh + r)\\ &= (1 - \frac{r}{h})\beta_{a,q} + \frac{r}{h}\beta_{a,q+1}\\ &= \beta_{a,q} + \frac{1}{h}(\beta_{a,q+1} - \beta_{a,q})(k - k^{(j)} - qh),\end{aligned}$$

where $k \in N_{k^{(j)},q}$, $r \in \{0, 1, \ldots, h-1\}$ and $q = 0, 1, \ldots, L-1$. Then we have

$$\begin{aligned}&\beta_a(k+1)\\ =&\beta_{a,q} + \frac{1}{h}(\beta_{a,q+1} - \beta_{a,q})(k + 1 - k^{(j)} - qh)\\ =&\beta_{a,q} + \frac{1}{h}(\beta_{a,q+1} - \beta_{a,q})(k - k^{(j)} - qh)\\ &+ \frac{1}{h}(\beta_{a,q+1} - \beta_{a,q})\\ =&(1 - \frac{r}{h})\beta_{a,q} + \frac{r}{h}\beta_{a,q+1} + \frac{1}{h}(\beta_{a,q+1} - \beta_{a,q}),\end{aligned}$$

where $k \in N_{k^{(j)},q}$.

Afterwards, for $k \in [k^{(j)} + \tau^*, k^{(j+1)})$, we establish the vector function $\beta_a(k) = \beta_{a,L}$, where $\beta_{a,L}$ is a constant vector. Hence, $\beta_a(k)$ with $a \in Z_\sigma$ is described as follows:

$$\beta_a(k) = \begin{cases}(1 - \frac{r}{h})\beta_{a,q} + \frac{r}{h}\beta_{a,q+1}, & k \in N_{k^{(j)},q},\\ \beta_{a,L}, \quad k \in [k^{(j)} + \tau^*, k^{(j+1)}),\end{cases}$$

where $r \in \{0, 1, \ldots, h-1\}$.

When $k \in N_{k^{(j)},q}$, we know that

$$\begin{aligned}&\Delta V_a(t_k)\\ =&V_a(t_{k+1}) - V_a(t_k)\\ =&\beta_a^{\mathrm{T}}(k+1)x(t_{k+1}) - \beta_a^{\mathrm{T}}(k)x(t_k)\\ =&\beta_a^{\mathrm{T}}(k+1)F_a x(t_k) - \beta_a^{\mathrm{T}}(k)x(t_k)\\ =&\{[(1 - \frac{r}{h})\beta_{a,q}^{\mathrm{T}}(F_a - I) + \frac{r}{h}\beta_{a,q+1}^{\mathrm{T}}(F_a - I) + \frac{1}{h}(\beta_{a,q+1}^{\mathrm{T}} - \beta_{a,q}^{\mathrm{T}})F_a\}x(t_k),\\ =&\{(1 - \frac{r}{h})[\beta_{a,q}^{\mathrm{T}}(F_a - I) + \frac{1}{h}(\beta_{a,q+1}^{\mathrm{T}} - \beta_{a,q}^{\mathrm{T}})F_a]\\ &+ \frac{r}{h}[\beta_{a,q+1}^{\mathrm{T}}(F_a - I) + \frac{1}{h}(\beta_{a,q+1}^{\mathrm{T}} - \beta_{a,q}^{\mathrm{T}})F_a]\}x(t_k).\end{aligned}$$

For $x(t_k) \neq \delta_{2^n}^{2^n}$, according to (5.40) and (5.41), we can obtain

$$\begin{aligned}\Delta V_a(t_k) &< (1-\frac{r}{h})\lambda_a\beta_{a,q}^{\mathrm{T}}x(t_k)+\frac{r}{h}\lambda_a\beta_{a,q+1}^{\mathrm{T}}x(t_k)\\&=\lambda_a[(1-\frac{r}{h})\beta_{a,q}^{\mathrm{T}}+\frac{r}{h}\beta_{a,q+1}^{\mathrm{T}}]x(t_k)\\&=\lambda_a\beta_a^{\mathrm{T}}(k)x(t_k)\\&=\lambda_a V_a(t_k),\end{aligned}$$

where $k \in \bigcup_{q=0}^{L-1} N_{k^{(j)},q} = [k^{(j)}, k^{(j)} + \tau^*)$. When $k \in [k^{(j)} + \tau^*, k^{(j+1)})$, for $x(t_k) \neq \delta_{2^n}^{2^n}$, from inequality (5.43), we have

$$\begin{aligned}\Delta V_a(t_k) &= V_a(t_{k+1}) - V_a(t_k)\\&=\beta_{a,L}^{\mathrm{T}}x(t_{k+1})-\beta_{a,L}^{\mathrm{T}}x(t_k)\\&=\beta_{a,L}^{\mathrm{T}}(F_a-I)x(t_k)\\&<\lambda_a\beta_{a,L}^{\mathrm{T}}x(t_k)\\&=\lambda_a V_a(t_k).\end{aligned}$$

For $x(t_k) = \delta_{2^n}^{2^n}$, we have $V_a(t_{k+1}) = V_a(t_k)$, where $k \in [k^{(j)}, k^{(j+1)})$.

Thus, for any $k \in [k^{(j)}, k^{(j+1)})$, we can obtain that $V_a(t_{k+1}) \leq (1+\lambda_a)V_a(t_k)$, which implies

$$\begin{aligned}V_a(t_k) &\leq (1+\lambda_a)V_a(t_{k-1})\\&\leq (1+\lambda_a)^2 V_a(t_{k-2})\\&\leq \cdots\\&\leq (1+\lambda_a)^{k-k^{(j)}}V_a(t_{k^{(j)}}), \quad k\in[k^{(j)},k^{(j+1)}).\end{aligned} \tag{5.47}$$

On the other hand, by inequality (5.44), one can obtain

$$V_b(t_{k^{(j+1)}}) \leq \mu_b V_a(t_{k^{(j+1)}}), \tag{5.48}$$

where $b = \sigma(t_{k^{(j+1)}}) \in Z_\sigma$ and $a \neq b$.

Then combined with inequalities (5.47), (5.48) and $k^{(i)} < n < k^{(i+1)}$, we have

$$\begin{aligned}
&V_{\sigma(t_n)}(t_n)\\
\leq&(1+\lambda_{\sigma(t_{k^{(i)}})})^{n-k^{(i)}}V_{\sigma(t_{k^{(i)}})}(t_{k^{(i)}})\\
\leq&(1+\lambda_{\sigma(t_{k^{(i)}})})^{n-k^{(i)}}\mu_{\sigma(t_{k^{(i)}})}V_{\sigma(t_{k^{(i)}-1})}(t_{k^{(i)}})\\
=&\mu_{\sigma(t_{k^{(i)}})}(1+\lambda_{\sigma(t_{k^{(i)}})})^{n-k^{(i)}}V_{\sigma(t_{k^{(i-1)}})}(t_{k^{(i)}})\\
\leq&\cdots\\
=&(1+\lambda_{\sigma(t_{k^{(i)}})})^{n-k^{(i)}}[\prod_{j=0}^{i-1}\mu_{\sigma(t_{k^{(j+1)}})}(1+\lambda_{\sigma(t_{k^{(j)}})})^{k^{(j+1)}-k^{(j)}}]V_{\sigma(t_{k^{(0)}})}(t_{k^{(0)}})\\
<&\frac{1}{\mu_{\sigma(t_{k^{(i+1)}})}}[\prod_{j=0}^{i}\mu_{\sigma(t_{k^{(j+1)}})}(1+\lambda_{\sigma(t_{k^{(j)}})})^{k^{(j+1)}-k^{(j)}}]V_{\sigma(t_{k^{(0)}})}(t_{k^{(0)}}).
\end{aligned} \tag{5.49}$$

From (5.45), we can derive

$$\mu_b(1+\lambda_a)^{\tau_{\max}} < 1, \quad a \neq b,\ \forall a, b \in Z_\sigma.$$

Thus, when

$$\mu_1 = \max \mu_b,\ \rho_1 = \max(1+\lambda_a)^{\tau_{\max}}, \quad a, b \in Z_\sigma$$

we have $\rho = \mu_1\rho_1 < 1$. Let $\mu_2 = \min \mu_b,\ b \in Z_\sigma$. Then inequality (5.49) can be converted to

$$V_{\sigma(t_n)}(t_n) < \frac{1}{\mu_2}\rho^{i+1}V_{\sigma(t_{k^{(0)}})}(t_{k^{(0)}}) = \frac{1}{\mu_2}\rho^{i+1}V_{\sigma(0)}(0). \tag{5.50}$$

Therefore, in view of (5.38), (5.39) and (5.46), if $n \to \infty$, i.e., $i \to \infty$, one can conclude that $x(t_n) \to \delta_{2^n}^{2^n}$, which further implies that the switched BN (5.37) is globally stable at x_e. It means that $x(t) \to \delta_{2^n}^{2^n}$ as $t \to \infty$.

The proof is thus completed. □

Remark 5.6 For the global stability analysis of a switched BN

$$x(t+1) = F_{\sigma(t)}x(t)$$

the above method is still applicable by constructing the Lyapunov function:

$$V_{\sigma(t)}(t) = \beta_{\sigma(t)}^{\mathrm{T}}(t)x(t).$$

Remark 5.7 Because the transformed switched BN switches only at the sampling instant, but not at each sampling instant, the problem considered in this study is more complex than directly studying the global stability of switched BNs with all subsystems unstable. The construction of the Lyapunov function and the definition of dwell time are also different from those in [2, 6–9].

5.2.3 A Biological Example

A biological example is shown to demonstrate the validity of Theorem 5.7.

Consider the BCN model studied in [5], which is a reduced model for the lac operon in the bacterium Escherichia coli.

$$\begin{cases} x_1(t+1) = \neg u_1(t) \wedge (x_2(t) \vee x_3(t)), \\ x_2(t+1) = \neg u_1(t) \wedge u_2(t) \wedge x_1(t), \\ x_3(t+1) = \neg u_1(t) \wedge (u_2(t) \vee (u_3(t) \wedge x_1(t))), \end{cases}$$

Here, x_1, x_2 and x_3 are state variables, representing the lac mRNA, high-concentration lactose, and medium-concentration lactose, respectively; u_1, u_2 and u_3 are control inputs, representing the extracellular glucose, high-concentration extracellular lactose, and medium-concentration extracellular lactose, respectively.

Setting

$$x(t) = \ltimes_{i=1}^{3} x_i(t)$$

and

$$u(t) = \ltimes_{j=1}^{3} u_j(t),$$

we obtain

$$x(t+1) = Mu(t)x(t), \tag{5.51}$$

where

$$M = \delta_8[8\ 8\ 8\ 8\ 8\ 8\ 8\ 8\ 8\ 8\ 8\ 8\ 8\ 8\ 8\ 8\ 8\ 8\ 8\ 8\ 8\ 8\ 8\ 8\ 8\ 8\ 8\ 8\ 8\ 8\ 8\ 8\\ 1\ 1\ 1\ 5\ 3\ 3\ 3\ 7\ 1\ 1\ 1\ 5\ 3\ 3\ 3\ 7\ 3\ 3\ 3\ 7\ 4\ 4\ 4\ 8\ 4\ 4\ 4\ 8\ 4\ 4\ 4\ 8].$$

The ASDC for BCN (5.51) is given in the form (5.36) as

$$u(t) = Ex(t_k), \quad t_k \leq t < t_{k+1}, \tag{5.52}$$

where $E = \delta_8[1\ 1\ 4\ 7\ 4\ 6\ 8\ 8]$.

Consider a sampling period $h_k \in \{1, 2\}$. Then a switched BN can be obtained as follows:

$$x(t_{k+1}) = F_{\sigma(t_k)} x(t_k), \tag{5.53}$$

where $F_1 = \delta_8[8\ 8\ 8\ 7\ 8\ 3\ 4\ 8]$ and $F_2 = \delta_8[8\ 8\ 8\ 4\ 8\ 1\ 8\ 8]$.

By calculating the state transition for any initial state $x(0) \in \Delta_8$, the first subsystem has the following two attractors of length 1 and length 2, respectively, i.e., (δ_8^8) and (δ_8^4, δ_8^7). Similarly, it follows that the second subsystem has two point attractors, (δ_8^8) and (δ_8^4). Based on the definition of stability in [1], one has that both of the two subsystems are unstable. Let $L = 1$, $\lambda_1 = \lambda_2 = 0.16$, $\mu_1 = \mu_2 = 0.86$, $\tau_{\min} = 1$. By constraints (5.38), (5.39), (5.40), (5.41), (5.42), (5.43), (5.44) and (5.45), we obtain the following feasible solution:

$$\begin{aligned}
&\tau_{\max} = 1,\\
&\beta_{1,0}^{\mathrm{T}} = (130, 132, 100, 146, 137, 100, 148, 0),\\
&\beta_{1,1}^{\mathrm{T}} = (176, 170, 100, 171, 172, 154, 169, 0),\\
&\beta_{2,0}^{\mathrm{T}} = (100, 146, 86, 147, 146, 132, 145, 0),\\
&\beta_{2,1}^{\mathrm{T}} = (152, 154, 117, 170, 160, 176, 173, 0).
\end{aligned}$$

According to Theorem 5.7, we know that the considered switched BN (5.37) can be globally stabilized by the switching signal $\sigma(t_k) \in \{1, 2\}$ and that BCN (5.51) under ASDC (5.36) is globally stable at δ_8^8.

Choose the initial state as $x(0) = \delta_8^7$. Figures 5.7, 5.8, and 5.9 show the corresponding state trajectory, controller $u(t)$, and the switching signal $\sigma(t_k)$. In Fig. 5.9, we can see that sampling instants are $t_0 = 0, t_1 = 1, t_2 = 3, t_3 = 4, t_4 = 6, t_5 = 7, t_6 = 9$ and the switching time sequence is $t_0, t_1, t_2, t_3, t_4, t_5$ satisfying $k^{(j+1)} - k^{(j)} = 1$.

5.3 Stabilization of Aperiodic Sampled-Data Boolean Control Networks: A Delay Approach

In this section, a novel method for the global stochastic stability analysis of aperiodic sampled-data Boolean control networks (BCNs) is introduced. Here, the sampling instants of aperiodic sampled-data control (ASDC) are uncertain and only the activation frequencies of the sampling interval are known. Using the semitensor product of matrices, a BCN under ASDC can be transformed into a Boolean network (BN) with stochastic delays. Specifically, the ASDC is represented as a delayed control. Here, the time-varying delay is a random variable generated by a Markov chain and its transition probability matrix can be obtained by the

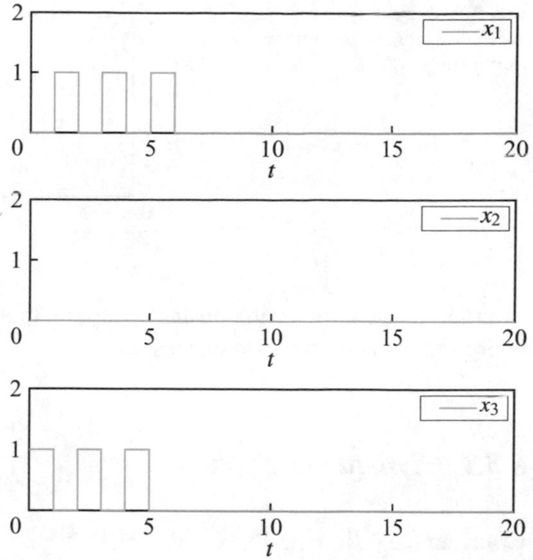

Fig. 5.7 State trajectory of (5.51) under the initial state $x(0) = \delta_8^7$

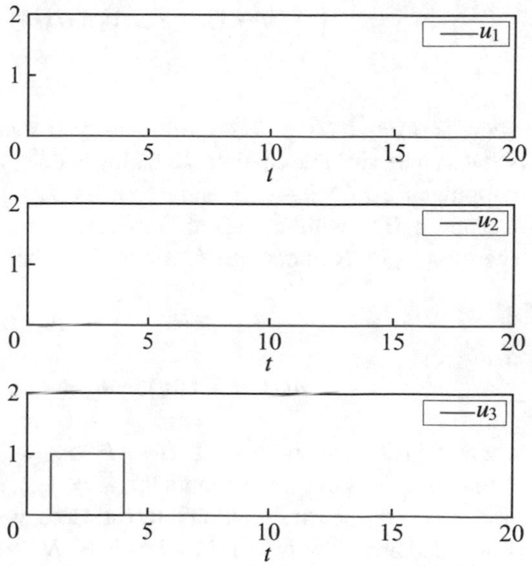

Fig. 5.8 Trajectory of controller $u(t)$ under initial state $x(0) = \delta_8^7$

activation frequencies of the sampling interval. Notably, the value of the time-varying delay is less than the upper bound of the sampling interval and when its present value is given, there are only two possible values that can be taken at the next moment. Subsequently, by using the Lyapunov function and augmented method, a sufficient condition for the global stochastic stability of BCNs under ASDC is provided. In particular, the aforementioned results are applicable to the

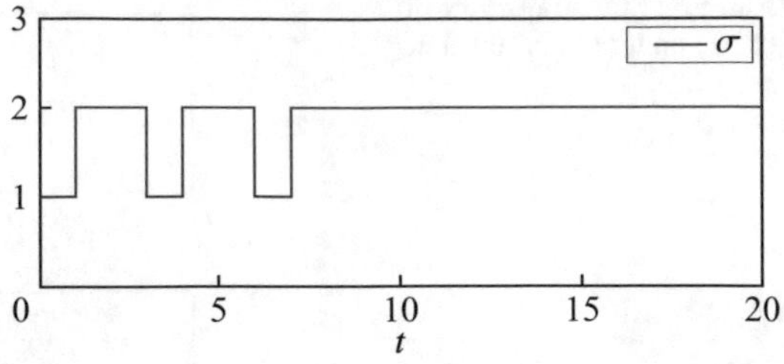

Fig. 5.9 Trajectory of switching signal $\sigma(t_k)$ under initial state $x(0) = \delta_8^7$

sampled-data control with constant sampling interval. Finally, a numerical example is presented to demonstrate our results.

5.3.1 System Description

Consider the following BCN under ASDC:

$$\begin{cases} X(k+1) = f(X(k), U(k)), \\ U(k) = e(X(\theta_i)), \quad \theta_i \le k < \theta_{i+1},\ i \in \mathbb{Z}_+, \end{cases}$$

where $X(k)$ and $U(k)$ is the n-dimensional state variable and the m-dimensional ASDC input variable at time k, taking values in $\mathcal{D}^n$ and $\mathcal{D}^m$, respectively. The mappings $f : \mathcal{D}^{n+m} \to \mathcal{D}^n$ and $e : \mathcal{D}^n \to \mathcal{D}^m$ are logical functions.

Then a BN with a logical form can be converted into an algebraic form. Therefore, a BCN under ASDC can be described as follows:

$$x(k+1) = Lu(k)x(k), \tag{5.54}$$

$$u(k) = Ex(\theta_i), \quad \theta_i \le k < \theta_{i+1},\ i \in \mathbb{Z}_+, \tag{5.55}$$

where $x(k) \in \Delta_{2^n}$ and $u(k) \in \Delta_{2^m}$. $L \in \mathcal{L}_{2^n \times 2^{n+m}}$ and $E \in \mathcal{L}_{2^m \times 2^n}$. It is worth noting that the sampling instants $\theta_i,\ i = 1, 2, \ldots$ are uncertain. $\theta_0 = 0$. Denote $h_i \triangleq \theta_{i+1} - \theta_i$ the ith sampling interval. Here, we assume that the sampling interval is bounded above by $N+1$, i.e., $1 \le h_i \le N+1,\ i = 0, 1, \ldots$.

5.3.2 Convert a Aperiodic Sampled-Data Control into a Delayed Control

The ASDC can be represented as a delayed control as follows:

$$u(k) = Ex(k - \tau_k), \quad \theta_i \le k < \theta_{i+1},\ i \in \mathbb{Z}_+, \tag{5.56}$$

where $\tau_k \triangleq k - \theta_i$ is a random variable. Since the value of τ_{k+1} is only related to τ_k, we assume that τ_k follows a Markov chain taking values in a finite set $\mathcal{M} = \{0, 1, \ldots, N\}$. The transition probability matrix of τ_k is $\Pi = [\pi_{ab}]$, where $\pi_{ab} = \mathbf{P}(\tau_{k+1} = b|\tau_k = a)$, $a, b \in \mathcal{M}$. Then we determine the transition probability matrix of τ_k. On the one hand, when $\tau_k = j \neq N$, $j \in \mathcal{M}$, if $\theta_{i+1} - \theta_i > j + 1$, then $\theta_i \leq k+1 < \theta_{i+1}$ and $\tau_{k+1} = k+1-\theta_i = \tau_k+1 = j+1$; if $\theta_{i+1}-\theta_i = j+1$, then $k + 1 = \theta_{i+1}$ and $\tau_{k+1} = k + 1 - \theta_{i+1} = 0$. Especially, when $\tau_k = N$, then $\tau_{k+1} \equiv 0$ $(h_i \leq N + 1)$. Therefore, the transition probability matrix Π can be expressed as follows:

$$\Pi = \begin{bmatrix} \pi_0 & 1-\pi_0 & 0 & \cdots 0 & 0 \\ \pi_1 & 0 & 1-\pi_1 & \cdots 0 & 0 \\ \vdots & \vdots & \vdots & \ddots \vdots & \vdots \\ \pi_{N-1} & 0 & 0 & \cdots 0 & 1-\pi_{N-1} \\ 1 & 0 & 0 & \cdots 0 & 0 \end{bmatrix}, \tag{5.57}$$

where $\pi_j \in (0, 1)$, $j = 0, 1, \ldots, N - 1$.

Remark 5.8 Since $\tau_k = k - \theta_i$, $\theta_i \leq k < \theta_{i+1}$, we have $\pi_j = \mathbf{P}(\tau_{k+1} = 0|\tau_k = j) = \mathbf{P}(\theta_{i+1} = k + 1|\theta_i = k - j)$ and $1 - \pi_j = \mathbf{P}(\tau_{k+1} = j + 1|\tau_k = j) = \mathbf{P}(\theta_{i+1} > k + 1|\theta_i = k - j)$, which means that the transition probability matrix of τ_k is also the transition probability matrix of θ_i.

On the other hand, when $\theta_i \leq k < \theta_{i+1}$, $\theta_{i+1} - \theta_i = j$, then $\tau_k \in \{0, 1, \ldots, j - 1\}$ in this sampling interval j, which means that the value of τ_k is related to the sampling interval. Thus, if we know the activation frequencies of the sampling interval, the corresponding transition probability matrix can be obtained. To be specific, we assume that p_j, $j = 1, 2, \ldots, N + 1$ is the activation frequency of the sampling interval j, where

$$p_j = \frac{\text{the total running time of the sampling interval } j}{\text{the whole interval}},$$

and $\sum_{j=1}^{N+1} p_j = 1$. Then the corresponding transition probability matrix Π can be expressed by p_j, $j = 1, 2, \ldots, N + 1$ as follows:

$$\Pi = \begin{bmatrix} \frac{p_1}{\sum_{j=1}^{N+1} \frac{p_j}{j}} & \frac{\sum_{j=2}^{N+1} \frac{p_j}{j}}{\sum_{j=1}^{N+1} \frac{p_j}{j}} & 0 & \cdots & 0 \\ \frac{\frac{p_2}{2}}{\sum_{j=2}^{N+1} \frac{p_j}{j}} & 0 & \frac{\sum_{j=3}^{N+1} \frac{p_j}{j}}{\sum_{j=2}^{N+1} \frac{p_j}{j}} & \cdots & 0 \\ \vdots & \vdots & \vdots & \ddots & \vdots \\ \frac{\frac{p_N}{N}}{\frac{p_N}{N} + \frac{p_{N+1}}{N+1}} & 0 & 0 & \cdots & \frac{\frac{p_{N+1}}{N+1}}{\frac{p_N}{N} + \frac{p_{N+1}}{N+1}} \\ 1 & 0 & 0 & \cdots & 0 \end{bmatrix}. \tag{5.58}$$

Given the sampling interval $h_i \in \{1, 2, 3\}$ and the activation frequencies of the sampling interval $p_1 = \frac{1}{6}$, $p_2 = \frac{1}{2}$, $p_3 = \frac{1}{3}$. Based on the above analysis, the corresponding transition probability matrix Π can be expressed as follows:

$$\Pi = \begin{bmatrix} \frac{6}{19} & \frac{13}{19} & 0 \\ \frac{9}{13} & 0 & \frac{4}{13} \\ 1 & 0 & 0 \end{bmatrix}.$$

Remark 5.9 If the specific sampling sequence (θ_i, $i = 1, 2, \ldots$) is unknown and only the activation frequencies of the sampling interval, i.e., p_j, $j = 1, 2, \ldots, N+1$ are known, it seems difficult to study the global stochastic stability of BCNs under ASDC by using the existing results [6, 10–12]. While, based on the aforementioned analysis, one can convert this ASDC into delayed control, and according to Eq. (5.58), the transition probability matrix of time-varying delay τ_k can be obtained. Then, by studying the stochastic stability of the transformed BN with a time-varying delay τ_k, we can get some results about global stochastic stability of the original BCN under ASDC (see Theorem 5.8 for details).

Remark 5.10 In particular, if the sampling interval is constant, the aforementioned analysis is still applicable. For example, if $h_i \equiv 2$, then the corresponding transition matrix Π is

$$\Pi = \begin{bmatrix} 0 & 1 \\ 1 & 0 \end{bmatrix}.$$

Remark 5.11 The aforementioned transformation method is not applicable if the specific sampling sequence (θ_i, $i = 1, 2, \ldots$) is known and its sampling interval is not constant, because the value of the time-varying delay τ_k at each time is determined in this case.

5.3.3 Global Stability

Consider BCN (5.54) and ASDC (5.55). For $\theta_i \leq k < \theta_{i+1}$, we have

$$\begin{cases} x(\theta_i + 1) = LEW_{[2^n]}\Phi_n x(\theta_i), \\ x(\theta_i + 2) = (LEW_{[2^n]})^2(\Phi_n)^2 x(\theta_i), \\ \vdots \\ x(\theta_{i+1}) = (LEW_{[2^n]})^{\theta_{i+1}-\theta_i}(\Phi_n)^{\theta_{i+1}-\theta_i} x(\theta_i), \end{cases}$$

i.e.,

$$x(k+1) = (LEW_{[2^n]})^{k+1-\theta_i}(\Phi_n)^{k+1-\theta_i}x(\theta_i), \quad \theta_i \leq k < \theta_{i+1}. \tag{5.59}$$

Definition 5.6 BCN (5.54) under ASDC (5.55) is said to be globally stochastically stable at $\delta_{2^n}^{2^n}$, if for any initial value $x(0) \in \Delta_{2^n}$ and $\theta_0 = 0$, the trajectory $x(k)$ of system (5.59) satisfies

$$\lim_{k\to\infty} \mathbb{E}\{x(k)|x(0), \theta_0 = 0\} = \delta_{2^n}^{2^n}. \tag{5.60}$$

Remark 5.12 In this section, we only consider that BCN (5.54) is globally stochastically stable at $\delta_{2^n}^{2^n}$. According to the coordinate transformation in [1], the corresponding result can be generalized to the global stochastic stability of BCN (5.54) to any $x_e \in \Delta_{2^n}$.

By $\tau_k = k - \theta_i$, $\theta_i \leq k < \theta_{i+1}$, one can get that

$$x(k+1) = (LEW_{[2^n]})^{\tau_k+1}(\Phi_n)^{\tau_k+1}x(k-\tau_k). \tag{5.61}$$

Remark 5.13 No matter which sampling sequence is, we always have

$$x(1) = Lu(0)x(0) = LEW_{[2^n]}\Phi_n x(0)$$

which means $\tau_0 \equiv 0$.

Theorem 5.8 *BCN (5.54) under ASDC (5.55) is globally stochastically stable at $\delta_{2^n}^{2^n}$, if and only if for any initial value $x(0) \in \Delta_{2^n}$ and $\tau_0 = 0$, the trajectory $x(k)$ of transformed system (5.61) satisfies*

$$\lim_{k\to\infty} \mathbb{E}\{x(k)|x(0), \tau_0 = 0\} = \delta_{2^n}^{2^n}. \tag{5.62}$$

Proof By Definition 5.6, we just need to prove that (5.60) is equivalent to (5.62). Since $\theta_0 = \tau_0 = 0$,

$$\begin{aligned} x(k+1) =& (LEW_{[2^n]})^{k+1-\theta_i}(\Phi_n)^{k+1-\theta_i}x(\theta_i) \\ =& (LEW_{[2^n]})^{\tau_k+1}(\Phi_n)^{\tau_k+1}x(k-\tau_k), \end{aligned}$$

and according to Remark 5.8 the transition probability matrix of τ_k is also the transition probability matrix of θ_i, we have (5.60) is equivalent to (5.62). □

Specially, if the ASDC (5.55) degenerates to the SDSFC, where the sampling interval h_i is constant. Without loss of generality, assume $h_i \equiv N + 1$, then the transition probability matrix of τ_k can be expressed as below:

$$\Pi = \begin{bmatrix} 0\,1\,0 \cdots 0\,0 \\ 0\,0\,1 \cdots 0\,0 \\ \vdots\,\vdots\,\vdots\,\ddots\,\vdots\,\vdots \\ 0\,0\,0 \cdots 0\,1 \\ 1\,0\,0 \cdots 0\,0 \end{bmatrix}.$$

Since $\tau_0 \equiv 0$, the time-varying delay τ_k of this case is deterministic. And we can obtain the following result.

Theorem 5.9 *BCN (5.54) under SDSFC (5.55) is globally stable at $\delta_{2^n}^{2^n}$, if and only if for any initial value $x(0) \in \Delta_{2^n}$ and $\tau_0 = 0$, the trajectory $x(k)$ of transformed system (5.61) satisfies*

$$\lim_{k\to\infty} \mathbb{E}\{x(k)|x(0), \tau_0 = 0\} = \delta_{2^n}^{2^n}.$$

Consider the system (5.61) and define the augmented state vector

$$\mathbf{X}(k) = [x^{\mathrm{T}}(k)\ x^{\mathrm{T}}(k-1)\ \ldots\ x^{\mathrm{T}}(k-N)]^{\mathrm{T}},$$

one has

$$\begin{aligned} x(k+1) =& (LEW_{[2^n]})^{\tau_k+1}(\Phi_n)^{\tau_k+1}R(\tau_k)\mathbf{X}(k) \\ \triangleq& A(\tau_k)\mathbf{X}(k), \end{aligned} \tag{5.63}$$

where $R(\tau_k) = \underbrace{[0\,0\ \ldots\ I\ \ldots\ 0]}_{\text{the } (\tau_k+1)-\text{th block is } I}$ and $A(\tau_k) = (LEW_{[2^n]})^{\tau_k+1}(\Phi_n)^{\tau_k+1}R(\tau_k)$.
Then, we have a new augmented system

$$\mathbf{X}(k+1) = G(\tau_k)\mathbf{X}(k), \tag{5.64}$$

where

$$G(\tau_k) = \begin{bmatrix} A(\tau_k) \\ B \end{bmatrix} \in M_{(N+1)2^n \times (N+1)2^n},$$

and

$$B = \begin{bmatrix} I_{2^n} & 0 & \cdots & 0 & 0 \\ 0 & I_{2^n} & \cdots & 0 & 0 \\ \vdots & \vdots & \ddots & \vdots & \vdots \\ 0 & 0 & \cdots & I_{2^n} & 0 \end{bmatrix} \in M_{N2^n \times (N+1)2^n}.$$

Remark 5.14 If there exists a $k > 0$, such that $x(k), x(k-1), \ldots, x(k-N)$ are all equal to $\delta_{2^n}^{2^n}$ (because $h_i \leq N+1,\ i = 0, 1, \ldots$), then $x(t) \equiv \delta_{2^n}^{2^n},\ t \geq k - N$.

When $x(k), x(k-1), \ldots, x(k-N)$ are all equal to $\delta_{2^n}^{2^n}$, one has

$$\mathbf{X}(k) = ((\delta_{2^n}^{2^n})^{\mathrm{T}}, (\delta_{2^n}^{2^n})^{\mathrm{T}}, \ldots, (\delta_{2^n}^{2^n})^{\mathrm{T}})^{\mathrm{T}}.$$

For simplicity, we define $((\delta_{2^n}^{2^n})^{\mathrm{T}}, (\delta_{2^n}^{2^n})^{\mathrm{T}}, \ldots, (\delta_{2^n}^{2^n})^{\mathrm{T}})^{\mathrm{T}}$ as Y. In the following, if there is no specification, then we write $Y = ((\delta_{2^n}^{2^n})^{\mathrm{T}}, (\delta_{2^n}^{2^n})^{\mathrm{T}}, \ldots, (\delta_{2^n}^{2^n})^{\mathrm{T}})^{\mathrm{T}}$.

Lemma 5.2 *BCN (5.54) under ASDC (5.55) is globally stochastically stable at $\delta_{2^n}^{2^n}$, if and only if for any $\mathbf{X}(0)$ and $\tau_0 = 0$,*

$$\lim_{k\to\infty} \mathbb{E}\{\mathbf{X}(k), k|\mathbf{X}(0), \tau_0 = 0\} = Y. \tag{5.65}$$

Proof (Necessity) By Theorem 5.8, one has that (5.62) holds. Because the value of τ_0 is determined, the trajectory of transformed system (5.61) does not depend on the values of $x(-1), x(-2), \ldots, x(-N)$, i.e., $\lim_{k\to\infty} \mathbb{E}\{x(k)|x(0), x(-1), \ldots, x(-N), \tau_0 = 0\} = \lim_{k\to\infty} \mathbb{E}\{x(k)|x(0), \tau_0 = 0\}$. Then, we can directly obtain that

$$\lim_{k\to\infty} \mathbb{E}\{\mathbf{X}(k), k|\mathbf{X}(0), \tau_0 = 0\} = Y.$$

(Sufficiency) For any $\mathbf{X}(0)$ and $\tau_0 = 0$, (5.65) implies

$$\lim_{k\to\infty} \mathbb{E}\{x(k)|x(0), x(-1), \ldots, x(-N), \tau_0 = 0\} = \delta_{2^n}^{2^n},$$

i.e., $\lim_{k\to\infty} \mathbb{E}\{x(k)|x(0), \tau_0 = 0\} = \delta_{2^n}^{2^n}$. □

Theorem 5.10 *Consider BCN (5.54) under ASDC (5.55). If there exist vectors $0 \leq \beta(i) \in \mathbb{R}^{(N+1)2^n},\ i \in \mathcal{M}$ and the following inequalities hold for all $i \in \mathcal{M}$:*

$$\left(\sum_{j\in\{0,i+1\}} \pi_{ij}\beta^{\mathrm{T}}(j)G(j) - \beta^{\mathrm{T}}(i)\right)\mathbf{X}(k) < 0, \quad \tau_{k-1} = i, \tag{5.66}$$

$$\left(\sum_{j\in\{0,i+1\}} \pi_{ij}\beta^{\mathrm{T}}(j)G(j) - \beta^{\mathrm{T}}(i)\right)Y = 0, \tag{5.67}$$

especially, when $i = N$, then $j = 0$, then BCN (5.54) under ASDC (5.55) is globally stochastically stable at $\delta_{2^n}^{2^n}$.

Proof Construct a Lyapunov function as

$$V(\mathbf{X}(k), k) \triangleq \beta^{\mathrm{T}}(\tau_{k-1})\mathbf{X}(k), \tag{5.68}$$

where the Lyapunov coefficients $\beta(\tau_{k-1}) \geq 0$ satisfy (5.66) and (5.67). And according to (5.67), we can further assume $\beta(\tau_{k-1})$ as follows:

$$\beta(\tau_{k-1}) = \left(\beta_1^{\mathrm{T}}(\tau_{k-1}), \ldots, \beta_{N+1}^{\mathrm{T}}(\tau_{k-1})\right)^{\mathrm{T}} \in \mathbb{R}^{(N+1)2^n}, \tag{5.69}$$

$$\beta_i(\tau_{k-1}) = \left(b_{i,1}(\tau_{k-1}), \ldots, b_{i,2^n-1}(\tau_{k-1}), 0\right)^{\mathrm{T}} \in \mathbb{R}^{2^n},$$

$$b_{i,j}(\tau_{k-1}) > 0,\ i = 1, 2, \ldots, (N+1),\ \ j = 1, 2, \ldots, 2^n - 1.$$

Let $\tau_{k-1} = i$ and $\tau_k = j$, the expectation of $\Delta V(\mathbf{X}(k), k)$ is

$$\begin{aligned}
&\mathbb{E}\{\Delta V(\mathbf{X}(k), k)\} \\
\triangleq&\mathbb{E}\{V(\mathbf{X}(k+1), k+1|\mathbf{X}(k), \tau_{k-1} = i)\} - V(\mathbf{X}(k), k) \\
=&\sum_{j=0}^{N} \pi_{ij}\beta^{\mathrm{T}}(j)\mathbf{X}(k+1) - \beta^{\mathrm{T}}(i)\mathbf{X}(k) \\
=&\left(\sum_{j=0}^{N} \pi_{ij}\beta^{\mathrm{T}}(j)G(j) - \beta^{\mathrm{T}}(i)\right)\mathbf{X}(k) \\
=&\left(\sum_{j\in\{0,i+1\}} \pi_{ij}\beta^{\mathrm{T}}(j)G(j) - \beta^{\mathrm{T}}(i)\right)\mathbf{X}(k),
\end{aligned}$$

in particular, when $i = N$, then $j = 0$.

If $\mathbf{X}(k) = Y$, then according to the construction of Lyapunov function and Remark 5.14, we have $\mathbb{E}\{V(\mathbf{X}(t), t)\} \equiv 0,\ t \geq k$. If $\mathbf{X}(k) \neq Y$, then according to (5.66), we have

$$\mathbb{E}\{\Delta V(\mathbf{X}(k), k)\} < 0. \tag{5.70}$$

Furthermore, we can find a scalar $\alpha \in \{\alpha : 0 < \alpha < 1\}$ satisfying

$$\mathbb{E}\{V(\mathbf{X}(k+1), k+1|\mathbf{X}(k), \tau_{k-1} = i)\} < \alpha V(\mathbf{X}(k), k). \tag{5.71}$$

Accordingly, we have

$$\begin{cases} \mathbb{E}\{V(\mathbf{X}(1), 1|\mathbf{X}(0), \tau_{-1})\} < \alpha V(\mathbf{X}(0), 0), \\ \mathbb{E}\{V(\mathbf{X}(2), 2|\mathbf{X}(1), \tau_0)\} < \alpha V(\mathbf{X}(1), 1), \\ \vdots \\ \mathbb{E}\{V(\mathbf{X}(k+1), k+1|\mathbf{X}(k), \tau_{k-1})\} < \alpha V(\mathbf{X}(k), k). \end{cases} \tag{5.72}$$

By induction based on Eq. (5.72), we have

$$\mathbb{E}\{V(\mathbf{X}(k+1), k+1|\mathbf{X}(0), \tau_{-1}, \tau_0)\} < \alpha^{k+1} V(\mathbf{X}(0), 0). \tag{5.73}$$

Clearly, we have

$$\begin{aligned} &\mathbb{E}\{V(\mathbf{X}(k+1), k+1|\mathbf{X}(0), \tau_{-1}, \tau_0 = 0)\} \\ <&\mathbb{E}\{V(\mathbf{X}(k+1), k+1|\mathbf{X}(0), \tau_{-1}, \tau_0)\} \\ <&\alpha^{k+1} V(\mathbf{X}(0), 0). \end{aligned}$$

Therefore,

$$\lim_{k\to\infty} \mathbb{E}\{V(\mathbf{X}(k), k|\mathbf{X}(0), \tau_{-1}, \tau_0 = 0)\} = 0,$$

for any $\mathbf{X}(0)$ and any $\tau_{-1} \in \mathcal{M}$. Then, we can conclude that

$$\lim_{k\to\infty} \mathbb{E}\{\mathbf{X}(k), k|\mathbf{X}(0), \tau_0 = 0\} = Y.$$

According to Lemma 5.2, the proof is completed. □

Remark 5.15 In fact, Theorem 5.10 gives a sufficient condition for the global stochastic stability of the BCN under ASDC, where the sampling instants are uncertain but the activation frequencies of the sampling interval are known. Here, we first convert BCN (5.54) under ASDC (5.55) into a system (5.61) with time-varying delay τ_k. Then, using the augmented method, the system (5.61) with a time-varying delay τ_k is transformed into a augmented system (5.64). Finally, Theorem 5.10 is obtained by the augmented system (5.64).

Remark 5.16 According to the construction of Lyapunov function in Eqs. (5.68) and (5.69), Eq. (5.67) can be guaranteed. Hence, from Theorem 5.10, the global stochastic stability problem is transformed into solving a linear programming problem (5.66) with $(N+1)^2(2^n-1)$ variables and $(N+1)^2(2^n-1)$ constraints. This linear programming problem (5.66) can be solved in $O((N+1)^2(2^n-1))$ time [13]. Hence, the computational complexity involved in Theorem 5.10 is polynomial in $(N+1)^2(2^n-1)$.

5.3.4 Biological Example

Now, an example is given to illustrate the validity of the aforementioned results.

Given a reduced model of the lac operon in the bacterium Escherichia coli [14], take ASDC into consideration, and then, the algebraic form of the model is as follows:

$$x(k+1) = Lu(k)x(k), \tag{5.74}$$

$$u(k) = Ex(\theta_i), \quad \theta_i \leq k < \theta_{i+1}, \ i \in \mathbb{Z}_+, \tag{5.75}$$

where $x(t) = \ltimes_{i=1}^{3} x_i(t), \ u(t) = \ltimes_{j=1}^{3} u_j(t), \ E = \delta_8[1\ 1\ 4\ 7\ 4\ 6\ 8\ 8]$ and

$$\begin{aligned} L = \delta_8[&8\ 8\ 8\ 8\ 8\ 8\ 8\ 8\ 8\ 8\ 8\ 8\ 8\ 8\ 8\ 8\ 8\ 8\ 8\ 8\ 8\ 8\ 8\ 8\ 8\ 8\ 8\ 8\ 8\ 8\ 8\ 8 \\ &1\ 1\ 1\ 5\ 3\ 3\ 3\ 7\ 1\ 1\ 1\ 5\ 3\ 3\ 3\ 7\ 3\ 3\ 3\ 7\ 4\ 4\ 4\ 8\ 4\ 4\ 4\ 8\ 4\ 4\ 4\ 8]. \end{aligned}$$

The ASDC can be represented as below:

$$u(k) = Ex(k - \tau_k), \quad \theta_i \leq k < \theta_{i+1}, \ i \in \mathbb{Z}_+.$$

Assume that the upper bound of the sampling interval is 3, and the activation frequencies of the sampling interval $p_1 = \frac{5}{57}, \ p_2 = \frac{16}{57}, \ p_3 = \frac{12}{19}$. Then the time-varying delay $\tau_k \in \mathcal{M} = \{0, 1, 2\}$ and the transition probability matrix of τ_k is obtained as below:

$$\Pi = \begin{bmatrix} 0.2 & 0.8 & 0 \\ 0.4 & 0 & 0.6 \\ 1 & 0 & 0 \end{bmatrix}.$$

Let $\mathbf{X}(k) = [x^{\mathrm{T}}(k)\ x^{\mathrm{T}}(k-1)\ x^{\mathrm{T}}(k-2)]^{\mathrm{T}}$, we can get $x(k+1) = A(\tau_k)\mathbf{X}(k)$, where

$$A(0) = \delta_8[8\ 8\ 8\ 7\ 8\ 3\ 4\ 8\ 0\ 0\ 0\ 0\ 0\ 0\ 0\ 0\ 0\ 0\ 0\ 0\ 0\ 0\ 0\ 0],$$

$$A(1) = \delta_8[0\ 0\ 0\ 0\ 0\ 0\ 0\ 0\ 8\ 8\ 8\ 4\ 8\ 1\ 8\ 8\ 0\ 0\ 0\ 0\ 0\ 0\ 0\ 0],$$

$$A(2) = \delta_8[0\ 0\ 0\ 0\ 0\ 0\ 0\ 0\ 0\ 0\ 0\ 0\ 0\ 0\ 0\ 0\ 8\ 8\ 8\ 7\ 8\ 1\ 8\ 8].$$

Then $\mathbf{X}(k+1) = G(\tau_k)\mathbf{X}(k)$, where

$$G(0) = \begin{bmatrix} A(0) \\ B \end{bmatrix}, G(1) = \begin{bmatrix} A(1) \\ B \end{bmatrix}, G(2) = \begin{bmatrix} A(2) \\ B \end{bmatrix},$$

and

$$B = \begin{bmatrix} I_8 & 0 & 0 \\ 0 & I_8 & 0 \end{bmatrix}.$$

By solving inequalities (5.66) and (5.67), one can obtain a feasible solution set as

$$\begin{aligned}
\beta(0)^{\mathrm{T}} &= (100, 12, 30, 126, 70, 100, 72, 0, 83, 11, 100, 150.52, \\
&\quad 60, 155.8, 17, 0, 10, 10, 10, 10, 10, 10, 40, 0), \\
\beta(1)^{\mathrm{T}} &= (130, 12, 48, 97, 55, 106, 100, 0, 11, 11, 11, 100, 53, \\
&\quad 11, 53, 0, 100, 10, 10, 87.4, 10, 61, 10, 0), \\
\beta(2)^{\mathrm{T}} &= (100, 12, 101, 224, 61, 187, 144, 0, 11, 11, 11, 11, 50, \\
&\quad 50, 41, 0, 10, 10, 10, 10, 80, 10, 60, 0).
\end{aligned}$$

According to Theorem 5.10, we know that the BCN (5.74) under ASDC (5.75) is globally stochastically stable at δ_8^8.

Remark 5.17 We can calculate

$$\mathbb{E}\{x(3)\} = \begin{bmatrix}
0 & 0 & 0 & 0 & 0 & 0.48 & 0 & 0 \\
0 & 0 & 0 & 0 & 0 & 0 & 0 & 0 \\
0 & 0 & 0 & 0 & 0 & 0 & 0 & 0 \\
0 & 0 & 0 & 0 & 0 & 0 & 0.2 & 0 \\
0 & 0 & 0 & 0 & 0 & 0 & 0 & 0 \\
0 & 0 & 0 & 0 & 0 & 0 & 0 & 0 \\
0 & 0 & 0 & 0.84 & 0 & 0 & 0 & 0 \\
1 & 1 & 1 & 0.16 & 1 & 0.52 & 0.8 & 1
\end{bmatrix} x(0),$$

$$\mathbb{E}\{x(4)\} = \begin{bmatrix}
0 & 0 & 0 & 0 & 0 & 0 & 0 & 0 \\
0 & 0 & 0 & 0 & 0 & 0 & 0 & 0 \\
0 & 0 & 0 & 0 & 0 & 0 & 0 & 0 \\
0 & 0 & 0 & 0.84 & 0 & 0 & 0 & 0 \\
0 & 0 & 0 & 0 & 0 & 0 & 0 & 0 \\
0 & 0 & 0 & 0 & 0 & 0 & 0 & 0 \\
0 & 0 & 0 & 0 & 0 & 0 & 0.168 & 0 \\
1 & 1 & 1 & 0.16 & 1 & 1 & 0.832 & 1
\end{bmatrix} x(0),$$

$$\mathbb{E}\{x(5)\} = \begin{bmatrix} 0 & 0 & 0 & 0 & 0 & 0 & 0 & 0 \\ 0 & 0 & 0 & 0 & 0 & 0 & 0 & 0 \\ 0 & 0 & 0 & 0 & 0 & 0 & 0 & 0 \\ 0 & 0 & 0 & 0 & 0 & 0 & 0.168 & 0 \\ 0 & 0 & 0 & 0 & 0 & 0 & 0 & 0 \\ 0 & 0 & 0 & 0 & 0 & 0 & 0 & 0 \\ 0 & 0 & 0 & 0.3984 & 0 & 0 & 0 & 0 \\ 1 & 1 & 1 & 0.6016 & 1 & 1 & 0.832 & 1 \end{bmatrix} x(0),$$

$$\mathbb{E}\{x(6)\} = \begin{bmatrix} 0 & 0 & 0 & 0 & 0 & 0 & 0 & 0 \\ 0 & 0 & 0 & 0 & 0 & 0 & 0 & 0 \\ 0 & 0 & 0 & 0 & 0 & 0 & 0 & 0 \\ 0 & 0 & 0 & 0.3984 & 0 & 0 & 0 & 0 \\ 0 & 0 & 0 & 0 & 0 & 0 & 0 & 0 \\ 0 & 0 & 0 & 0 & 0 & 0 & 0 & 0 \\ 0 & 0 & 0 & 0 & 0 & 0 & 0.07968 & 0 \\ 1 & 1 & 1 & 0.6016 & 1 & 1 & 0.92032 & 1 \end{bmatrix} x(0),$$

$$\vdots$$

Finally, one can obtain $\lim_{k\to\infty} \mathbb{E}\{x(k)|x(0), \theta_0 = 0\} = \delta_8^8$. Therefore, by Definition 5.6, we can verify that the BCN (5.74) under ASDC (5.75) is globally stochastically stable at δ_8^8.

5.4 Summary

In this chapter, we mainly discussed the stabilization of aperiodic sampled-data BCNs. This chapter was mainly divided into three parts. Global stability, stabilization and guaranteed cost analysis were studied for BCNs under ASDC in the first part. A BCN under ASDC is transformed into a switched BN, and further global stability of the BCN under ASDC can be obtained by studying the global stability of the transformed switched BN. However, in this part, the switched BN considered has not fewer than one stable subsystem. Then in the second part, we further studied the global stability of BCNs under ASDC, where all subsystems of the transformed switched BN can be unstable. Finally, in the third part, a delay method for the global stochastic stability analysis of aperiodic sampled-data BCNs was introduced. Here, the sampling instants are uncertain and only the activation frequencies of the sampling interval are known. Therefore, the stabilization problem can not be analyzed by using the above two methods.

The major contributions of the first part are given as below. (i) By converting the BCN under ASDC into a switched BN, the global stability of BCNs under ASDC is first studied. Here, the switched BN can only switch at sampling instants, while it does not mean that the switches occur at each sampling instant. (ii) For the switched BN containing both stable subsystems and unstable subsystems, we denote the activation frequencies of the stable subsystems and unstable subsystems, respectively. A sufficient condition for global stability BCNs under ASDC is derived and an upper bound of the cost function is determined. (iii) An algorithm is presented to construct ASDCs for global stabilization of BCNs.

Then, we have studied the global stability of BCNs under ASDCs. Using semi-tensor product, we converted a BCN under ASDC into a switched BN, whose subsystems are all unstable. Some results for global stability of BCNs under ASDC have been obtained by means of a discretized Lyapunov function and DT. The validity has been demonstrated by a biological example. The major contributions of the second part are given as below. (i) the proposed method here not only can solve the global stability problem of the BCN under ASDC when all subsystems of the transformed switched BN are unstable, but also adapt to work on the global stability of switched BNs containing all modes unstable. Thus, the problems left in [2] and [6] are solved. (ii) when compared with the direct research of the global stability of switched BNs with all unstable subsystems, the problem considered in this part is more complex. Although we transform this problem into studying the global stability of the transformed switched BN containing all modes unstable, the switching instant must be the sampling instant of the original BCN under ASDC. Therefore, the construction of Lyapunov function and the definition of dwell time are different from those in [2, 6–9].

Finally, we investigated the global stochastic stability of BCNs under ASDCs by a delay approach. Using the semi-tensor product, a BCN under ASDC could be converted into a BN with a time-varying delay. We regarded the time-varying delay as a random variable modeled by a Markov chain and its transition probability matrix could be expressed by the activation frequencies of the sampling interval. Then, a sufficient condition for global stochastic stability of BCNs under ASDC was obtained by using the Lyapunov function and augmented method. The obtained results were then illustrated by a biological example. We will study the stabilization of aperiodic sampled-data BCNs under noisy sampling interval in the future. The main contributions are as follows: (i) when the sampling instants are uncertain and only the activation frequencies of the sampling interval are known, by transforming the ASDC into delayed control, the global stochastic stability of the BCN under this ASDC is first considered in this article; (ii) for SDSFC (constant sampling interval), we can also convert it into delayed control, and then the global stability of the BCN under SDSFC can be studied by a delay approach.

In addition to the results about stabilization of aperiodic sampled-data BCNs mentioned in this chapter, we have also found some other studies on this. For example, the time-variant state feedback stabilization of constrained delayed BCNs under nonuniform sampled-data control was studied in [15]. Based on the controllability matrix, a necessary and sufficient condition was proposed for the nonuniform sampled-data reachability of constrained delayed BCNs. By virtue of the nonuniform sampled-data reachability, a new procedure was established to design time-variant state feedback sampled-data stabilizers via reachable set approach. In [16], model-free reinforcement learning based control was proposed in order to minimize model design efforts and regulate gene gene regulatory networks with high complexities. A deep Q-learning protocol was used to stabilize probabilistic Boolean control networks in an aperiodic control framework.

References

1. Cheng, D., Qi, H., Li, Z.: Analysis and Control of Boolean Networks: A Semi-Tensor Product Approach. Springer Science & Business Media, New York (2010)
2. Meng, M., Lam, J., Feng, J., et al.: Stability and guaranteed cost analysis of time-triggered Boolean networks. IEEE Trans. Neural Netw. Learn. Syst. **29**(8), 3893–3899 (2017)
3. Hespanda, J.P., Morse, A.S.: Stability of switched systems with average dwell-time. In: Proceedings of the 38th IEEE Conference on Decision and Control, pp. 2655–2660 (1999)
4. Fornasini, E., Valcher, M.E.: Optimal control of Boolean control networks. IEEE Trans. Autom. Control **59**(5), 1258–1270 (2013)
5. Li, H., Wang, Y., Liu, Z.: Simultaneous stabilization for a set of Boolean control networks. Syst. Control Lett. **62**(12), 1168–1174 (2013)
6. Lu, J., Sun, L., Liu, Y., et al.: Stabilization of Boolean control networks under aperiodic sampled-data control. SIAM J. Control Optim. **56**(6), 4385–4404 (2018)
7. Xiang, W., Xiao, J.: Stabilization of switched continuous-time systems with all modes unstable via dwell time switching. Automatica **50**(3), 940–945 (2014)
8. Feng, S., Wang, J., Zhao, J.: Stability and robust stability of switched positive linear systems with all modes unstable. IEEE/CAA J. Autom. Sin. **6**(1), 167–176 (2017)
9. Liu, Z., Zhang, X., Lu, X., et al.: Stabilization of positive switched delay systems with all modes unstable. Nonlinear Anal. Hybrid Syst **29**, 110–120 (2018)
10. Liu, Y., Tong, L., Lou, J., et al.: Sampled-data control for the synchronization of Boolean control networks. IEEE Trans. Cybern. **49**(2), 726–732 (2018)
11. Yu, Y., Feng, J., Wang, B., et al.: Sampled-data controllability and stabilizability of Boolean control networks: nonuniform sampling. J. Frankl. Inst. **355**(12), 5324–5335 (2018)
12. Zhu, S., Liu, Y., Lou, J., et al.: Sampled-data state feedback control for the set stabilization of Boolean control networks. IEEE Trans. Syst. Man Cybern. Syst. **50**(4), 1580–1589 (2018)
13. Megiddo, N.: Linear programming in linear time when the dimension is fixed. J. ACM (JACM) **31**(1), 114–127 (1984)
14. Veliz-Cuba, A., Stigler, B.: Boolean models can explain bistability in the *lac* operon. J. Comput. Biol. **18**(6), 783–794 (2011)
15. Kong, X., Li, H.: Time-variant feedback stabilization of constrained delayed Boolean networks under nonuniform sampled-data control. Int. J. Control Autom. Syst. **19**(5), 1819–1827 (2021)
16. Bajaria, P., Yerudkar, A., Del Vecchio, C.: Aperiodic sampled-data stabilization of probabilistic Boolean control networks: deep Q-learning approach with relaxed Bellman operator. In: 2021 European Control Conference (ECC), pp. 836–841 (2021)

Part IV
Event-Triggered Control

Chapter 6
Event-Triggered Control for Logical Control Networks

Abstract In this chapter, we consider the logical control networks under event-triggered control, and mainly study the stabilization, disturbance decoupling problem, and output regulation problem of logical control networks under event-triggered control.

6.1 Stabilization of Logical Control Networks: An Event-Triggered Control Approach

In this section, we investigate the global stabilization problem of k-valued logical control networks via event-triggered control, where the control inputs only work at several certain individual states. Compared with traditional state feedback control, the designed event-triggered control approach not only shortens the transient period of logical networks but also decreases the number of controller executions. The content of this paper is divided into two parts. In the first part, a necessary and sufficient criterion is derived for the event-triggered stabilization of k-valued logical control networks, and a construction procedure is developed to design all time-optimal event-triggered stabilizers. In the second part, the switching-cost-optimal event-triggered stabilizer is designed to minimize the number of controller executions. A labeled digraph is obtained based on the dynamic of the overall system. Utilizing this digraph, we formulate a universal and unified procedure called the minimal spanning in-tree algorithm to minimize the triggering event set. Furthermore, we illustrate the effectiveness of obtained results through several numerical examples.

6.1.1 Dynamics of k-Valued Logical Control Networks Under Event-Triggered Controllers

The k-valued logical control network under event-triggered control, presented as follows, consists of an inherent non-control k-valued logical network (6.1a), an

Y. Liu et al., *Sampled-data Control of Logical Networks*,
https://doi.org/10.1007/978-981-19-8261-3_6

alternative k-valued logical control networks (6.1b), and a triggering event set $\Lambda \subseteq \mathcal{D}_k^n$ standing for certain individual states where the control inputs are triggered:

$$\begin{cases} x_1(t+1) = f_1(x_1(t), \ldots, x_n(t)), \\ x_2(t+1) = f_2(x_1(t), \ldots, x_n(t)), \\ \quad \vdots \\ x_n(t+1) = f_n(x_1(t), \ldots, x_n(t)), \end{cases} \tag{6.1a}$$

$$\begin{cases} x_1(t+1) = f_1'(x_1(t), \ldots, x_n(t), u_1(t), \ldots, u_m(t)), \\ x_2(t+1) = f_2'(x_1(t), \ldots, x_n(t), u_1(t), \ldots, u_m(t)), \\ \quad \vdots \\ x_n(t+1) = f_n'(x_1(t), \ldots, x_n(t), u_1(t), \ldots, u_m(t)), \end{cases} \tag{6.1b}$$

where $f_i : \mathcal{D}_k^n \to \mathcal{D}_k$ and $f_i' : \mathcal{D}_k^{n+m} \to \mathcal{D}_k$, $i = 1, \ldots, n$ are logical functions, and $x_i \in \mathcal{D}_k$, $u_j \in \mathcal{D}_k$, $j = 1, 2, \ldots, m$ are states and control inputs, respectively.

The event-triggered control mechanism is essentially an intermittent control strategy. In particular, when the dynamic of inherent system (6.1a) evolves desirably, the system is maintained in the form of (6.1a) and the control inputs are not triggered. Otherwise, it means that the system's state locates in the set Λ, k-valued logical control network (6.1b) works, and the control inputs are considered.

Then, we present the equivalent algebraic expression of the event-triggered controlled k-valued logical control network. To facilitate the analysis, let $x(t) = \Gamma_n(x_1(t), x_2(t), \ldots, x_n(t)) \in \Delta_N$ and $u(t)=\Gamma_m(u_1(t), u_2(t), \ldots, u_m(t)) \in \Delta_M$, where $N = k^n$ and $M = k^m$. k-valued logical network (6.1a) and k-valued logical control network (6.1b) can be algebraically represented as follows:

$$x_i(t+1) = M_i x(t), \quad i \in [1, n], \tag{6.2a}$$

$$x_j(t+1) = M_j' u(t) x(t), \quad j \in [1, n], \tag{6.2b}$$

where $M_i \in \mathcal{L}_{k\times N}$ and $M_j' \in \mathcal{L}_{k\times MN}$. Then, the equations in (6.2a) and (6.2b) are further multiplied to show that

$$x(t+1) = Lx(t), \tag{6.3a}$$

$$x(t+1) = L' u(t) x(t), \tag{6.3b}$$

where L=$M_1 * M_2 * \cdots * M_n \in \mathcal{L}_{N\times N}$ is the inherent transition matrix, and $L' = M_1' * M_2' * \cdots * M_n' \in \mathcal{L}_{N\times MN}$ is the transition matrix of alternative subsystem (6.1b), where $*$ is the Khatri-Rao product [1].

Therefore, if we define $\Gamma(\Lambda) := \{\Gamma_n(\mathbf{x}) : \mathbf{x} \in \Lambda\}$, the overall dynamic of the k-valued logical control network with event-triggered control can be synoptically described as

$$x(t+1) = \begin{cases} Lx(t), & x(t) \in \Delta_N \backslash \Gamma(\Lambda), \\ L'u(t)x(t), & x(t) \in \Gamma(\Lambda). \end{cases} \tag{6.4}$$

Or equivalently, the above dynamic can be given as

$$x(t+1) = \left[L\ L'\right]\widetilde{u}(t)x(t) := \widetilde{L}\widetilde{u}(t)x(t), \tag{6.5}$$

where the novel control $\widetilde{u}(t) \in \Delta_{M+1}$ is constructed from $u(t)$ as follows: (1) If $x(t) \in \Delta_N \backslash \Gamma(\Lambda)$, then $\widetilde{u}(t) := \delta_{M+1}^1$. (2) If $x(t) \in \Gamma(\Lambda)$, then one obtains that $\widetilde{u}(t) := [0, u(t)^{\mathrm{T}}]^{\mathrm{T}}$.

The state trajectory of system (6.5) with $x(0; x_0, \widetilde{\mathbf{u}}) = x_0$ with respect to a certain control sequence $\widetilde{\mathbf{u}} : \{0, 1, 2, \ldots\} \to \Delta_{M+1}$ is recorded as $x(t; x_0, \widetilde{\mathbf{u}})$. Then, the concept of the global event-triggered stabilization for system (6.5) with respect to $x^* \in \Delta_N$ is presented, where x^* is supposed to be δ_N^r without loss of generality.

Definition 6.1 For a given state $\delta_N^r \in \Delta_N$, system (6.5) is said to be globally stabilizable to δ_N^r, i.e., δ_N^r-stabilization, if for every $x_0 \in \Delta_N$, a positive integer T and a control sequence $\widetilde{\mathbf{u}} : \{0, 1, 2, \ldots\} \to \Delta_{M+1}$ exist such that $t \geq T$ implies $x(t; x_0, \widetilde{\mathbf{u}}){=}\delta_N^r$.

Remark 6.1 Because system (6.5) contains the information of the triggering event set $\Gamma(\Lambda)$, the stabilization of system (6.5) also can be called the event-triggered stabilization of system (6.4). Without raising any confusion, we simply refer to "stabilization" in the following sections.

In this study, the control input $u(t)$ in (6.4) is considered as the feedback of state $x(t)$, that is,

$$u(t) = Gx(t) = \delta_M[\beta_1\ \beta_2\ \cdots\ \beta_N]x(t), \tag{6.6}$$

where $G \in \mathcal{L}_{M\times N}$ is called the state feedback matrix. In response to (6.6), $\widetilde{u}(t)$ also can be regarded as a special feedback of $x(t)$ with the "state feedback matrix" $\widetilde{G}$, namely $\widetilde{u}(t) = \widetilde{G}x(t)$. In details, $\widetilde{G} = \delta_{M+1}[\gamma_1\ \gamma_2\ \cdots\ \gamma_N]$ is built as

$$\gamma_j = \begin{cases} 1, & \delta_N^j \in \Delta_N \backslash \Gamma(\Lambda), \\ \beta_j + 1, & \delta_N^j \in \Gamma(\Lambda). \end{cases} \tag{6.7}$$

The Objective of this paper is to design the possible state feedback matrix $\widetilde{G} \in \mathcal{L}_{(M+1)\times N}$ such that k-valued logical control network (6.5) is globally stabilizable to δ_N^r under two classes of event-triggered controllers, that is, the time-optimal stabilizer and the switching-cost-optimal stabilizer. Here, the time-optimal stabilizer aims to minimize the transient period and the switching-cost-optimal stabilizer aims to minimize the cardinal number of triggering event set $|\Gamma(\Lambda)|$.

6.1.2 Design of the Time-Optimal Event-Triggered Controller

In this subsection, the time-optimal event-triggered stabilizers are designed. We consider a υ-step reachable set with respect to state δ_N^r that is defined as in [2].

$$\mathscr{R}_\upsilon(r) = \Big\{\delta_N^j \in \Delta_N : \text{A control sequence } \widetilde{\mathbf{u}}(0), \widetilde{\mathbf{u}}(1), \ldots, \widetilde{\mathbf{u}}(w-1) \in \Delta_{M+1} \\ \text{exists such that } x(w; \delta_N^j, \widetilde{\mathbf{u}}(0), \widetilde{\mathbf{u}}(1), \ldots, \widetilde{\mathbf{u}}(w-1)) = \delta_N^r\Big\}. \tag{6.8}$$

On the basis of $\mathscr{R}_\upsilon(r)$ defined above, the following theorem can be obtained; its proof is straightforward and omitted.

Theorem 6.1 *For a given state $\delta_N^r \in \Delta_N$, system (6.4) can be globally δ_N^r-stabilization by an event-triggered controller if and only if both of the following conditions are satisfied:*

1. $\delta_N^r \in \mathscr{R}_1(r)$;
2. *An integer $l \in [1, N-1]$ exists such that $\mathscr{R}_L(r) = \Delta_N$.*

Without any confusion, the minimal integer satisfying the second condition is denoted by l^*. Based on the assumption that conditions in Theorem 6.1 are satisfied, the implication is that $\mathscr{R}_{i+1}(r) \supseteq \mathscr{R}_i(r)$ for all $i \in [0, l^*-1]$, where $\mathscr{R}_0(r) = \{\delta_N^r\}$. Then, we aim to develop a constructive procedure for the "state feedback matrix" $\widetilde{G}$, under which the transient period of (6.5) is minimal.

To this end, we split Δ_N into mutually disjoint sets as

$$\Delta_N = (\mathscr{R}_{l^*}(r) \setminus \mathscr{R}_{l^*-1}(r)) \cup \cdots \cup (\mathscr{R}_2(r) \setminus \mathscr{R}_1(r)) \cup (\mathscr{R}_1(r) \setminus \mathscr{R}_0(r)) \cup \mathscr{R}_0(r). \tag{6.9}$$

To each $\delta_N^i \in \Delta_N$, a unique integer $l_i \in [1, l^*]$ can be found such that $\delta_N^i \in \mathscr{R}_{l_i}(r) \setminus \mathscr{R}_{l_i-1}(r)$. Define $\alpha_i = \tilde{L}\delta_{MN}^i$ for $i \in [1, MN]$. The 'state feedback matrix' $\widetilde{G} = \delta_{M+1}[\gamma_1\ \gamma_2\ \cdots\ \gamma_N]$ can be given by the following procedure:

1. If $\alpha_r = r$, let $\gamma_r = 1$. Otherwise, namely $\alpha_r \neq r$, let γ_r be a solution of $\alpha_{(\gamma_r-1)N+r} = r$;
2. For $i \in [1, N] \setminus \{r\}$, if $\delta_N^{\alpha_i} \in \mathscr{R}_{l_i-1}(r)$, let $\gamma_r = 1$; Otherwise, let γ_r be a solution of $\delta_N^{\alpha_{(\gamma_r-1)N+i}} \in \mathscr{R}_{l_i-1}(r)$.

Remark 6.2 Under the constructed controller, all states in Δ_N can reach δ_N^r after at most l^* steps. This time-optimal event-triggered stabilizer simultaneously reduces the control inputs as much as possible. If all time-optimal stabilizers are necessary, we only need to modify the above procedure trivially. Therefore, we ignore it here.

Once matrix $\widetilde{G}$ is obtained, the triggering event set $\Gamma(\Lambda)$ can immediately be calculated as $\{\delta_N^i : \gamma_i = 1\}$, and the initial state feedback matrix G is equal to

$G = \delta_M[\beta_1\ \beta_2\ \cdots\ \beta_N]$, where $\beta_i = \gamma_i - 1$ if $\gamma_i \neq 1$ and β_i can be arbitrarily selected in $[1, M]$ for $\gamma_i = 1$.

Let $L = \delta_4[1\ 1\ 3\ 4]$ and $L' = \delta_4[1\ 3\ 4\ 1\ 1\ 3\ 4\ 1]$. We construct a novel system in the form of (6.5) with state transition matrix:

$$\widetilde{L} = \delta_4[1\ 1\ 3\ 4\ 1\ 3\ 4\ 1\ 1\ 3\ 4\ 1]. \tag{6.10}$$

Let $r = 1$. We can easily calculate that $\mathscr{R}_1(1) = \{\delta_4^1, \delta_4^2, \delta_4^4\}$ and $\mathscr{R}_2(1) = \Delta_4$. As $\delta_4^1 \in \mathscr{R}_1(1)$ and $\mathscr{R}_2(1) = \Delta_4$, this system can be globally stabilizable to δ_4^1 under event-triggered control.

According to $\widetilde{L}$, let γ_i be as in the aforementioned procedure. One has $\gamma_1 = 1$, $\gamma_2 = 1$, $\gamma_3 = 2, 3$ and $\gamma_4 = 2, 3$. Correspondingly, we can take the triggering event set $\Gamma(\Lambda) = \{\delta_4^1, \delta_4^2\}$. β_1, β_2, β_3 and β_4 can be arbitrarily chosen from $\{1, 2\}$.

As observed from the preceding calculation, the transient period is 2. If we use the traditional state feedback control as in [2], it will be 3. Thus, we can formulate that the designed event-triggered controllers reduce the control cost and transient period more efficiently than the traditional state feedback controllers do.

6.1.3 Design of Switching-Cost-Optimal Event-Triggered Stabilizer

In this subsection, we assume that the conditions in Theorem 6.1 are satisfied, and we focus on designing the event-triggered stabilizer with optimal switching cost. That is, to minimizing the triggering set $\Gamma(\Lambda)$. In [3], an adjustment method has been formulated to minimize the triggering event set $\Gamma(\Lambda)$ in some special cases. However, this method is not capable of addressing certain generalized senses. Thus, a universal and unified approach to design the switching-cost-optimal stabilizer is still valuable and meaningful.

We consider a logical system (6.5) with transition matrices $L = \delta_8[4\ 2\ 1\ 3\ 6\ 5\ 8\ 5]$ and $L' = \delta_8[4\ 2\ 1\ 2\ 6\ 8\ 3\ 3\ 4\ 2\ 1\ 2\ 6\ 5\ 3\ 3]$. We can easily confirm that this network is globally δ_8^2-stabilization under event-triggered control by Theorem 6.1.

According to the approach purposed in [3], we first draw the attractors and basis[1] of k-valued logical network with L as in Fig. 6.1.

We can select a possible state in the attractor $\{\delta_8^1, \delta_8^3, \delta_8^4\}$ and $\{\delta_8^5, \delta_8^6\}$. Then, we add the feasible control inputs at these two states such that the overall system can be stablized to δ_8^2. That is, the minimal number of controller execution is equal to 2.

However, as the transition matrix L' of k-valued logical control network (6.3b) is determined, if δ_8^5 is selected, then it evolves to δ_8^6 under every control input. Otherwise, if δ_8^6 is selected, it evolves to δ_8^8 under $u = \delta_2^1$ or δ_2^2. Obviously, any

[1] Please refer to [4] for more details on attractors and the basis of k-valued logical networks.

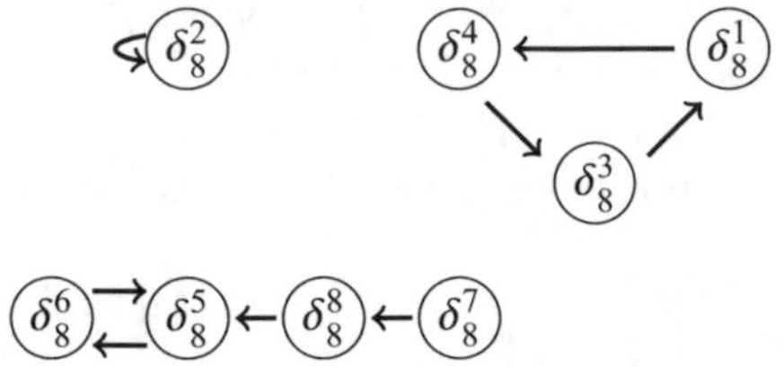

Fig. 6.1 State transition graph of k-valued logical network with respect to transition matrix $L = \delta_8[4\ 2\ 1\ 3\ 6\ 5\ 8\ 5]$

state in the attractor $\{\delta_8^5, \delta_8^6\}$ cannot reach δ_8^2 by this approach. Thus, this method to minimize the triggering event set $\Gamma(\Lambda)$ is not applicable to this example.

Remark 6.3 In fact, as the transition matrix L' is known, considering L unilaterally is infeasible when minimizing $\Gamma(\Lambda)$ as in [3]. In the following, based on the knowledge of graph theory, we present a universal and unified approach to minimize the triggering event set.

First of all, a labelled digraph $\mathcal{G}$ is derived for equivalent graphical description of the dynamic of k-valued logical control network (6.5). The labelled digraph $\mathcal{G}$ is an ordered pair (V, A) consisting of a set of vertices $V := [1, N]$ and a set of directed arcs A. For every arc $(i, j) \in A$, vertices i and j are respectively named as the starting vertex and the ending one of arc (i, j).

For k-valued logical network (6.3a), because $L \in \mathcal{L}_{N\times N}$ is a Boolean matrix, it can be associated with a labelled diagraph $\mathcal{G}_0 = (V, A_0)$. Here, A_0 is a real line arc set, where $\mathcal{G}_0$ has a real line arc (i, j) joining i to j if and only if $[L]_{ji} = 1$. For k-valued logical control network (6.3b), L' is partitioned into $[L_1'\ L_2' \cdots L_M']$, where L_μ', $\mu \in [1, M]$ are control-dependent transition matrices. Similar to the construction of $\mathcal{G}_0$, the labelled digraph for L_μ, denoted by $\mathcal{G}_\mu$, is associated with an order pair (V, A_μ), where A_μ is a dashed line arc set when a dashed line arc $(i, j) \in A_\mu$ if and only if $\left[L_\mu'\right]_{ji} = 1$. Furthermore, by uniting these labelled digraphs $\mathcal{G}_0$ and $\mathcal{G}_\mu$, $\mu \in [1, M]$, we can obtain the overall labelled digraph $\mathcal{G} = (V, A)$ as

$$\mathcal{G} = \bigcup_{\mu=0}^{M} \mathcal{G}_\mu = \left(V, \bigcup_{\mu=0}^{M} A_\mu\right).$$

Remark 6.4 In fact, the arc set A consists of some real line and dashed line arcs corresponding to the dynamics of (6.3a) and (6.3b), respectively. For convenience, there are denoted by an identical set A without distinction from notation. From the construction of $\mathcal{G}$, we can easily find that more than one arc may exist in the same direction with the same starting and ending vertices. Operating on the labelled digraph $\mathcal{G}$ may cause unnecessary issues and high time complexity. Therefore, we construct pretreatment for $\mathcal{G}$ before presenting the algorithm.

To facilitate the analysis, some pretreatment is operated on the labelled digraph $\mathcal{G}$. The labelled digraph after pretreatment is also denoted by $\mathcal{G} := (V, A)$

for convenience, where A represents the arc set of the labelled digraph after pretreatment. The pretreatment is listed as follows.

1. Delete all self loops.
2. For all ordered pair $(i, j) \in [1, N] \times [1, N]$ and $i \neq j$, we retain the arc with minimal weight joining i to j and delete the others. If two such arcs exist, we select the arbitrary one.
3. Assign each dashed line arc joining i to j by a control set

$$u_{(i,j)} := \left\{\mu \mid [L_\mu]_{ji} = 1,\ \mu \in [1, M]\right\}.$$

As mentioned in [5], the stabilization problem of k-valued logical control network can be equivalently described by the existence of spanning in-tree with the designated vertex r, which is call the root of tree. Thus, an approach to find the switching-cost-optimal event-triggered stabilizer is exactly to find a spanning in-tree at root r with the minimal number of dashed line arcs in labelled graph $\mathcal{G}$.

To this end, weights N and 1 are respectively assigned to each dashed line arc and real line. Let $w(u, v)$ denote the weight on every arc (i, j), and then let $\mathcal{G} := (V, A, W)$ denote the labelled digraph $\mathcal{G}$ with weight, where W is a set of weight $w(i, j)$ for all $(i, j) \in A$. The spanning in-tree at root r with the minimal sum of weight is called the minimal spanning in-tree of labelled digraph $\mathcal{G}$. In the graph theory, an effective algorithm has been proposed to find the minimal spanning in-tree; it is called Edmonds's Algorithm [6]. Moreover, a universal and unified procedure is firstly derived for the switching-cost-optimal event-triggering stabilizer.

Algorithm 22 Minimal spanning in-tree algorithm

Step 1: Initialize $i := 0,\ V_0 := V,\ E_0 := A$ and $W_0 := W$. Designate vertex r as the root.
Step 2: Calculate $J_1 = \{(v, \theta(v)) \mid v \in V \setminus \{r\}\}$, where an order pair $(v, \theta(v))$ is the minimal weight arc among all $(v, j) \in E_0$.
Step 3: Check whether directed cycles exists in (V_i, J_{i+1}). If do, then proceed to Step 4. Otherwise, proceed to Step 7.
Step 4: Contract every cycle $\mathcal{C}$ into one new vertex to obtain a new diagraph $(V_{i+1}, E_{i+1}, W_{i+1})$; the weight set W_{i+1} is updated from W_i as follows: then $i := i + 1$ and go to Step 5.

- If (u, v) is an arc joining cycle $\mathcal{C}$, its weight is kept unchanged.
- If (u, v) is an arc away cycle $\mathcal{C}$, its weight is reassigned as $w(u, v) - w(v, \theta(v))$.
- The weights of the other arcs are kept unchanged.

Step 5: Perform pretreatment for the novel labeled digraph (V_i, E_i, W_i).
Step 6: Calculate $J_{i+1} = \{(v, \theta(v)) \mid v \in V \setminus \{r\}\}$, where an order pair $(v, \theta(v))$ is the minimal weight arc among all $(v, j) \in E_i$. Then, return to Step 3.
Step 7: Expand the contracted cycles formed during the above phase in reverse order of their contraction and remove one arc from each cycle to form a spanning in-tree.

Remark 6.5 The time complexity of Algorithm 22 is $O(HN)$, where $N = k^n$ and $H = |A|$.

The return minimal spanning in-tree in Algorithm 22 is denoted by $\mathscr{G}^0 = (V, A^0, W^0)$, where $A^0 \subseteq A$ and $W_0 \subseteq W$. Once $\mathscr{G}^0$ is obtained, the corresponding event-triggered controllers can be constructed immediately. For every arc set $D \subseteq A$, $\lfloor D \rfloor$ consists of the starting vertex of each arc in D.

In the following, we consider a logical system (6.5) with transition matrices

$$L = \delta_8[4\ 2\ 1\ 3\ 6\ 5\ 8\ 5]$$

and

$$L' = \delta_8[4\ 2\ 1\ 2\ 6\ 8\ 3\ 3\ 4\ 2\ 1\ 2\ 6\ 5\ 3\ 3].$$

Firstly, weights 8 and 1 are respectively assigned to the dashed line and the real line arcs. The labelled digraph after pretreatment is presented as Fig. 6.2.

Then, J_1 is calculated by Step 2 of Algorithm 22 as Fig. 6.3.

As (V_0, J_1) has cycles $\mathscr{C}_1$={1, 3, 4} and $\mathscr{C}_2$={5, 6}, we proceed to Step 4 of Algorithm 22. As drawn in Fig. 6.4, cycles $\mathscr{C}_1$ and $\mathscr{C}_2$ are contracted into novel vertices U and V, respectively.

Consequently, Fig. 6.5 is obtained by repeating Step 5 and 6 of Algorithm 22. A still cycle in (V_1, J_2) still exists. Thus, vertices V and 8 are further recontracted and

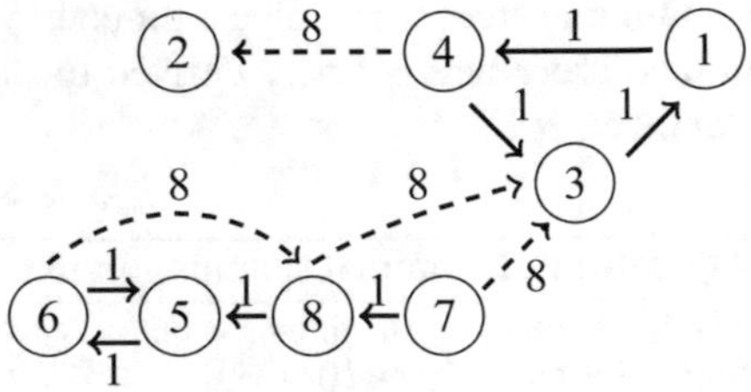

Fig. 6.2 Labelled digraph after pretreatment (V_0, E_0, W_0)

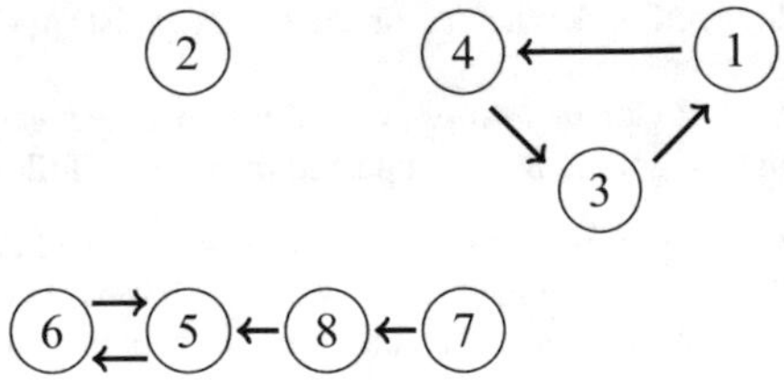

Fig. 6.3 Calculate set J_1=$\{(v, \theta(v)) \mid v \in [1, 8]\}$ by Step 2 in Algorithm 22. That is, $\theta(1) = 4,\ \theta(3) = 1,\ \theta(4) = 3,\ \theta(5) = 6,\ \theta(6) = 5,\ \theta(7) = 8$ and $\theta(8) = 5$

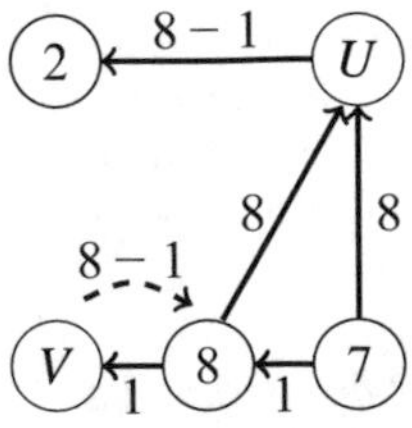

Fig. 6.4 New constructed weighted directed graph (V_1, E_1, W_1). Based on Algorithm 22, $w(U, 2) = 8 - 1$ and $w(V, 8) = 8 - 1$. The weights of the other arcs remain unchanged

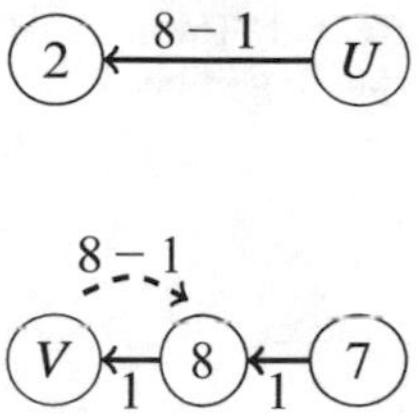

Fig. 6.5 Find the set J_2 in Figure 6.4, where $J_2 = \{(V, 8), (8, V), (7, 8), (U, 2)\}$

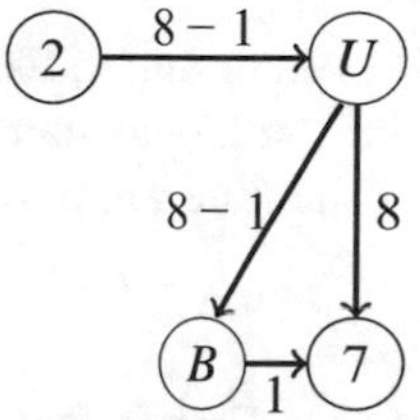

Fig. 6.6 The vertices V and 8 are contracted into a novel vertex B. Let $w(U, B) = 8 - 1$ and the weights of the other arcs be unchanged

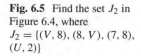

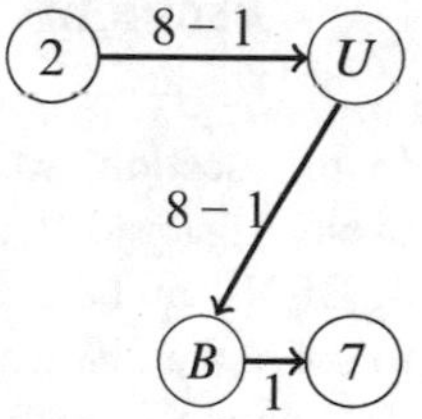

Fig. 6.7 Find the set J_3 here, where $J_3 = \{(2, U), (U, B), (B, 7)\}$

the weights of arcs are updated by Step 4 in Fig. 6.6. By repeating Step 5 and 6, (V_2, J_3) is obtained without any cycle as shown in Fig. 6.7.

Finally, using Step 7 of Algorithm 22, we expand the contracted cycles formed during the preceeding phase in reverse order of their contraction and remove one arc from each cycle to form a spanning in-tree. Therefore, arcs (3, 4), (5, 6) and (5, 8) are removed. The obtained minimal spanning in-tree $\mathscr{G}^0$ is presented as that in Fig. 6.8.

Based on the minimal spanning in-tree $\mathscr{G}^0$, using Algorithm 23, the corresponding triggering event set is designed as $\Gamma(\Lambda) = \{\delta_8^4, \delta_8^6, \delta_8^8\}$, and the possible state feedback matrices are $G = \delta_8[* * * * 1 * *]$, where $*$ is 1 or 2.

Algorithm 23 Corresponding event-triggered controller design from minimal spanning in-tree.

Step 1: Construct the triggering event set $\Gamma(\Lambda)$. If $[L]_{rr} = 1$, then $\Gamma(\Lambda) = \{\delta_N^i : i \in \lfloor A^0 \backslash A_0 \rfloor\}$. Otherwise, $\Gamma(\Lambda) = \{\delta_N^i : i \in \lfloor A^0 \backslash A_0 \rfloor \cup \{r\}\}$.
Step 2: Determine the state feedback matrix G. Let β_r be randomly selected in Δ_M if $r \notin \Gamma(\Lambda)$; else, $\beta_r = u_{(r,r)}$. For every $j \in [1, N] \backslash \{r\}$, if $j \in \Gamma(\Lambda)$, a unique integer $t_j \in [1, N]$ satisfies $(t_j, j) \in A^0$. Let β_j be an arbitrary integer in $u_{(t_j, j)}$. Otherwise, let β_j be an arbitrary integer in $[1, M]$. The feasible state feedback matrix can be designed as $G = \delta_M[\beta_1\ \beta_2\ \cdots\ \beta_N]$.

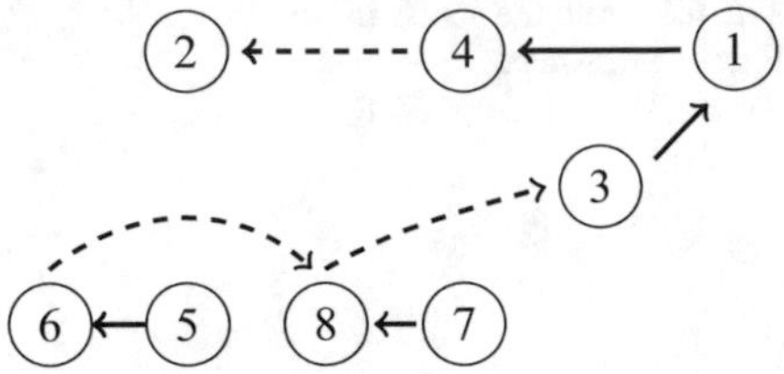

Fig. 6.8 The minimal spanning in-tree $\mathscr{G}^0$

According to Fig. 6.8, the number of control execution is equal to 3. If we utilize the traditional state feedback control, the number of control execution would be 7 in the transient period because all states need to be controlled.

6.2 Event-Triggered Control for the Disturbance Decoupling Problem

In this section, we investigate the disturbance decoupling problem of BCNs by event-triggered control. Using the semi-tensor product of matrices, algebraic forms of BCNs can be achieved, based on which, event-triggered controllers are designed to solve the disturbance decoupling problem of BCNs. In addition, the disturbance decoupling problem of Boolean partial control networks is also derived by event-triggered control. Finally, two illustrative examples demonstrate the effectiveness of proposed methods.

6.2.1 Definition of Disturbance Decoupling Problem

Consider a BCN with n nodes, m control inputs, p outputs, and q disturbance inputs as

$$x_\lambda(t+1) = f_\lambda(X(t), U(t), \bar{\xi}(t)), \tag{6.11a}$$

$$x_\beta(t+1) = f_\beta(X(t), U(t), \bar{\xi}(t)), \tag{6.11b}$$

$$y_j(t) = h_j(X^1(t)), \quad t \in \mathbb{Z}_+, \tag{6.11c}$$

where $X(t) = (x_1(t), x_2(t), \ldots, x_n(t))$, $X^1(t) = (x_1(t), x_2(t) \ldots, x_r(t))$, $U(t) = (u_1(t), u_2(t) \ldots, u_m(t))$ and $\bar{\xi}(t) = (\xi_1(t), \xi_2(t), \ldots, \xi_q(t))$, $f_\lambda, f_\beta : \mathcal{D}^{n+q+m} \to \mathcal{D}$, and $h_j : \mathcal{D}^r \to \mathcal{D}$ are logical functions with $\lambda \in [1, r], \beta \in [r+1, n]$, $j \in [1, p]$.

For each f_i and h_j, there exist their unique structure matrices M_{f_i} and M_{h_j}, $i \in [1, n]$, $j \in [1, p]$.

Let $x^1(t) = \ltimes_{i=1}^{r} x_i(t)$, $x^2(t) = \ltimes_{i=r+1}^{n} x_i(t)$, $\xi(t) = \ltimes_{i=1}^{q} \xi_i(t)$, $u(t) = \ltimes_{i=1}^{m} u_i(t)$ and $y(t) = \ltimes_{j=1}^{p} y_j(t)$, then, one can obtain the following algebraic form:

$$\begin{cases} x^1(t+1) = \tilde{L}u(t)x(t)\xi(t), \\ x^2(t+1) = \tilde{L}_2 u(t)x(t)\xi(t), \\ y(t) = Hx^1(t), \end{cases} \tag{6.12}$$

where $\tilde{L} = M_{f_1} * M_{f_2} * \cdots * M_{f_r} \in \mathcal{L}_{2^r \times 2^{n+m+q}}$, $\tilde{L}_2 = M_{f_{r+1}} * M_{f_{r+2}} * \cdots * M_{f_n} \in \mathcal{L}_{2^{n-r} \times 2^{n+m+q}}$, $H = M_{h_1} * M_{h_2} * \cdots * M_{h_p} \in \mathcal{L}_{2^p \times 2^r}$, and $*$ is the Khatri-Rao product [1].

Now, we give the definition of disturbance decoupling problem for BCNs.

Definition 6.2 Consider system (6.11a), (6.11b), and (6.11c). The disturbance decoupling problem is solvable, if for every initial state $x^1(0) \in \Delta_{2^r}$, there exists a state feedback time-variant controller in the form of

$$\begin{cases} u_1(t) = g_t^1(X(t)), \\ u_2(t) = g_t^2(X(t)), \\ \vdots \\ u_m(t) = g_t^m(X(t)), \end{cases} \tag{6.13}$$

where g_t^i, $i \in [1, m]$ are time-variant Boolean functions, such that the closed-loop system consisting of (6.11a), (6.11b), (6.11c) and (6.13) can be expressed by

$$\begin{cases} x_i(t+1) = \hat{f}_i(X^1(t)), \\ x_k(t+1) = \hat{f}_k(X(t), \bar{\xi}(t)), \\ y_j(t) = h_j(X^1(t)), \end{cases} \tag{6.14}$$

where $\hat{f}_i : \mathcal{D}^r \to \mathcal{D}$, $i \in [1, r]$ and $\hat{f}_k : \mathcal{D}^{n+q} \to \mathcal{D}$, $k \in [r+1, n]$ are logical functions.

For each g_t^i, $i \in [1, m]$, there exists its unique structure matrix $M_{g_t^i}$, then $u(t) = K_t(x^1(0))x(t)$ with $K_t(x^1(0)) = \ltimes_{i=1}^{m} M_{g_t^i} \in \mathcal{L}_{2^m \times 2^n}$ called time-variant event-triggered gain matrix. Thus, the disturbance decoupling problem of system (6.11a), (6.11b), (6.11c) is solvable by (6.13) if and only if $x^2(t)$ and $\xi(t)$ are redundant variables in (6.11a). The following lemma shows how to obtain redundant variables.

Lemma 6.1 ([7]) *Given an integer $r \le n$, let M_G be the structure matrix of a given logical mapping $G = (g_1(x_1 \cdots x_n), \ldots, g_k(x_1 \cdots x_n)) : \mathcal{D}^n \to \mathcal{D}^k$. Split M_G into 2^r equal blocks as $M_G = [\mathrm{Blk}_1(M_G) \cdots \mathrm{Blk}_{2^r}(M_G)]$. For the given logical* *mapping G,* $x_{r+1}, \cdots, x_n$ *are redundant variables if and only if for any* $\mu \in \Omega_r$,

$$\operatorname{rank}\big(\mathrm{Blk}_\mu(M_G)\big) = 1,$$

where $Blk_\mu(M_G) \in \mathcal{L}_{2^k \times 2^{n-r}}$.

6.2.2 Event-Triggered Control of Boolean Control Networks

It should be pointed out that controllers are in change every moment in (6.13), which implies that the cost will be large. Motivated by this, the event-triggered control is provided to solve the disturbance decoupling problem. A triggering mechanism that decides when the control input has to be updated is proposed, and a feedback controller that determines the control input is designed. First, event-triggered control formulation is presented in the sequel.

Consider system (6.12), given $x^1(0) = \delta_{2^r}^{\lambda}$, and suppose that the initial control $u(t) = K_0(x^1(0))x(t)$. Triggering time t' is defined if $\operatorname{rank}(\mathrm{Blk}_i(\tilde{L}u(t'))) \neq 1$ assumed by $x^1(t') = \delta_{2^r}^{i}$. The sequence of triggering time can be denoted by: $t_1 < t_2 < \cdots < t_\rho$, which corresponds to a sequence of control updates

$$u(t_1), \ldots, u(t_\rho).$$

Since Δ_{2^r} is a finite set, then $\rho \leq 2^r$. Between event-triggered executions the controller is assumed invariant and depends on the last event-triggered gain matrix update, that is,

$$u(t) = u(t_\iota) = K_{t_\iota}(x^1(0))x(t), \text{ for all } t \in [t_\iota, t_{\iota+1}), \tag{6.15}$$

where $\iota \in [1, \rho]$, and when $\iota = \rho$, let $t_{\rho+1} = \infty$. Suppose that $K_{t_{\nu-1}}(x^1(0)) = \delta_{2^m}[v_1^{t_{\nu-1}}\ v_2^{t_{\nu-1}}\ \cdots\ v_{2^n}^{t_{\nu-1}}]$, $\nu \in [2, \rho+1]$, then

$$x^1(t+1) = \tilde{L}K_{t_{\nu-1}}(x^1(0))\Phi_n x^1(t)x^2(t)\xi(t), \tag{6.16}$$

where $t \in [t_{\nu-1}, t_\nu)$, and $\Phi_n = \mathrm{Diag}\{\delta_{2^n}^1, \delta_{2^n}^2, \ldots, \delta_{2^n}^{2^n}\} \in \mathcal{L}_{2^{2n} \times 2^n}$ is the power reducing matrix.

Split $\tilde{L}$ into 2^m equal blocks as

$$\tilde{L} = [\mathrm{Blk}_1(\tilde{L})\ \mathrm{Blk}_2(\tilde{L})\ \cdots\ \mathrm{Blk}_{2^m}(\tilde{L})],$$

where $\mathrm{Blk}_i(\tilde{L}) \in \mathcal{L}_{2^r \times 2^{n+q}}$, $i \in \Omega_m$. For any $i \in \Omega_m$, split $\mathrm{Blk}_i(\tilde{L})$ into 2^n equal blocks as

$$\mathrm{Blk}_i(\tilde{L}) = [\tilde{L}_{i,1}\ \tilde{L}_{i,2}\ \cdots\ \tilde{L}_{i,2^n}],$$

where $\tilde{L}_{i,j} \in \mathcal{L}_{2^r \times 2^q}$, $j \in \Omega_n$. Then, it follows from (6.16) that

$$\begin{aligned} x^1(t+1) =& [\tilde{L}_{v_1^{t_{\nu-1}},1} \ \tilde{L}_{v_2^{t_{\nu-1}},2} \ \cdots \ \tilde{L}_{v_{2^n}^{t_{\nu-1}},2^n}]x^1(t)x^2(t)\xi(t), \\ =& \bar{L}_{t_{\nu-1}} x^1(t)x^2(t)\xi(t), \end{aligned} \tag{6.17}$$

where $\bar{L}_{t_{\nu-1}} = [\tilde{L}_{v_1^{t_{\nu-1}},1} \ \tilde{L}_{v_2^{t_{\nu-1}},2} \ \cdots \ \tilde{L}_{v_{2^n}^{t_{\nu-1}},2^n}]$. Further partition $\bar{L}_{t_{\nu-1}}$ into 2^r equal blocks as

$$\bar{L}_{t_{\nu-1}} = [\mathrm{Blk}_1(\bar{L}_{t_{\nu-1}}) \ \mathrm{Blk}_2(\bar{L}_{t_{\nu-1}}) \ \cdots \ \mathrm{Blk}_{2^r}(\bar{L}_{t_{\nu-1}})], \tag{6.18}$$

where $\mathrm{Blk}_i(\bar{L}_{t_{\nu-1}}) \in \mathcal{L}_{2^r \times 2^{n+q-r}}$, $i \in \Omega_r$. It follows from event-triggered control formulation that for any $t \in [t_{\nu-1}, t_\nu)$, $\mathrm{rank}(\mathrm{Blk}_j(\bar{L}_{t_{\nu-1}})) = 1$ assumed by $x^1(t) = \delta_{2^r}^j$ by $K_{t_{\nu-1}}(x^1(0))$.

Now, assume that $x^1(t_\nu) = \delta_{2^r}^{\lambda_\nu}$, if $\mathrm{rank}(\mathrm{Blk}_{\lambda_\nu}(\bar{L}_{t_{\nu-1}})) \neq 1$, then t_ν is next triggering time after triggering time $t_{\nu-1}$, which implies that $\mathrm{Blk}_{\lambda_\nu}(\bar{L}_{t_{\nu-1}})$ should be changed, i.e., $K_{t_\nu}(x^1(0))$ needs to be designed. Therefore, it is learned from (6.17) and (6.18) that $v^{t_{\nu-1}}_{(\lambda_\nu-1)2^{n-r}+1}, v^{t_{\nu-1}}_{(\lambda_\nu-1)2^{n-r}+2}, \ldots, v^{t_{\nu-1}}_{\lambda_\nu 2^{n-r}}$ should be altered such that $\mathrm{rank}(\mathrm{Blk}_{\lambda_\nu}(\bar{L}_{t_\nu})) = 1$.

In order to obtain $v^{t_{\nu-1}}_{(\lambda_\nu-1)2^{n-r}+i}$, $i \in \Omega_{n-r}$. For any $i \in \Omega_{n-r}$, we define the following sets:

$$\Lambda((\lambda_\nu-1)2^{n-r}+i) = \{\delta_{2^r}^{k_w} : \text{there exists } j \in \Omega_m \text{ such that } \tilde{L}_{j,(\lambda_\nu-1)2^{n-r}+i} = \delta_{2^r}[k_w \ \cdots \ k_w]\}.$$

For each set $\Lambda((\lambda_\nu - 1)2^{n-r} + i)$, $i \in \Omega_{n-r}$, we can construct a set, denoted by $\Lambda((\lambda_\nu - 1)2^{n-r} + i)^{k_w}$, as

$$\Lambda((\lambda_\nu - 1)2^{n-r} + i)^{k_w} = \left\{\delta_{2^m}^j \in \Delta_{2^m} : \tilde{L}_{j,(\lambda_\nu-1)2^{n-r}+i} = \delta_{2^r}[k_w \ \cdots \ k_w]\right\}.$$

Based on the above analysis, one can conclude that the following result.

Theorem 6.2 *Given an initial state $x^1(0) \in \Delta_{2^r}$, assume that the initial controller $u(t) = K_0(x^1(0))x(t)$ and the triggering times $t_1 < \cdots < t_\rho$, $\rho \leq 2^r$. Then for any $\iota \in [1, \rho]$, $t \in [t_\iota, t_{\iota+1})$, assume $x^1(t) = \delta_{2^r}^{\lambda_\iota}$, $\lambda_\iota \in \Omega_r$ and the corresponding $u(t) = K_{t_\iota}(x^1(0))x(t)$, then rank($Blk_{\lambda_\iota}(\tilde{L}u(t))$) = 1 if and only if*

$$\Xi_{\lambda_\iota} := \Lambda((\lambda_\iota - 1)2^{n-r} + 1) \cap \Lambda((\lambda_\iota - 1)2^{n-r} + 2) \cap \cdots \cap \Lambda(\lambda_\iota 2^{n-r+1}) \neq \varnothing. \tag{6.19}$$

Additionally, if (6.19) holds, then $K_{t_\iota}(x^1(0))$ can be designed as

$$K_{t_\iota}(x^1(0)) = \delta_{2^m}[v_1^{t_\iota} \ v_2^{t_\iota} \ \cdots \ v_{2^n}^{t_\iota}], \tag{6.20}$$

with $v_{\mu}^{t_\iota} = v_{\mu}^{t_{\iota-1}}$, $\mu \in [1, 2^n] \backslash [(\lambda_\iota - 1)2^{n-r} + 1, \lambda_\iota 2^{n-r}]$,

$$\delta_{2^m}^{v^{t_\iota}_{(\lambda_\iota - 1)2^{n-r}+i}} \in \bigcup_{\delta_{2^r}^{k_w} \in \Xi_{\lambda_\iota}} \Lambda((\lambda_\iota - 1)2^{n-r} + i)^{k_w},$$

where $i \in \Omega_{n-r}$.

Proof The necessity is trivial by Lemma 6.1. As for the sufficiency, if (6.19) holds, and suppose that for any $\iota \in [1, \rho]$, $K_{t_\iota}(x^1(0)) = \delta_{2^m}[v_1^{t_\iota}\ v_2^{t_\iota}\ \cdots\ v_{2^n}^{t_\iota}]$ with $\delta_{2^m}^{v_i^{t_\iota}}$ given by (6.20), $i \in \Omega_n$, then one can get that

$$\begin{aligned} x^1(t+1) &= [\tilde{L}_{v_1^{t_\iota},1}\ \cdots\ \tilde{L}_{v_{2^n}^{t_\iota},2^n}]x^1(t)x^2(t)\xi(t), \\ &= \bar{L}_{t_\iota} x^1(t)x^2(t)\xi(t), \end{aligned} \tag{6.21}$$

where $\bar{L}_{t_\iota} = [\tilde{L}_{v_1^{t_\iota},1}\ \cdots\ \tilde{L}_{v_{2^n}^{t_\iota},2^n}]$. Partition $\bar{L}_{t_\iota}$ into 2^r parts denoted by $\mathrm{Blk}_i(\bar{L}_{t_\iota})$, $i \in \Omega_r$. Since $\Xi_{\lambda_\iota} \neq \varnothing$, for any $\iota \in [1, \rho]$, then $\mathrm{rank}(\mathrm{Blk}_{\lambda_\iota}(\bar{L}_{t_v})) = 1$. Therefore, for any $\iota \in [1, \rho]$, $t \in [t_\iota, t_{\iota+1})$, $x^1(t+1)$ is independent of $x^2(t)$ and $\xi(t)$, which completes the proof. □

The following corollary comes from Theorem 6.2 immediately.

Corollary 6.1 *The disturbance decoupling problem of system (6.12) is solvable if for any initial states* $x^1(0) \in \Delta_{2^r}$, *condition (6.19) in Theorem 6.2 holds.*

Based on Theorem 6.2, we present an algorithm to design event-triggered controllers for the disturbance decoupling problem of system (6.12).

Algorithm 24 Event-triggered controllers designing depends on initial states for disturbance decoupling problem

---

Step 1. Given an initial state $x^1(0) = \delta_{2^r}^{\lambda}$, and the corresponding initial event-triggered gain matrix is assumed by $K_0(x^1(0))$.
Step 2. The first triggering time t_1 is ensured by $K_0(x^1(0))$, then $K_{t_1}(x^1(0))$ is designed according to (6.20). Similarly, the second triggering time t_2 can be also obtained based on $K_{t_1}(x^1(0))$, then $K_{t_2}(x^1(0))$ is designed. In this way, $x^1(t+1)$ is independent of $x^2(t)$ and $\xi(t)$ with initial state $x^1(0) = \delta_{2^r}^{\lambda}$.

It is found that the obtained controllers from Algorithm 24 depend on initial states. In the following, we aim to design a uniform controller which is independent of initial states. Moreover, the event-triggered gain matrix will be time-invariant for easy applications. Actually, the controller is called global time-invariant state feedback controller. To get global time-invariant state feedback gain matrix assumed K, we only need to modify Algorithm 24 as follows.

Therefore, we have the following result.

Algorithm 25 Global time-invariant state feedback controllers designing for disturbance decoupling problem.

Step 1. Assume that the sequence of triggering time has been obtained by Step 2 in Algorithm 24, denoted by $t_1 < \cdots < t_\rho,\ \rho \le 2^r$ and $K_{t_\rho}(x^1(0))$ has also been obtained by Algorithm 24.
Step 2. If for any initial state $\hat{x}^1(0) \in \Delta_{2^r} \backslash x^1(0)$ assumed by $\hat{x}^1(0) = \delta_{2^r}^i$ such that $\mathrm{rank}(\mathrm{Blk}_i(\tilde{L}u(t))) = 1$ with $u(t) = K_{t_\rho}(x^1(0))x(t)$, then global time-invariant state feedback gain matrix $K = K_{t_\rho}(x^1(0))$. If there exists an initial state $x'^1(0) = \delta_{2^r}^{\lambda'}$ such that $\mathrm{rank}(\mathrm{Blk}_{\lambda'}(\tilde{L}u(t))) \neq 1$ with $K_0(x'^1(0)) = K_{t_\rho}(x^1(0))$, then, the first triggering time will be $t_1 = 0$ when the initial state and controller are $x'^1(0)$ and $u(t) = K_0(x'^1(0))x(t)$, respectively.
Step 3. It is learned from (6.17) and (6.18) that the $(\lambda' - 1)2^{n-r} + 1$-th column, $(\lambda' - 1)2^{n-r} + 2$-th column, $\cdots$, $\lambda' 2^{n-r}$-th column of $K_0(x'^1(0))$ need to be altered, which can be determined by (6.20). The next process can be done as Step 2 in Algorithm 24, then $x^1(t+1)$ is independent of $x^2(t)$ and $\xi(t)$ with initial state $x'^1(0)$.
Step 4. Repeat Step 2 until for any initial state, $x^1(t+1)$ is independent of $x^2(t)$ and $\xi(t)$.

Theorem 6.3 *The disturbance decoupling problem of system (6.12) is solvable by global time-invariant state feedback control $u(t) = Kx(t),\ K \in \mathcal{L}_{2^m \times 2^n}$ designed as Algorithm 25.*

Consider the following BCN [8]:

$$\begin{cases} x_1(t+1) = x_2(t) \wedge [(x_3(t) \to \xi(t)) \vee u_1(t)] \\ x_2(t+1) = x_2(t) \wedge \xi(t) \wedge u_2(t) \\ x_3(t+1) = (x_2(t) \leftrightarrow \xi(t)) \wedge u_1(t) \\ y(t) = x_1(t) \wedge x_2(t). \end{cases} \tag{6.22}$$

Let $x^1 = x_1(t)x_2(t),\ u(t) = u_1(t)u_2(t)$. By simple computation, we have

$$\begin{aligned} \tilde{L} = \delta_{2^2}[&1\,2\,1\,2\,3\,4\,3\,4\,1\,1\,1\,1\,3\,3\,3\,3\,2\,2\,2\,2\,4\,4\,4\,4\,1\,1\,1\,1\,3\,3\,3\,3 \\ &1\,4\,1\,2\,3\,4\,3\,4\,1\,3\,1\,1\,3\,3\,3\,3\,2\,4\,2\,2\,4\,4\,4\,4\,1\,3\,1\,1\,3\,3\,3\,3], \end{aligned} \tag{6.23}$$

and

$$H = \delta_2[1\,2\,2\,2].$$

Assume that $x^1(0) = \delta_{2^2}^4$, and $K_0(\delta_{2^2}^4) = \delta_{2^2}[2\,2\,3\,4\,3\,2\,1\,3]$, then it follows from (6.17) that

$$\begin{aligned} x^1(t+1) &= \delta_{2^2}[2\,2\,2\,2\,3\,4\,4\,4\,1\,3\,1\,1\,3\,3\,3\,3]x^1(t)x^2(t)\xi(t) \\ &= \bar{L}x^1(t)x^2(t)\xi(t). \end{aligned} \tag{6.24}$$

When $x^1(0) = \delta_{2^2}^4$, $\mathrm{rank}(\mathrm{Blk}_4(\bar{L})) = 1$, which means that $x^1(1)$ is independent of $x^2(0)$ and $\xi(0)$. Now $x^1(1) = \delta_{2^2}^3$, since $\mathrm{rank}(\mathrm{Blk}_3(\bar{L})) \neq 1$, then by event-triggered control formulation, the first triggering time $t_1 = 1$. Therefore, v_5^1 and v_6^1 need to be altered, while others unchange. From (6.23), we have $\Lambda(5) = \Lambda(6) = \{\delta_{22}^1\}$, then $\Lambda(5)^1 = \{\delta_{22}^1, \delta_{22}^2\}$, and $\Lambda(6)^1 = \{\delta_{22}^1, \delta_{22}^2, \delta_{22}^3, \delta_{22}^4\}$. Since $\Lambda(5)^1 \cap \Lambda(6)^1 = \{\delta_{22}^1, \delta_{22}^2\}$, then by Theorem 6.2, $K_1(\delta_{22}^4)$ can be designed by $v_5^1 = v_6^1 = 1$. Therefore, $x^1(2) = \delta_{22}^1$.

Similarly, since $\text{rank}(\text{Blk}_1(\bar{L})) = 1$, then $x^1(3)$ is independent of $x^2(2)$ and $\xi(2)$. When $x^1(3) = \delta_{22}^2$, $\text{rank}(\text{Blk}_2(\bar{L})) \neq 1$ leads to $t_2 = 3$. One gets $\Lambda(3) = \Lambda(4) = \{\delta_{22}^4\}$, and $\Lambda(3)^4 = \Lambda(4)^4 = \{\delta_{22}^2, \delta_{22}^4\}$. Letting $v_3^3 = 2$ and $v_4^3 = 4$, then $K_3(\delta_{22}^4)$ can also be obtained. Moreover, it is found that $\rho = 2$ by $K_3(\delta_{22}^4)$. Therefore, we can conclude that when $x^1(0) = \delta_{22}^4$, the disturbance decoupling problem of system (6.22) is solvable by $K_0(\delta_{22}^4)$, $K_1(\delta_{22}^4)$, and $K_3(\delta_{22}^4)$ designed from the above. In addition, since there does not exist $x'^1(0) \in \Delta_{22}$ assumed by $x'^1(0) = \delta_{22}^i$ such that $\text{rank}(\text{Blk}_i(\tilde{L}u(t))) \neq 1$ with $u(t) = K_3(\delta_{22}^4)x(t)$, then one has $K = K_3(\delta_{22}^4) = \delta_{22}[2\ 2\ 2\ 4\ 1\ 1\ 1\ 3]$ by Algorithm 25. As a result, the disturbance decoupling problem of system (6.22) is solvable by global time-invariant state feedback control $u(t) = Kx(t)$ with $K = \delta_{22}[2\ 2\ 2\ 4\ 1\ 1\ 1\ 3]$.

Finally, we can verify our result by numerical simulation. Consider disturbance sequences

$$\xi^1 = \{\xi(0), \ldots, \xi(5)\} = \left\{\delta_2^1, \delta_2^2, \delta_2^2, \delta_2^2, \delta_2^1, \delta_2^1\right\}$$

and

$$\xi^2 = \left\{\delta_2^2, \delta_2^1, \delta_2^2, \delta_2^1, \delta_2^2, \delta_2^2\right\},$$

respectively. The first subgraph of Fig. 6.9 displays $x^1(t)$, which is affected by $\xi^1(t)$ and $\xi^2(t)$, respectively, with initial state $x(0) = \delta_4^4$ and $K_0(\delta_{22}^4)$. The second subgraph presents the evolution of $x^1(t)$ under the designed event-triggered controllers $K_0(\delta_{22}^4)$, $K_1(\delta_{22}^4)$, and $K_3(\delta_{22}^4)$, which shows that $x^1(t+1)$ is independent of $\xi^1(t)$ and $\xi^2(t)$. Actually, $x^1(t)$ under the designed controller is not affected by any disturbance sequence.

6.2.3 Event-Triggered Control of Boolean Partial Control Networks

We consider the following BN:

$$\begin{cases} x_i(t+1) = f_i(X(t),\ \bar{\xi}(t)), \\ y_j(t) = h_j(X^1(t)). \end{cases} \tag{6.25}$$

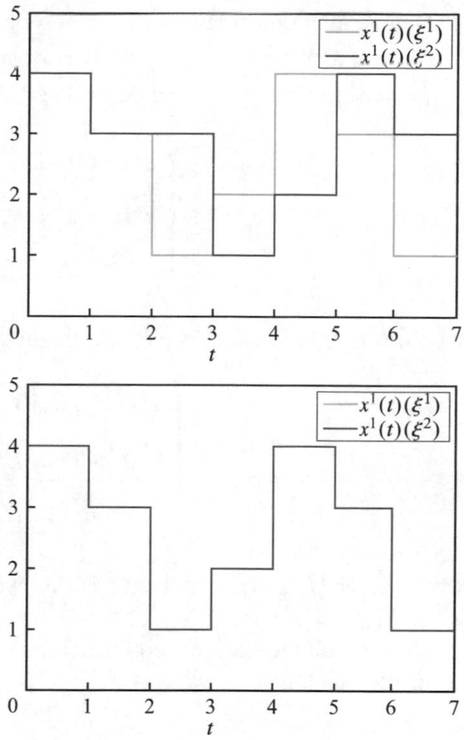

Fig. 6.9 $x^1(t)(\xi^1)$ and $x^1(t)(\xi^2)$ represent the trajectory of $x^1(t)$, which is affected by ξ^1 and ξ^2 respectively, with the same initial state $x^1(0) = \delta_{22}^4$ and the corresponding initial event-triggered gain matrix $K_0(\delta_{22}^4)$. The number in x-axis corresponds to the time t, and j in y-axis of the graph corresponds to the $x^1(t) = \delta_{22}^j$, $j \in [1, 4]$, respectively

By the similar procedure given in (6.12), system (6.25) can be converted into the following form:

$$\begin{cases} x^1(t+1) = Lx^1(t)x^2(t)\xi(t), & (6.26a) \\ x^2(t+1) = L_2x^1(t)x^2(t)\xi(t), & (6.26b) \\ y(t) = Hx^1(t), & (6.26c) \end{cases}$$

where $L \in \mathcal{L}_{2^r \times 2^{n+q}}$, $L_2 \in \mathcal{L}_{2^{n-r} \times 2^{n+q}}$ and $H \in \mathcal{L}_{2^p \times 2^r}$.

In order to make the disturbance decoupling problem of system (6.25) solvable, we consider another kind of event-triggered control [9]. Assume that some states are affected by intermittent external controls, i.e., the controller only works at some certain states. In detail, assume that there exist the trajectories of some states $x^1(t)$ depend on $x^2(t)$ and $\xi(t)$, then event-triggered controllers need to be added to BN (6.25) such that the trajectories are independent of $x^2(t)$ and $\xi(t)$. In addition, it is learned from (6.16) and (6.17) that we only design the controllers in dynamic system of $x^1(t)$. In other words, the controllers in dynamic system of $x^2(t)$ are redundant, which implies that we just need to consider Boolean networks with partial control inputs called Boolean partial control networks.

Denote the control signal depending on $x^1(t)$ by $f_{x^1(t)}$, then $f_{x^1(t)}$ takes value in Δ_2, and assume that $f_{x^1(t)} = \delta_2^1$ corresponds to the non-control BN (6.25) and $f_{x^1(t)} = \delta_2^2$ to Boolean partial control networks as follows:

$$\begin{cases} x^1(t+1) = \tilde{L}u(t)x(t)\xi(t), \\ x^2(t+1) = L^2x(t)\xi(t), \\ \quad\quad y(t) = Hx^1(t), \end{cases} \tag{6.27}$$

where $\tilde{L} \in \mathcal{L}_{2^r \times 2^{n+m+q}}$. As a result, the overall dynamics become

$$\begin{cases} x^1(t+1) = \hat{L}f_{x^1(t)}x(t)\xi(t), \\ x^2(t+1) = L^2x(t)\xi(t), \\ \quad\quad y(t) = Hx^1(t), \end{cases} \tag{6.28}$$

where $\hat{L} = [L \ \tilde{L}u(t)]$, and $u(t)$ is the event-triggered control which only works when $f_{x^1(t)} = \delta_2^2$.

Now, we aim to design a controller $u(t) = Kx(t)$ and find events that trigger the controller, such that

$$f_{x^1(t)} = \begin{cases} \delta_2^1 & \text{for } x^1(t) \in \Upsilon, \\ \delta_2^2 & \text{for } x^1(t) \in \Delta_{2^r} \backslash \Upsilon. \end{cases}$$

Therefore, $\Upsilon \subseteq \Delta_{2^r}$ needs to be properly chosen.

From the above analysis, we can conclude that if $\delta_{2^r}^i \in \Upsilon$, then $f_{x^1(t)} = \delta_2^1$, which means that if $x^1(t) = \delta_{2^r}^i$, then $x^1(t+1)$ is unaffected by $x^2(t)$ and $\xi(t)$, and $\text{rank}(\text{Blk}_i(L)) = 1$. In other words, if $\delta_{2^r}^i \notin \Upsilon$, Lemma 6.1 fails to be satisfied. Then, controllers need to be added to system (6.25) such that $x^1(t+1)$ is independent of $x^2(t)$ and $\xi(t)$. Therefore, we choose system (6.27) and $f_{x^1(t)} = \delta_2^2$ in this situation.

From the above analysis, set Υ can be obtained by $\Upsilon = \{\delta_{2^r}^i, i \in \Omega_r : \text{rank}(\text{Blk}_i(L)) = 1\}$. Then, for any $x^1(t) \in \Upsilon$, $x^1(t+1)$ is unaffected by $x^2(t)$ and $\xi(t)$. While for $x^1(t) \in \Delta_{2^r} \backslash \Upsilon$, system (6.27) will be chosen and the controllers should be designed to guarantee that $x^1(t+1)$ is independent of $x^2(t)$ and $\xi(t)$. Assume that the desired controller $u(t) = Kx(t)$ with $K = \delta_{2^m}[v_1 \ \cdots \ v_{2^n}]$, then we have the following result.

Theorem 6.4 *Denote $\Delta_{2^r}\backslash\Upsilon$ by $\{\delta_{2^r}^{\alpha_1}, \delta_{2^r}^{\alpha_2}, \ldots, \delta_{2^r}^{\alpha_{\iota'}}\}$, $\iota' \leq 2^r$. For any $\delta_{2^r}^{\alpha_i} \in \{\delta_{2^r}^{\alpha_1}, \delta_{2^r}^{\alpha_2}, \ldots, \delta_{2^r}^{\alpha_{\iota'}}\}$, $i \in [1, \iota']$, $rank(Blk_{\alpha_i}(\tilde{L}u(t))) = 1$, if and only if*

$$\Xi_{\alpha_i} := \Lambda((\alpha_i - 1)2^{n-r} + 1) \cap \Lambda((\alpha_i - 1)2^{n-r} + 2) \cap \cdots \cap \Lambda(\alpha_i 2^{n-r+1}) \neq \varnothing. \tag{6.29}$$

Additionally, if (6.29) holds, then the event-triggered gain matrix can be designed as

$$K = \delta_{2^m}[v_1\ v_2\ \cdots\ v_{2^n}], \tag{6.30}$$

with $v_j \in \Omega_m$, $j \in [1, 2^n]\backslash[(\alpha_i - 1)2^{n-r} + 1, \alpha_i 2^{n-r}]$, $i \in [1, \iota']$, and $\delta_{2^m}^{v_\lambda} \in \bigcup_{\delta_{2^r}^{k_w} \in \Xi_{\alpha_i}} \Lambda(\lambda)^{k_w}$, $\lambda \in \Omega_n$, where $\alpha_i \in \Omega_r$ is the unique integer satisfies that $\lambda = (\alpha_i - 1)2^{n-r} + \tau$, $\tau \in \Omega_{n-r}$.

Proof From the above analysis, it is easy to see that the necessity holds, and we only prove the sufficiency. Suppose $K = \delta_{2^m}[v_1\ v_2\ \cdots\ v_{2^n}]$ as (6.30), then one can get from $u(t) = Kx(t)$ that

$$x^1(t+1) = [\tilde{L}_{v_1,1}\ \tilde{L}_{v_2,2}\ \cdots\ \tilde{L}_{v_{2^n},2^n}]x^1(t)x^2(t)\xi(t) = \bar{L}x^1(t)x^2(t)\xi(t). \tag{6.31}$$

Partition $\bar{L}$ into 2^r parts denoted by $\text{Blk}_v(\bar{L})$, $v \in \Omega_r$. Since for any $i \in [1, \iota']$, $\Xi_{\alpha_i} \neq \varnothing$, which implies that for any given Ξ_{α_i}, $\text{rank}(Blk_{\alpha_i}(\bar{L})) = 1$. □

The following corollary is easily obtained from Theorem 6.4.

Corollary 6.2 *The disturbance decoupling problem of system (6.28) is solvable by event-triggered control $u(t) = Kx(t)$ given by (6.30).*

Remark 6.6 Corollary 6.1 is obtained from the aspect of BCNs, where controllers always exist in systems, and event-triggered gain matrices are constantly updated. While, the event-triggered gain matrix given in Corollary 6.2 are invariant, and only when the states belong to Υ, the controlled system (6.27) is chosen and the event-triggered control is active. Therefore, the latter is quite meaningful to reduce the control times and cost.

Consider the following apoptosis network [10]: Fas ligand that belongs to the tumor necrosis factor families initiates a pro-apoptotic pathway. Inversely, Fas ligand also knows how to trigger an anti-apoptotic pathway. As a result, Apoptosis maintains a stable levels. However, Fas ligand activity is increased rapidly when the interferon is added, which means that apoptosis is excessive and it leads to diabetes, neurodegenerative diseases, AIDS, etc. Then, Fas ligand can be regarded as a disturbance. Therefore, a simple cell apoptosis Boolean model can be expressed as follows,

$$\begin{cases} x_1(t+1) = \neg x_2(t) \wedge \xi(t), \\ x_2(t+1) = \neg x_1(t) \wedge x_3(t), \\ x_3(t+1) = x_2(t) \vee \xi(t), \end{cases} \tag{6.32}$$

where inhibitor of apoptosis proteins (IAP) is denoted by x_1, active caspase 3 (C3a) by x_2, and active caspase 8 (C8a) by x_3; Fas ligand is regarded as disturbance input ξ.

Now, we study the disturbance decoupling problem of system (6.32), and consider the following output equations:

$$\begin{cases} y_1(t) = x_1(t) \rightarrow x_2(t), \\ y_2(t) = x_1(t) \leftrightarrow x_2(t). \end{cases} \tag{6.33}$$

Let $x^1(t) = \ltimes_{i=1}^{2} x_i(t)$, then one has the following algebraic form:

$$x^1(t+1) = Lx^1(t)x_3(t)\xi(t) = \delta_{2^2}[4\ 4\ 4\ 4\ 2\ 3\ 2\ 4\ 3\ 3\ 4\ 4\ 1\ 3\ 2\ 4]x^1(t)x_3(t)\xi(t). \tag{6.34}$$

For this example, one can easily obtain that $\Upsilon = \{\delta_{2^2}^1\}$. Assume that the control system of (6.32) is given as follows:

$$\begin{cases} x_1(t+1) = u_1(t) \leftrightarrow [\neg x_2(t) \wedge \xi(t)], \\ x_2(t+1) = u_2(t) \vee [\neg x_1(t) \wedge x_3(t)], \\ x_3(t+1) = x_2(t) \vee \xi(t), \end{cases} \tag{6.35}$$

and

$$\begin{aligned} \tilde{L} = \delta_{2^2}[&4\ 4\ 4\ 4\ 4\ 4\ 4\ 4\ 1\ 1\ 4\ 4\ 1\ 1\ 4\ 4\ 4\ 4\ 4\ 4\ 4\ 4\ 4\ 4\ 4\ 4\ 4\ 4\ 4\ 4\ 4\ 4 \\ &4\ 4\ 4\ 4\ 1\ 1\ 4\ 4\ 1\ 1\ 4\ 4\ 4\ 4\ 4\ 4\ 4\ 4\ 4\ 4\ 4\ 4\ 4\ 4\ 4\ 4\ 4\ 4\ 4\ 4\ 4\ 4]. \end{aligned}$$

It is easy to obtain that $\Lambda(3) = \Lambda(5) = \Lambda(7) = \{\delta_{2^2}^1, \delta_{2^2}^4\}$, and $\Lambda(4) = \Lambda(6) = \Lambda(8) = \{\delta_{2^2}^4\}$, which means that $\Xi_2 = \Xi_3 = \Xi_4 = \{\delta_{2^2}^4\}$. Moreover, $\Lambda(3)^4 = \{\delta_{2^2}^1, \delta_{2^2}^2, \delta_{2^2}^4\}$, $\Lambda(5)^4 = \{\delta_{2^2}^2, \delta_{2^2}^4\}$, $\Lambda(7)^4 = \{\delta_{2^2}^2, \delta_{2^2}^3, \delta_{2^2}^4\}$, and $\Lambda(4)^4 = \Lambda(6)^4 = \Lambda(8)^4 = \Delta_{2^2}$. Therefore, the disturbance decoupling problem of system (6.32) is solvable from Corollary 6.2. Let $v_1 = v_6 = 1$, $v_4 = v_8 = 2$, $v_2 = v_7 = 3$, and

$v_3 = v_5 = 4$. As a result, the overall dynamics becomes

$$\begin{cases} x^1(t+1) = \hat{L} f_{x^1(t)} x(t)\xi(t), \\ x_3(t+1) = x_2(t) \vee \xi(t), \\ \quad y_1(t) = x_1(t) \to x_2(t), \\ \quad y_2(t) = x_1(t) \leftrightarrow x_2(t), \end{cases}$$

and

$$f_{x^1(t)} = \begin{cases} \delta_2^1 \text{ for } x^1(t) \in \{\delta_{22}^1\}, \\ \delta_2^2 \text{ for } x^1(t) \in \{\delta_{22}^2, \delta_{22}^3, \delta_{22}^4\}, \end{cases}$$

where $\hat{L} = [L\ \tilde{L}u_1(t)u_2(t)]$, $u_1 = [x_1 \wedge (x_2 \leftrightarrow x_3)] \vee [\neg(x_1 \vee x_3)]$, and $u_2 = [x_1 \wedge x_2] \vee [\neg(x_1 \vee (x_2 \leftrightarrow x_3))]$.

6.3 Event-Triggered Control for Output Regulation of Probabilistic Logical Systems with Delays

In this section, we investigate the output regulation problem of probabilistic k-valued logical systems with delays by an intermittent control scheme. Two types of event-triggered control are designed via semi-tensor product of matrices. According to the algebraic state-space representation of probabilistic k-valued systems with delays, the problem is transformed into the existence of solutions of algebraic equations. Then we obtain the sufficient and necessary condition for the output regulation problem of probabilistic k-valued logical systems with delays. Two types of approaches are given to design the event-triggered control laws.

6.3.1 Problem Formulation

We probe the following probabilistic logical systems with n nodes, m inputs, and τ time state delays:

$$x_i(t+1) = f_i^{k_i}(u_1(t), \ldots, u_m(t), x_1(t), \ldots, x_n(t), x_1(t-1), \ldots, x_n(t-1), \ldots, x_1(t-\tau), \ldots, x_n(t-\tau)), \quad i = 1, \ldots, n, 1 \le k_i \le l_i, \tag{6.36}$$

$$y_j(t) = h_j(x_1(t), \ldots, x_n(t), x_1(t-1), \ldots, x_n(t-1), \ldots, x_1(t-\tau), \ldots, x_n(t-\tau)), \tag{6.37}$$

where $i = 1, \ldots, n,\ 1 \leq k_i \leq l_i,\ j = 1, \ldots, p$, for $t \geq 0, x_1(t), x_2(t), \ldots, x_n(t) \in \mathcal{D}_k$ are states and $y_1(t), \ldots, y_p(t)$ are outputs, $u_1(t), u_2(t), \ldots, u_m(t) \in \mathcal{D}_k$ are control inputs, and $f_i : D_k^{m+n(\tau+1)} \to D_k\ (i = 1, 2, \ldots, n)$ are opted from a finite logical function set $F_i = \{f_i^1, f_i^2, \ldots, f_i^{l_i}\}$ with probability p_i^j, where $\Sigma_{j=1}^{l_i} p_i^j = 1$ holds for any $i = 1, 2, \ldots, n$ and $h_j : D_k^{n(\tau+1)} \to D_k\ (j = 1, 2, \ldots, p)$ are logical functions. Ultimately, there are $\kappa = \Pi_{i=1}^n l_i$ models.

Let $\Sigma_\lambda = \{f_1^{\lambda_1}, f_2^{\lambda_2}, \ldots, f_n^{\lambda_n}\}$ be the λ-th of system (6.36) and (6.37), and the probability of the λ-th model is selected as

$$\mathbf{P}(f_i = f_i^{\lambda_i}) = p_i^j, \quad j = 1, 2, \ldots, l_i$$

Then the probability of Σ_λ is

$$P_\lambda = \Pi_{i=1}^n p_i^j.$$

Define $x(t) = \ltimes_{i=1}^n x_i(t)$ and $u(t) = \ltimes_{j=1}^m u_j(t)$. The algebraic form of system (6.36) and (6.37) is converted as

$$\begin{aligned} x_i(t+1) &= M_i u(t)x(t)x(t-1)\cdots x(t-\tau), \quad i = 1, 2, \ldots, n, \\ y_j(t) &= H_j x(t)x(t-1)\cdots x(t-\tau), \quad j = 1, 2, \ldots, p, \end{aligned}$$

where M_i and H_j are the structure matrices of each f_i and h_j respectively, and M_i are chosen from the set $\{M_i^1, M_i^2, \ldots, M_i^{l_i}\}$ with the probability $\{p_i^1, p_i^2, \ldots, p_i^{l_i}\}$.

Denote $\mathbb{E}x(t)$ as the mathematical expectation of state at time t, then the evolution of state expectation is described as

$$\begin{cases} \mathbb{E}x_i(t+1) = \mathbb{E}[M_i u(t)x(t)x(t-1)\cdots x(t-\tau)] = \tilde{M}_i u(t)\mathbb{E}x(t)\cdots \mathbb{E}x(t-\tau), \\ \qquad \mathbb{E}y_j(t+1) = H_j \mathbb{E}x(t)\mathbb{E}x(t-1)\cdots \mathbb{E}x(t-\tau), \end{cases} \tag{6.38}$$

where $\tilde{M}_i = \Sigma_{j=1}^{l_i} p_i^j M_i^j,\ i = 1, 2, \ldots, n$ and $j = 1, 2, \ldots, p$.

The dynamics of the logical reference networks with n_1 nodes are proposed as

$$\begin{cases} \hat{x}_i(t+1) = \hat{f}_i(\hat{x}_1(t), \ldots, \hat{x}_{n_1}(t)), \quad i = 1, \ldots, n_1, \\ \quad \hat{y}_j(t) = \hat{h}_j(\hat{x}_1(t), \ldots, \hat{x}_{n_1}(t)), \quad j = 1, \ldots, p, \end{cases} \tag{6.39}$$

for $t \geq 0$, where $\hat{x}_1(t), \hat{x}_2(t), \ldots, \hat{x}_{n_1}(t) \in \mathcal{D}_k$ and $\hat{y}_1(t), \ldots, \hat{y}_p(t)$ are states, $\hat{f}_i\ i = 1, 2, \ldots, n_1$ and $\hat{h}_j,\ j = 1, 2, \ldots, p$ are logical functions.

Let $\hat{x}(t) = \ltimes_{i=1}^{n_1} \hat{x}_i(t)$. The algebraic form of system (6.39) can be acquired as follows,

$$\begin{cases} \hat{x}_i(t+1) = \hat{M}_i \hat{x}(t), & i = 1, 2, \dots, n_1, \\ \hat{y}_j(t) = \hat{H}_j \hat{x}(t), & j = 1, 2, \dots, p, \end{cases} \tag{6.40}$$

where $\hat{M}_i \in \mathcal{L}_{k \times k^{n_1}}$ and $\hat{H}_j \in \mathcal{L}_{k \times k^{n_1}}$ are structure matrix of each $\hat{f}_i$ and $\hat{h}_j$.

Multiply the equations in (6.38) and (6.40) together leads to the following equations,

$$\begin{cases} \mathbb{E}x(t+1) = Lu(t)\mathbb{E}x(t)\mathbb{E}x(t-1)\cdots\mathbb{E}x(t-\tau), \\ \mathbb{E}y(t) = H\mathbb{E}x(t)\mathbb{E}x(t-1)\cdots\mathbb{E}x(t-\tau), \end{cases} \tag{6.41}$$

$$\begin{cases} \hat{x}(t+1) = \hat{L}\hat{x}(t), \\ \hat{y}(t) = \hat{H}\hat{x}(t), \end{cases} \tag{6.42}$$

where $L \in \mathcal{L}_{k^n \times k^{m+n(\tau+1)}}$, $\hat{L} \in \mathcal{L}_{k^{n_1} \times k^{n_1}}$, $H \in \mathcal{L}_{k^p \times k^{n(\tau+1)}}$, $\hat{H} \in \mathcal{L}_{k^p \times k^{n_1}}$ and $\mathrm{Col}_k(L) = \ltimes_{j=1}^{n} \mathrm{Col}_k(\tilde{M}_j)$.

Definition 6.3 The state $x(t) \in \Delta_{k^n}$ is said to be globally stabilized to the fixed point $\delta_{k^{n+n_1}}^a$, if there exists control input $u(t)$ and an integer $T > 0$, for random initial state sequence $x(0), x(-1), \dots, x(-\tau) \in \Delta_{k^{n+n_1}}$, such that $x(t) = \delta_{k^{n+n_1}}^a$, $t \geq T$.

The output regulation problem aims to conduct the state feedback control laws of system:

$$U_i(t) = g_i(\hat{x}_1(t), \dots, \hat{x}_{n_1}(t), x_1(t), \dots, x_n(t), \dots, x_1(t-\tau), \dots, x_n(t-\tau)), \tag{6.43}$$

where g_i, $i = 1, \dots, m$ denote logical functions. Furthermore, there exists an integer $T > 0$, such that

$$y(t; x(0), x(-1), \dots, x(-\tau), u(t)) = \hat{y}(t; \hat{x}(0)), \quad t > T. \tag{6.44}$$

holds for $\forall x(t) \in \Delta_{k^n}$ and $\hat{x}(t) \in \Delta_{k^{n_1}}$.

Definition 6.4 The output regulation problem is solvable if we can find a control in the form of (6.43) and an integer $T > 0$, such that

$$y(t; x(0), x(-1), \dots, x(-\tau), u(t)) = \hat{y}(t; \hat{x}(0)), \quad t > T. \tag{6.45}$$

holds for $\forall x(t) \in \Delta_{k^n}$ and $\hat{x}(t) \in \Delta_{k^{n_1}}$.

6.3.2 The Existence of Solutions of the Output Regulation Problem

In this section, we present the analysis process of the output regulation problem with probabilistic-delay systems, combined with the algorithm of output-state set S and a sufficient and necessary condition to conduct the output regulation based primarily on the existence of state feedback control laws.

Let $w_0(t) = \mathbb{E}x(t)$, $w_1(t) = \mathbb{E}x(t-1)$, $w_2(t) = \mathbb{E}x(t-2), \ldots, w_\tau(t) = \mathbb{E}x(t-\tau)$ and $w(t) = \ltimes_{i=0}^{\tau} w_i(t)$. Then (6.41) can be written as

$$\begin{cases} \mathbb{E}x(t+1) = Lu(t)w(t), \\ \mathbb{E}y(t) = Hw(t), \end{cases} \tag{6.46}$$

$$\begin{aligned} w(t+1) &= \ltimes_{i=0}^{\tau} w_i(t+1) \\ &= \mathbb{E}x(t+1)\mathbb{E}x(t)\cdots\mathbb{E}x(t-\tau+1) \\ &= Lu(t)\mathbb{E}x(t)\cdots\mathbb{E}x(t-\tau)\mathbb{E}x(t)\cdots\mathbb{E}x(t-\tau+1) \\ &= Lu(t)W_{[k^{n\tau},k^{n(\tau+1)}]}\Phi_{n\tau}w(t) \\ &= L(I_{k^m} \otimes W_{[k^{n\tau},k^{n(\tau+1)}]}\Phi_{n\tau})u(t)w(t) \\ &\triangleq Ku(t)w(t), \end{aligned}$$

where $K = L(I_{k^m} \otimes W_{[k^{n\tau},k^{n(\tau+1)}]}\Phi_{n\tau})$.

We premeditate a type of event-triggered control to present the output regulation problem. Suppose that intermittent external controls influence several states, i.e., control only affects several states of the system. That is called the event-triggered control. A triggering mechanism is proposed, determining the time for control input to be updated.

Let Π be the control sets of the states. Utilizing semi-tensor product, we are able to obtain that

$$w(t+1) = \begin{cases} K^1 \ltimes u(t) \ltimes w(t) & w(t) \in \Psi, \\ K^2 \ltimes w(t) & w(t) \in \Delta_{k^{n+n_1}} \backslash \Psi, \end{cases} \tag{6.47}$$

where $w(t)$ is state, $u(t)$ is the event-triggered control input which operates when $w(t) \in \Psi$, $K^1 \in \mathcal{L}_{k^{n(\tau+1)} \times k^{m+n(\tau+1)}}$ and $K^2 \in \mathcal{L}_{k^{n(\tau+1)} \times k^{n(\tau+1)}}$. Moreover, we consider no control action as a special control method. Then we use $\delta_{k^m+1}^{k^m+1}$ (which is a formal definition) to represent the control $u_0(t)$. Let

$$\tilde{u}(t) = \begin{cases} u(t) & w(t) \in \Psi, \\ u_0(t) & w(t) \in \Delta_{k^{n+n_1}} \backslash \Psi. \end{cases} \tag{6.48}$$

Thus, we can get

$$w(t+1) = [K^1\ K^2]\tilde{u}(t)w(t). \tag{6.49}$$

The objective aims to investigate the output regulation problem of probabilistic k-valued logical system, we combine the two systems in the form of semi-tensor product.

Let $z(t) = w(t)\hat{x}(t)$ and $s(t) = Ey(t)\hat{y}(t)$, then augmented system can be acquired as follows,

$$\begin{aligned} z(t+1) &= w(t+1)\hat{x}(t+1) \\ &= [K^1\ K^2]\tilde{u}(t)w(t)\hat{L}\hat{x}(t) \\ &= [K^1\ K^2](I_{k^{n(\tau+1)}\times(k^m+1)} \otimes \hat{L})\tilde{u}(t)w(t)\hat{x}(t) \\ &= [K^1\ K^2](I_{k^{n(\tau+1)}\times(k^m+1)} \otimes \hat{L})\tilde{u}(t)z(t) \\ &\triangleq [\tilde{L}\ \bar{L}]\tilde{u}(t)z(t), \end{aligned} \tag{6.50}$$

$$\begin{aligned} s(t) &= Ey(t)\hat{y}(t) \\ &= Hw(t)\hat{H}\hat{x}(t) \\ &= H(I_{k^{n(\tau+1)}} \otimes \hat{H})w(t)\hat{x}(t) \\ &= H(I_{k^{n(\tau+1)}} \otimes \hat{H})z(t) \\ &\triangleq \tilde{H}z(t), \end{aligned} \tag{6.51}$$

where $\tilde{L} = K^1(I_{k^{n(\tau+1)}\times(k^m+1)} \otimes \hat{L})$, $\bar{L} = K^2(I_{k^{n(\tau+1)}\times(k^m+1)} \otimes \hat{L})$, and $\tilde{H} = H(I_{k^{n(\tau+1)}} \otimes \hat{H})$. This part aims to conduct the state feedback event-triggered control

$$\tilde{u}(t) = Gz(t), \tag{6.52}$$

where $G \in \mathcal{L}_{(k^m+1)\times k^{n(\tau+1)+n_1}}$ is the state feedback control matrix. If the output regulation problem is solvable, then the system will stay in the nonempty set $S \subseteq \Omega :\triangleq \bigcup_{j=1}^{k^p} \Omega_j$ and the set Ω_j is defined as

$$\Omega_j = \{\delta^i_{k^{n(\tau+1)+n_1}} : \mathrm{Col}_i(\tilde{H}) = \delta^j_{k^p} \ltimes \delta^j_{k^p}\}.$$

According to the corresponding analysis above, we draw the algorithm for the output-state set S listed as follows:

Algorithm 26 An algorithm for output-state set S

1: Input $\tilde{H},\ \tilde{L},\ \bar{L}$
2: **for** each integer $1 \leq j \leq k^p$ **do**
3: calculate $\Omega_j = \{\delta^i_{k^{n(\tau+1)+n_1}} : \mathrm{Col}_i(\tilde{H}) = \delta^j_{k^p} \ltimes \delta^j_{k^p}\}$
4: **end for**
5: $\Omega := \bigcup_{j=1}^{k^p} \Omega_j$
6: Randomly select a non-empty subset of Ω and denote the set as S

Theorem 6.5 *The output regulation problem is solvable, if and only if there exists an integer* $1 \leq T \leq k^{n(\tau+1)+n_1}$, *such that*

$$R^T = \delta_{k^{n(\tau+1)+n_1}} \underbrace{[j_1 \ \cdots \ j_s]}_{k^{n(\tau+1)+n_1}},$$

where $R = [\tilde{L}\ \bar{L}]G\Phi_{n(\tau+1)+n_1}$ *and* $\delta^{j_k}_{k^{n(\tau+1)+n_1}} \in S,\ 1 \leq k \leq s.$

Proof For any state in set S, logical control networks converge to it, which means that the control sequence needs to meet the conditions as follows,

$$\mathbb{E}y(t) = \hat{y}(t).$$

The recursive relationship of logical systems are as follows:

$$\mathbb{E}y(t) = Hw(t) = HKu(t-1)w(t-1),$$
$$\hat{y}(t) = \hat{H}\hat{x}(t).$$

Under the event-triggered control, we can draw the corresponding equation,

$$\begin{aligned}
&\mathbb{E}y(t) - \hat{y}(t)\\
&= HKu(t-1)w(t-1) - \hat{H}\hat{L}\hat{x}(t-1)\\
&= HKGw(t-1)\hat{x}(t-1)w(t-1) - \hat{H}\hat{L}\hat{x}(t-1)\\
&= HKGW_{[k^{n_1},k^{n(\tau+1)}]}\Phi_{n(\tau+1)}w(t-1)\hat{x}(t-1) - \hat{H}\hat{L}P_{n(\tau+1)}w(t-1)\hat{x}(t-1)\\
&= (HKGW_{[k^{n_1},k^{n(\tau+1)}]}\Phi_{n(\tau+1)} - \hat{H}\hat{L}P_{n(\tau+1)})w(t-1)\hat{x}(t-1),
\end{aligned}$$

where $P_{n(\tau+1)}w(t-1) = I_{n(\tau+1)}$. We introduce this symbol $P_{n(\tau+1)}$ just to unify the equation form.

Denote $W = HKGW_{[k^{n_1},k^{n(\tau+1)}]}\Phi_{n(\tau+1)} - \hat{H}\hat{L}P_{n(\tau+1)}$, then $\mathbb{E}y(t) - \hat{y}(t) = Ww(t-1)\hat{x}(t-1) = WR^{t-1}w(0)\hat{x}(0)$.

Therefore, output regulation problem is solvable, which is equivalent to the equations $WR^{t-1} = 0$ and

$$R^T = \delta_{k^{n(\tau+1)+n_1}} \underbrace{[j_1 \ \cdots \ j_s]}_{k^{n(\tau+1)+n_1}}.$$

□

6.3.3 Event-Triggered Control Design I

Owing to the augmented system (6.50), a feasible method is given to conduct the state feedback event-triggered control.

First, divide the matrix $\tilde{L}$ into the same k^m partitions:

$$\tilde{L} = [\mathrm{Blk}_1(\tilde{L}) \ \mathrm{Blk}_2(\tilde{L}) \ \cdots \ \mathrm{Blk}_{k^m}(\tilde{L})]$$

where $\mathrm{Blk}_i(\tilde{L}),\ i = 1, \ldots, k^m$ is the i-th partition, and every columns of which are probabilistic vectors.

Definition 6.5 State Z is reachable from Z_0 at time $s > 0$ if a sequence of controls $\{U(t), t = 0, 1, \ldots\}$ can be found such that the trajectory of (6.50) with initial value Z_0 and controls $\{U(t), t = 0, 1, \ldots\}$ will reach Z at time s.

Then, we will give some definitions of reachable sets listed as follows:

$$S_0 = \{\delta^j_{k^{n(\tau+1)+n_1}} : \exists 1 \leq T \leq k^{n(\tau+1)+n_1}, \text{ s.t. } \tilde{L}^T \delta^j_{k^{n(\tau+1)+n_1}} \in S\},$$

$$S_0' = \{\delta^j_{k^{n(\tau+1)+n_1}} \in Q : \exists \delta^{v_j}_{k^m} \text{s.t. } \mathrm{Col}_j(\mathrm{Blk}_{v_j}(\tilde{L})) \bigcap S \neq \varnothing\}.$$

where Q is $\Delta_{k^{n(\tau+1)+n_1}} \setminus S_0$.

The construction of the above two sets classifies the initial state, and divides the target state into $u(t)$ control state and $u_0(t)$ control state.

$$\begin{cases} S_i = \bigcup_{T=0}^{k^{n(\tau+1)+n_1}} R_T(S_{i-1}'), \\ S_i' = \{\delta^j_{k^{n(\tau+1)+n_1}} \in \Delta_{k^{n(\tau+1)+n_1}} \setminus \bigcup_{k=0}^{i-1} \tilde{S}_k : \exists \delta^{v_j}_{k^m} \text{s.t.} \mathrm{Col}_j(\mathrm{Blk}_{v_j}(\tilde{L})) \bigcap \tilde{S}_{i-1} \neq \varnothing\}, \end{cases} \tag{6.53}$$

where $i \geq 1,\ \tilde{S}_0 = S_0 \cup S_0',\ R_T(S_i') = \{\delta^j_{k^{n(\tau+1)+n_1}} : \tilde{L}^T \delta^j_{k^{n(\tau+1)+n_1}} \in S_i'\}$, and $\tilde{S}_i = S_i' \cup S_i$.

Theorem 6.6 *The output regulation problem is solvable, if and only if the following conditions are satisfied:*

(I) $\Omega \neq \varnothing$ *and* $\tilde{S}_0 \neq \varnothing$*;*
(II) There exist an integer $0 \leq T \leq k^{n(\tau+1)+n_1} - 1$ *and a nonempty set* $S \subseteq \Omega$*, such that*

$$\bigcup_{i=0}^{T} \tilde{S}_i = \Delta_{k^{n(\tau+1)+n_1}}. \tag{6.54}$$

Proof (Necessity) Note that the output regulation problem is solvable, it is easy to see that $\Omega \neq \varnothing$. By definition, the states in the nonempty subset of Ω (defined as set S) are globally stabilized from arbitrary initial state $z(0) = \delta^{j}_{k^{n(\tau+1)+n_1}}$ at time T, if and only if there exist an integer T and control $u(t)$, such that

$$z(T) = \delta^{j_k}_{k^{n(\tau+1)+n_1}} \in S.$$

It is liable to examine that $\delta^{j}_{k^{n(\tau+1)+n_1}} \in \bigcup_{i=0}^{T} \tilde{S}(j_k)$. Thus, we only prove the situation that $\delta^{j_k}_{k^{n(\tau+1)+n_1}}$ is fixed. When $T = 0$, we can see that

$$z(0) = \delta^{j_k}_{k^{n(\tau+1)+n_1}},$$

i.e., $\delta^{j}_{k^{n(\tau+1)+n_1}} \in \tilde{S}_0(j_k)$. When $T = k - 1$, we assume that

$$z(k-1) = \delta^{j_k}_{k^{n(\tau+1)+n_1}},$$

i.e., $\delta^{j}_{k^{n(\tau+1)+n_1}} \in \bigcup_{i=0}^{k-1} \tilde{S}(j_k)$, thus

$$\delta^{j}_{k^{n(\tau+1)+n_1}} \in \bigcup_{i=0}^{k} \tilde{S}_i(j_k).$$

Assume that γ is the minimum value that satisfies the above conditions. Then we have proved that $\bigcup_{i=0}^{T} \tilde{S}_i = \Delta_{k^{n(\tau+1)+n_1}}$. Then we need to prove $T \leq k^{n(\tau+1)+n_1} - 1$, which is equivalent to

$$\left| \bigcup_{i=0}^{\gamma} \tilde{S}_i \right| \geq \gamma + 1. \tag{6.55}$$

If the above formula does not hold, then $|\bigcup_{i=0}^{\gamma}\tilde{S}_i| \leq \gamma$. There exists $0 \leq i \leq k$ such that $\tilde{S}_i = \varnothing$, i.e., $S_l = \varnothing$ and $S_i^{'} = \varnothing$. Thus, we can draw the conclusion that $\bigcup_{i=0}^{T}\tilde{S}_i \neq \Delta_{k^{n(\tau+1)+n_1}}$. We adopt the mathematical induction to prove (6.55).

When $k = \gamma = 1$, if $|\tilde{S}_0 \cup \tilde{S}_1| < 2$, then we can see that $|\tilde{S}_0 \cup \tilde{S}_1| = 1$, $\tilde{S}_k = \varnothing$ for $k \geq 1$. There exits a contradiction with the condition $\bigcup_{i=0}^{\gamma}\tilde{S}_l = \Delta_{k^{n(\tau+1)+n_1}}$.

Assume that all integers $k \leq \gamma - 1$ satisfy the above condition $|\bigcup_{i=0}^{\gamma-1}\tilde{S}_i| \geq \gamma$. If

$$\left|\bigcup_{i=0}^{\gamma}\tilde{S}_i\right| = \left|\bigcup_{i=0}^{\gamma-1}\tilde{S}_i \cup \tilde{S}_\gamma\right| \leq \gamma,$$

then we have $\tilde{S}_\gamma = \varnothing$. There exits a contradiction with the assumption that γ is the minimum value that satisfies the above conditions. Therefore,

$$\left|\bigcup_{i=0}^{\gamma}\tilde{S}_i\right| \geq \gamma + 1$$

is true for any $k \leq \gamma$.

(Sufficiency) Note that the equation (6.54) means that all states in set S are globally stabilizable. Let $x^* = \delta_{k^{n(\tau+1)+n_1}}^{j} \in \Delta_{k^{n(\tau+1)+n_1}}$. We can obtain the sets $S_i^{'}$ and S_i respectively. For $1 \leq j \leq k^{n(\tau+1)+n_1}$, we can see that

$$\mathrm{Col}_j(G) = \begin{cases} \delta_{k^m+1}^{k^m+1}, & \delta_{k^{n(\tau+1)+n_1}}^{j} \in \bigcup_{i=0}^{T} S_i, \\ \delta_{k^m+1}^{u_j}, & \delta_{k^{n(\tau+1)+n_1}}^{j} \in \bigcup_{i=0}^{T} S_i^{'}, \end{cases} \tag{6.56}$$

where $u_j = \max\{\mathrm{Col}_{u_j}(\mathrm{Blk}_j(\tilde{L})) \subset \tilde{S}_i\}$.

For any state $x^* \in \Delta_{k^{n(\tau+1)+n_1}} = \bigcup_{i=0}^{T}\tilde{S}_i$, there are four cases: (i) $x^* \in S_0$; (ii) $x^* \in S_0^{'}$; (iii) $x^* \in S_i$; (iv) $x^* \in S_i^{'}$.

- Case 1: If $x^* \in S_0$, then there exists an integer T such that $\bar{L}^T x^* \in S$.
- Case 2: If $x^* \in S_0^{'}$, then there exists a control u such that

$$\begin{aligned} Lux^* &= LW_{[k^{n(\tau+1)+n_1}, k^m]}x^*u \\ &= \mathrm{Blk}_j(L^\star)u \\ &= \mathrm{Col}_{u_j}(\mathrm{Blk}_j(L^\star)) \in S, \end{aligned}$$

where $L^\star = LW_{[k^{n(\tau+1)+n_1}, k^m]}$.

- Case 3: If $x^* \in S_i$, then there exists an integer $T^{'}$ such that $\bar{L}^{T^{'}} x^* \in S_i^{'}$.

- Case 4: If $x^* \in S_i'$, then there exists a control input u such that

$$Lux^* = LW_{[k^{n(\tau+1)+n_1},k^m]}x^*u$$
$$= \text{Blk}_j(L^\star)u$$
$$= \text{Col}_{u_j}(\text{Blk}_j(L^\star)) \in \tilde{S}_{i-1}.$$

According to the comprehensive classification, the output regulation problem is solvable.   □

Assuming that the conditions in Theorem 6.6 are satisfied, we can obtain the state feedback matrix in the output regulation problem. The algorithm is presented to find the event-triggered state feedback matrix G:

Algorithm 27 An algorithm for event-triggered state feedback matrix

---

1: Input: S, $\tilde{L}$, $\bar{L}$
2: Initialize: $S_0 := \varnothing$, $S_0' := \varnothing$
3: **for** $j := 1$ to $k^{n(\tau+1)+n_1}$ **do**
4:     **for** $T := 1$ to $k^{n(\tau+1)+n_1}$ **do**
5:         **if** $\bar{L}^T \delta_{k^{n(\tau+1)+n_1}}^j \in S$ **then** $S_0 \leftarrow \delta_{k^{n(\tau+1)+n_1}}^j$
6:         **end if**
7:     **end for**
8: **end for**
9: **for** $j := 1$ to $k^{n(\tau+1)+n_1}$ **do**
10:     **for** $v_j := 1$ to k^m **do**
11:         **while** $\delta_{k^{n(\tau+1)+n_1}}^j \in \Delta_{k^{n(\tau+1)+n_1}} \backslash S_0$ and $\text{Col}_j(\text{Blk}_{v_j}(\tilde{L})) \bigcap S \neq \varnothing$ **do** $S_0' \leftarrow \delta_{k^{n(\tau+1)+n_1}}^j$
12:         **end while**
13:     **end for**
14: **end for**
15: $i := 1$
16: **while** $\bigcup_{k=0}^{i-1} \tilde{S}_k \subsetneq \Delta_{k^{n(\tau+1)+n_1}}$ **do**
17:     **for** $T := 1$ to $k^{n(\tau+1)+n_1}$ **do**
18:         **for** $j := 1$ to $k^{n(\tau+1)+n_1}$ **do**
19:             **if** $\bar{L}^T \delta_{k^{n(\tau+1)+n_1}}^j \in S_{i-1}'$ **then**
20:                 $S_i \leftarrow \delta_{k^{n(\tau+1)+n_1}}^j$
21:             **end if**
22:         **end for**
23:         **for** $v_j := 1$ to k^m **do**
24:             **if** $\text{Col}_j(\text{Blk}_{v_j}(\tilde{L})) \bigcap \tilde{S}_{i-1} \neq \varnothing$ and $\delta_{k^{n(\tau+1)+n_1}}^j \in \Delta_{k^{n(\tau+1)+n_1}} \backslash \bigcup_{k=0}^{i-1} \tilde{S}_k$ **then**
25:                 $S_i' \leftarrow \delta_{k^{n(\tau+1)+n_1}}^j$
26:             **end if**
27:         **end for**
28:     **end for**
29:     $\tilde{S}_i := S_i' \cup S_i$
30:     $i := i+1$
31: **end while**

Remark 6.7

1. According to Algorithm 27, the design of the event-triggered state feedback matrix is

$$\mathrm{Col}_j(G) = \delta_{k^m+1}[k^m+1\ j_1\ \cdots\ j_2\ k^m+1\ \cdots\ j_r\ k^m+1],$$

where

$$\mathrm{Col}_j(G) = \begin{cases} \delta_{k^m+1}^{u_j}, & u_j \in j_1, j_2, \cdots, j_r, \\ \delta_{k^m+1}^{k^m+1}, & \text{otherwises.} \end{cases}$$

2. The algorithm complexity for conducting the output-state set S is $O(k^p)$. Based on Algorithm 27, the computational complexity of searching for event-triggered state feedback matrix is calculated as follows: it needs $k^{2(n(\tau+1)+n_1)}$, $k^{n(\tau+1)+n_1+m}$ steps to obtain $S_0, S_0^{'}, \ldots$, and the total steps are $k^{2(n(\tau+1)+n_1)} + k^{n(\tau+1)+n_1+m} + 1 + k^{n(\tau+1)+n_1}[k^{n(\tau+1)+n_1}(k^{n(\tau+1)+n_1+m} + k^{2(n(\tau+1)+n_1)}) + 2]$. Thus, the algorithm complexity for conducting the event-triggered state feedback matrix is $O[k^{3(n(\tau+1)+n_1)+m} + k^{4(n(\tau+1)+n_1)}]$.

6.3.4 Event-Triggered Control Design II

In this part, we utilize another event-triggered control for the probabilistic k-valued logical systems with delays.

Consider the system (6.41), we can draw the conclusion that

$$\begin{aligned} z(t+1) &= w(t+1)\hat{x}(t+1) \\ &= Ku(t)w(t)\hat{L}\hat{x}(t) \\ &= K(I_{k^{m+n(\tau+1)}} \otimes \hat{L})u(t)w(t)\hat{x}(t) \\ &= K(I_{k^{m+n(\tau+1)}} \otimes \hat{L})u(t)z(t) \\ &\triangleq Mu(t)z(t), \end{aligned}$$

where $M = K(I_{k^{m+n(\tau+1)}} \otimes \hat{L})$.

In this problem, we use the time-variable controller for the reason that time-variable controller is closer to the actual situation. So we use the time-variable controller $u_i(t) = \Xi_i(t, z(0))z(t)$. Then control is able to transform as

$$u(t) = \Xi(t, z(0))z(t)$$

and

$$\mathrm{Col}_j(\Xi(t, z(0))) = \ltimes_{i=1}^{k^{n(\tau+1)+n_1}} \mathrm{Col}_j(\Xi_i(t, z(0))).$$

Then we can see that

$$z(t+1) = Mu(t)z(t) = \ltimes_{i=t}^{0}(M\Xi(i, z(0))\Phi_{k^{n(\tau+1)+n_1}})z(0).$$

Denote the k-step reachable set of S by $\Upsilon_k(S)$, then the structure of $\Upsilon_k(S)$ is given as follows:

$$\begin{aligned}
&\Upsilon_1(S) = \left\{\delta^{\beta}_{k^{n(\tau+1)+n_1}} : \exists\delta^{j_\beta}_{k^m}, \text{ s.t. } \mathrm{Blk}_{j_\beta}(M)\cdot\delta^{\beta}_{k^{n(\tau+1)+n_1}} \in S\right\},\\
&\Upsilon_{k+1}(S) =\\
&\left\{\delta^{\beta}_{k^{n(\tau+1)+n_1}} : \exists\delta^{j_\beta}_{k^m}, \text{ s.t. } \left[\mathrm{Blk}_{j_\beta}(M)\cdot\delta^{\beta}_{k^{n(\tau+1)+n_1}}\right] \circ \left(\mathbf{1}_{k^{n(\tau+1)+n_1}} - \sum_{a\in\Upsilon_k(S)} a\right) = \mathbf{0}\right\},
\end{aligned} \tag{6.57}$$

where $\delta^{j_\beta}_{k^m}$ is the value of control input. And $\Upsilon(S) = \bigcup_{k=1}^{n(\tau+1)+n_1}\Upsilon_k(S)$ represents the all reachable states without control.

The operator $\circ$ above is defined as: $X \circ Y = (p_1 \wedge q_1, p_2 \wedge q_2, \ldots, p_n \wedge q_n)^{\mathrm{T}}$, where $X = (p_1, p_2, \ldots, p_n)^{\mathrm{T}}$, $Y = (q_1, q_2, \ldots, q_n)^{\mathrm{T}}$, $p_i \wedge q_i = 1$ if and only if $p_i q_i > 0$, otherwise, $p_i \wedge q_i = 0$.

For arbitrary initial state $z(0) = \delta^{\beta}_{k^{n(\tau+1)+n_1}}$, the structure of $z(t+1)$ is given as follows,

$$\begin{aligned}
\Gamma(t+1) &= \{z(t+1) : z(t+1) = Mu(t)z(t)\}\\
&= \mathrm{Blk}_{\beta}(\ltimes_{i=t}^{0}(M\Xi(i, z(0))\Phi_{k^{n(\tau+1)+n_1}})).
\end{aligned} \tag{6.58}$$

Theorem 6.7 *The output regulation problem is solvable, if and only if* $z(0) = \delta^{\beta}_{k^{n(\tau+1)+n_1}} \in \Upsilon_1(S)$ *and* $S \subseteq \Upsilon_1(S)$.

Proof We only need to prove the systems are globally stabilized to the set S under the control design above.

Given control $\delta^{j_\beta}_{k^m}$, we assume the first time t_1 to update the control when

$$\hat{\Gamma}(t_1) = \{z(t_1) \in \Gamma(t_1) \subseteq S : M\Xi(0, z(0))\Phi_{k^{n(\tau+1)+n_1}} z(t_1) \notin S\}.$$

Then we update the control as $\Xi(t_1, z(0))$. Let

$$\Xi(t_1, z(0)) = \delta_{k^m}[\eta_1^1\ \eta_2^1\ \cdots\ \eta^1_{k^{n(\tau+1)+n_1}}], \tag{6.59}$$

where η^1_λ satisfies the following conditions

$$\eta^1_\lambda = \begin{cases} \{j_\beta\} & \lambda = \beta, \\ \{1, 2, \dots, k^m\} & \text{otherwise}, \end{cases}$$

where the elements in

$$Blk_\beta(M\Xi(t_1, z(0))\Phi_{k^{n(\tau+1)+n_1}} \ltimes^0_{i=t_1-1} (M\Xi(0, z(0))\Phi_{k^{n(\tau+1)+n_1}}))$$

are included in set S.

In general, similarly we can draw the other controllers as

$$\Xi(t_l, z(0)) = \delta_{k^m}[\eta^l_1 \; \eta^l_2 \; \cdots \; \eta^l_{k^{n(\tau+1)+n_1}}],$$

where η^l_λ satisfies the following conditions

$$\eta^l_\lambda = \begin{cases} \{j_\beta\} & \lambda = \beta, \\ \{1, 2, \dots, k^m\} & \text{otherwise}. \end{cases}$$

When $\hat{\Gamma}(t_s) \neq \varnothing$ and $\hat{\Gamma}(t_{s+1}) = \varnothing$, the systems achieve set stabilization, and the output of k-valued probabilistic logical systems with delays is same with the reference systems.

Similarly, we can get a complete control sequence which satisfies the probabilistic logic systems with time delays, and the reference systems maintain output synchronization after a certain time T. □

Assuming that the conditions in Theorem 6.7 are satisfied, we can obtain the state feedback matrix on the output regulation problem. The detailed procedures are listed as follows:

Algorithm 28 An algorithm for event-triggered state feedback matrix

step 1: Calculate the set $\hat{\Gamma}(t_1)$ and $\Xi(t_1, z(0))$.
step 2: For each $i = 2, \dots, s$, calculate $\hat{\Gamma}(t_s)$ and $\Xi(t_s, z(0))$.
step 3: If $\hat{\Gamma}(t_s) \neq \varnothing$ and $\hat{\Gamma}(t_{s+1}) = \varnothing$, stop; If not, go to step 2.
step 4: The event-triggered state feedback control can be designed as, $u(t) = \Xi(t, z(0))$, $t \in [t_i, t_{i+1})$.

Remark 6.8 When it comes to the definition of the update laws, it indicates that the updating time interval exists a lower bound $\tau^\star > 0$ for $t_{k+1} - t_k$, $\forall k \in \mathcal{N}$. We define the minimal inter-event time [11] as the upper bound of the updating time interval $\tau^\star$, which means $inf_{k\in\mathcal{N}}\{t_{k+1} - t_k\}$. If $\tau^\star = 0$, then the updates of event-triggered control operate so fast as to be unable to perform in computer simulation, which is called the Zeno behavior. Moreover, the Zeno behavior only occurs in continuous-time controllers. In the worst case, every time interval requires updating controller, and the minimal inter-event time is 1.

6.3.5 Examples

A gene regulatory example for studying metastatic melanoma in [12] is given as a specific evidence to support the wide applications of output regulation. The following network contains four genes: S100P ,WNT5A, STC3 and pirin, while the first two genes are considered as states x_1 and x_2 and the rest are considered as controls $u1$ and $u2$ respectively. In the case of outputs, S100P and WNT5A are expressed as y_1 and y_2. Finally, set $x_1, x_2, u_1, u_2, y_1, y_2 \in \mathcal{D} = \{0, 1\}$. In the real circumstance, time delay is usually accompanied by the expression of gene S100P. Thus, the purpose is to observe the behaviour of gene S100P.

$$\begin{cases} x_1(t+1) = f_1(x_1(t-1), x_2(t), u_1(t), u_2(t)), \\ x_2(t+1) = f_2(x_1(t-1), x_2(t), u_1(t), u_2(t)), \\ y(t) = x_1(t-1), \end{cases} \tag{6.60}$$

where $f_1^1 = u_1(t) \wedge [u_2(t) \wedge (x_1(t-1) \rightarrow x_2(t)) \vee (\neg u_2(t) \wedge (x_1(t-1) \wedge x_2(t)))] \vee [\neg u_1(t) \wedge u_2(t) \wedge (x_1(t-1) \leftrightarrow x_2(t))]$, $f_1^2 = u_1(t) \wedge (\neg x_1(t-1) \wedge x_2(t)) \vee u_2(t) \wedge (\neg x_2(t))$, and $f_2 = \neg u_1(t) \wedge u_2(t) \wedge x_2(t) \vee (\neg x_1(t-1)) \vee x_2(t)$.

Assume that $f_1 \in \{f_1^1, f_1^2\}$ with probability $\mathbf{P}(f_1 = f_1^1) = \mathbf{P}(f_1 = f_1^2) = 0.5$, with the corresponding structure matrices,

$$\begin{aligned} M_1^1 = \delta_2[&1\ 1\ 1\ 1\ 2\ 2\ 1\ 1\ 1\ 1\ 1\ 1\ 2\ 2\ 1\ 1\ 1\ 1\ 1\ 1\ 2\ 2 \\ &2\ 2\ 1\ 1\ 1\ 1\ 2\ 2\ 2\ 2\ 1\ 1\ 2\ 2\ 2\ 2\ 1\ 1\ 1\ 1\ 2\ 2 \\ &2\ 2\ 1\ 1\ 2\ 2\ 2\ 2\ 2\ 2\ 2\ 2\ 2\ 2\ 2\ 2\ 2\ 2\ 2\ 2]; \\ M_1^2 = \delta_2[&1\ 1\ 1\ 1\ 1\ 1\ 1\ 1\ 1\ 1\ 1\ 1\ 1\ 1\ 1\ 1\ 2\ 2\ 1\ 1\ 2\ 2 \\ &2\ 2\ 2\ 2\ 1\ 1\ 2\ 2\ 2\ 2\ 1\ 1\ 2\ 2\ 1\ 1\ 1\ 1\ 1\ 1\ 2\ 2 \\ &1\ 1\ 1\ 1\ 2\ 2\ 2\ 2\ 2\ 2\ 2\ 2\ 2\ 2\ 2\ 2\ 2\ 2\ 2\ 2]; \\ M_2 = \delta_2[&1\ 1\ 2\ 2\ 1\ 1\ 2\ 2\ 1\ 1\ 2\ 2\ 1\ 1\ 2\ 2\ 2\ 2\ 2\ 2\ 2\ 2 \\ &1\ 1\ 2\ 2\ 2\ 2\ 2\ 2\ 1\ 1\ 1\ 1\ 1\ 1\ 1\ 1\ 2\ 2\ 1\ 1\ 1\ 1 \\ &1\ 1\ 2\ 2\ 2\ 2\ 1\ 1\ 2\ 2\ 2\ 2\ 2\ 2\ 1\ 1\ 2\ 2\ 2\ 2], \end{aligned}$$

where L is

$$
\begin{aligned}
&[\delta_4^1, \delta_4^1, \delta_4^2, \delta_4^2, \frac{1}{2}\delta_4^1 + \frac{1}{2}\delta_4^3, \frac{1}{2}\delta_4^1 + \frac{1}{2}\delta_4^3, \delta_4^2, \delta_4^2, \delta_4^1, \delta_4^1, \delta_4^2, \delta_4^2,\\
&\frac{1}{2}\delta_4^1 + \frac{1}{2}\delta_4^3, \frac{1}{2}\delta_4^1 + \frac{1}{2}\delta_4^3, \delta_4^2, \delta_4^2, \frac{1}{2}\delta_4^2 + \frac{1}{2}\delta_4^4, \frac{1}{2}\delta_4^2 + \frac{1}{2}\delta_4^4, \delta_4^2, \delta_4^2,\\
&\delta_4^4, \delta_4^4, \delta_4^3, \delta_4^3, \frac{1}{2}\delta_4^2 + \frac{1}{2}\delta_4^4, \frac{1}{2}\delta_4^2 + \frac{1}{2}\delta_4^4, \delta_4^3, \delta_4^3, \delta_4^4, \delta_4^4, \delta_4^3, \delta_4^3,\\
&\delta_4^1, \delta_4^1, \delta_4^3, \delta_4^3, \frac{1}{2}\delta_4^1 + \frac{1}{2}\delta_4^3, \frac{1}{2}\delta_4^1 + \frac{1}{2}\delta_4^3, \delta_4^2, \delta_4^2, \delta_4^1, \delta_4^1, \delta_4^3, \delta_4^3,\\
&\frac{1}{2}\delta_4^1 + \frac{1}{2}\delta_4^3, \frac{1}{2}\delta_4^1 + \frac{1}{2}\delta_4^3, \delta_4^3, \delta_4^3, \delta_4^4, \delta_4^4, \delta_4^3, \delta_4^3, \delta_4^4, \delta_4^4, \delta_4^4, \delta_4^4, \delta_4^4,\\
&\delta_4^4, \delta_4^3, \delta_4^3, \delta_4^4, \delta_4^4, \delta_4^4, \delta_4^4].
\end{aligned}
$$

The dynamics of the logical reference system are:

$$
\begin{cases}
\hat{x}_1(t+1) = (\hat{x}_1(t) \vee \hat{x}_2(t)) \vee (\hat{x}_1(t) \wedge \hat{x}_2(t)) \vee \\
(\hat{x}_1(t) \leftrightarrow \hat{x}_2(t)) \\
\hat{x}_2(t+1) = \hat{x}_2(t) \vee (\neg \hat{x}_1(t)), \\
\hat{y}(t) = \hat{x}_1(t).
\end{cases}
\tag{6.61}
$$

where $\hat{L} = \delta_4[1\ 2\ 1\ 1]$, $\hat{H} = \delta_2[1\ 1\ 2\ 2]$, and $H = \delta_2[1\ 1\ 2\ 2\ 1\ 1\ 2\ 2\ 1\ 1\ 2\ 2\ 1\ 1\ 2\ 2]$.

Claims: According to (6.50), we can get $\tilde{H} = H(I_{k^{n(\tau+1)}} \otimes \hat{H})$ and

$$
\Omega_1 = \delta_{2^6}[1\ 2\ 5\ 6\ 17\ 18\ 21\ 22\ 33\ 34\ 37\ 38\ 49\ 50\ 53\ 54],
$$

$$
\Omega_2 = \delta_{2^6}[11\ 12\ 15\ 16\ 27\ 28\ 31\ 32\ 43\ 44\ 47\ 48\ 59\ 60\ 63\ 64].
$$

According to the second event-triggered control method, the control design process is as follows,

Step 1. Given $S = \{\delta_{2^6}^1, \delta_{2^6}^2\}$.

Step 2. On the basis of control design of Algorithm 28 and Eq. (6.58), we can calculate that the state set which can reach $\delta_{2^6}^1$ under control δ_4^1 is $\Upsilon_1(\{\delta_{2^6}^1\}) = \delta_{2^6}[1\ 3\ 4\ 5\ 7\ 8\ 33\ 35\ 36\ 37\ 39\ 40]$, and the state set which can reach $\delta_{2^6}^{17}$ under control δ_4^1 is $\Upsilon_2(\{\delta_{2^6}^2\}) = \delta_{2^6}[2\ 6\ 18\ 22\ 34\ 38\ 50\ 54]$. Then $\Gamma(0) = \Upsilon_1(\{\delta_{2^6}^1\}) \cup \Upsilon_2(\{\delta_{2^6}^2\})$. Similarly, we can calculate that $\Gamma(1) = \delta_{2^6}[10\ 14\ 17\ 19\ 20\ 21\ 23\ 24\ 26\ 30\ 42\ 46\ 49\ 51\ 52\ 53\ 55\ 56\ 58\ 62]$. Then $\hat{\Gamma}(1) = \delta_{2^6}[9\ 11\ 12\ 13\ 15\ 16\ 25\ 27\ 28\ 29\ 31\ 32\ 34\ 38\ 41\ 43\ 44\ 45\ 47\ 48\ 57\ 59\ 60\ 61\ 63\ 64]$, which can be transformed into $\Gamma(0) \cup \Gamma(1)$ under the control δ_4^2.

Step 3. The corresponding control matrix of the output regulation problem raised in the example is $u(1) = [\eta_1\ \eta_2\ \cdots\ \eta_{26}]$, where $\{\eta_9, \eta_{11}, \eta_{12}, \eta_{13}, \eta_{15}, \eta_{16}, \eta_{25}, \eta_{27}, \eta_{28}, \eta_{29}, \eta_{31}, \eta_{32}, \eta_{34}, \eta_{38}, \eta_{41}, \eta_{43}, \eta_{44}, \eta_{45}, \eta_{47}, \eta_{48}, \eta_{57}, \eta_{59}, \eta_{60}, \eta_{61}, \eta_{63}, \eta_{64}\}$ are chosen from $\{\delta_4^2\}$ and others are chosen from $\{\delta_4^1\}$.

Consider the following probabilistic two-valued logical systems with delays:

$$\begin{cases} x_1(t+1) = f_1(t, u(t), x_1(t), x_2(t-1)), \\ x_2(t+1) = f_2(t, u(t), x_1(t-1), x_2(t-1)), \\ y(t) = g(x_1(t), x_2(t), x_1(t-1), x_2(t-1)), \end{cases} \tag{6.62}$$

where $\mathbf{P}(f_1(t) = f_1^1) = \mathbf{P}(f_1(t) = f_1^2) = \frac{1}{2}$, $\mathbf{P}(f_2(t) = f_2^1) = \mathbf{P}(f_2(t) = f_2^2) = \mathbf{P}(f_2(t) = f_2^3) = \frac{1}{3}$, $f_1^1, f_1^2, f_2^1, f_2^2, f_2^3$ are logical control functions, and $x_1(t), x_2(t) \in \mathcal{D} = \{0, 1\}$, $u(t) \in \mathcal{D}^2 = \{0, 1, 2, 3\}$.

Then the structure matrix is $L = (A\ B\ A\ B\ A\ B\ A\ B)$, where $A = (\frac{1}{3}\delta_4^1 + \frac{1}{6}\delta_4^2 + \frac{1}{3}\delta_4^3 + \frac{1}{6}\delta_4^4, \frac{1}{2}\delta_4^1 + \frac{1}{2}\delta_4^3, \frac{1}{3}\delta_4^1 + \frac{2}{3}\delta_4^2, \delta_4^1, \frac{1}{2}\delta_4^2 + \frac{1}{2}\delta_4^4, \frac{1}{3}\delta_4^1 + \frac{1}{6}\delta_4^2 + \frac{1}{3}\delta_4^3 + \frac{1}{6}\delta_4^4, \delta_4^2, \frac{2}{3}\delta_4^1 + \frac{1}{3}\delta_4^2)$, $B = (\frac{1}{2}\delta_4^1 + \frac{1}{2}\delta_4^3, \frac{1}{2}\delta_4^1 + \frac{1}{2}\delta_4^3, \frac{2}{3}\delta_4^1 + \frac{1}{3}\delta_4^2, \delta_4^1, \frac{1}{6}\delta_4^1 + \frac{1}{3}\delta_4^2 + \frac{1}{6}\delta_4^3 + \frac{1}{3}\delta_4^4, \frac{1}{3}\delta_4^1 + \frac{1}{6}\delta_4^2 + \frac{1}{3}\delta_4^3 + \frac{1}{6}\delta_4^4, \frac{1}{3}\delta_4^1 + \frac{2}{3}\delta_4^2, \frac{2}{3}\delta_4^1 + \frac{1}{3}\delta_4^2)$.

The dynamics of the logical reference system are:

$$\begin{cases} \hat{x}_1(t+1) = \hat{x}_1(t) \wedge \hat{x}_2(t)) \vee \neg\hat{x}_1(t), \\ \hat{x}_2(t+1) = \hat{x}_1(t) \vee (\neg\hat{x}_1(t) \wedge \neg\hat{x}_2(t)), \\ \hat{y}(t) = \hat{x}_1(t) \vee (\neg\hat{x}_1(t) \wedge \neg\hat{x}_2(t)). \end{cases} \tag{6.63}$$

Our aim is to design the event-triggered control for (6.62) and (6.63). Let $x(t) = \ltimes_{i=1}^2 x_i(t)$ and $u(t) = \ltimes_{i=1}^2 u_i(t)$. We can obtain $\hat{L} = \delta_4[1\ 3\ 2\ 1]$, $\hat{H} = \delta_2[1\ 1\ 2\ 1]$, and $H = \delta_2[1\ 1\ 2\ 2\ 1\ 2\ 1\ 2\ 1\ 1\ 2\ 1\ 2\ 1\ 1\ 1]$.

Claims: According to (6.50), we can get $\tilde{H} = H(I_{k^{n(\tau+1)}} \otimes \hat{H})$ and $\Omega_1 = \delta_{26}[1\ 2\ 4\ 5\ 6\ 8\ 17\ 18\ 20\ 25\ 26\ 28\ 33\ 34\ 36\ 37\ 38\ 40\ 41\ 42\ 44\ 53\ 54\ 56\ 57\ 58\ 60\ 61\ 62\ 64]$
$\Omega_2 = \delta_{26}[11\ 15\ 23\ 31\ 47\ 51]$.

According to the first event-triggered control method, the control design process is as follows :

Step 1. Given $S = \{\delta_{26}^1, \delta_{26}^{23}\}$. In other words, the fixed points in $w(t)$ are $\{\delta_{16}^1, \delta_{16}^6\}$, $K_{16\times 64} = (A_1\ B_1\ A_2\ B_2\ A_3\ B_3\ A_4\ B_4)$, where $\mathrm{Col}_1(A_i) = \frac{1}{3}\delta_{16}^i + \frac{1}{6}\delta_{16}^{i+4} + \frac{1}{3}\delta_{16}^{i+8} + \frac{1}{6}\delta_{16}^{i+12}$, $\mathrm{Col}_2(A_i) = \frac{1}{2}\delta_{16}^i + 0\delta_{16}^{i+4} + \frac{1}{2}\delta_{16}^{i+8} + 0\delta_{16}^{i+12}, \ldots, \mathrm{Col}_8(A_i) = \frac{2}{3}\delta_{16}^i + \frac{1}{3}\delta_{16}^{i+4} + 0\delta_{16}^{i+8} + 0\delta_{16}^{i+12}$, $\mathrm{Col}_1(B_i) = \frac{1}{2}\delta_{16}^i + 0\delta_{16}^{i+4} + \frac{1}{2}\delta_{16}^{i+8} + 0\delta_{16}^{i+12}$, $\mathrm{Col}_2(B_i) = \frac{1}{2}\delta_{16}^i + 0\delta_{16}^{i+4} + \frac{1}{2}\delta_{16}^{i+8} + 0\delta_{16}^{i+12}, \ldots, \mathrm{Col}_8(B_i) = \frac{2}{3}\delta_{16}^i + \frac{1}{3}\delta_{16}^{i+4} + 0\delta_{16}^{i+8} + 0\delta_{16}^{i+12}$.

Step 2. On the basis of control design of Remark 6.7 and Eq. (6.53), we can calculate that

$$S_0 = \{\delta_{16}^1, \delta_{16}^3, \delta_{16}^6, \delta_{16}^8, \delta_{16}^{11}, \delta_{16}^{13}, \delta_{16}^{14}, \delta_{16}^{15}, \delta_{16}^{16}\},$$

$$S_0^{'} = \{\delta_{16}^2, \delta_{16}^4, \delta_{16}^5, \delta_{16}^7, \delta_{16}^9, \delta_{16}^{10}, \delta_{16}^{12}\}.$$

It is obvious to see that $S_0 \cup S_0^{'} = \Delta_{16}$.

Step 3. The corresponding control of $S_0^{'}$ is $u = \delta_4[1\ 1\ 2\ 2\ 1\ 1\ 1]$.

Step 4. The event-triggered state feedback gain matrix can be determined as $G = \delta_5[5\ 1\ 5\ 1\ 2\ 5\ 2\ 5\ 1\ 1\ 5\ 1\ 5\ 5\ 5\ 5]$.

6.4 Summary

The global stabilization problem of k-valued logical control networks was first addressed via the time-optimal event-triggered controller and switching-cost-optimal event-triggered controller. The main contributions of this study are listed as follows:

- In the first part of the study, the time-optimal event-triggered stabilizer is designed. Through semi-tensor product technique, the algebraic framework of the k-valued logical control network under event-triggered control is established; it consists of a network inherent transition matrix, an alternative network transition matrix, and a triggering event set. Similar to the time-optimal state feedback stabilizer in [2], a necessary and sufficient criterion is derived for the event-triggered stabilization. Furthermore, a constructive procedure is developed to design all time optimal event-triggered stabilizers.
- In the second part, the switching-cost-optimal stabilizer, which is event-triggered and has a minimal number of controller executions, is designed. The labeled digraph is constructed to describe the dynamic behavior of the event-triggered controlled k-valued logical control network. Moreover, based on knowledge of graph theory, the number of controller executions is minimized through a universal procedure called minimal spanning in-tree algorithm. It deserves formulating that this algorithm can handle all circumstances and overcome the constraint of the method in [3].

Then we investigate the disturbance decoupling problem of BCNs by event-triggered control. We first convert the dynamics of BNs (BCNs) into an algebraic form by using semi-tensor product, then two kinds of event-triggered controllers are designed. The first one always exists systems and depends on initial states, then the designing algorithm is given, based on which the global time-invariant state feedback controllers are also designed. The second one is executed after the occurrence of some events, then the designing algorithm is also proposed.

Finally, we introduce two types of event-triggered control to resolve the output regulation problem. It is worth noting that the feedback controller and triggering mechanism are the essential elements of event-triggered control systems, which is different from other control methods. It is when the triggering mechanism operates that we update the control inputs. Thus, we can, in turn, obtain the feedback control laws. Therefore, one of the most critical topics in the study of event-triggered control systems is how to design the triggering mechanism. Different from the state feedback control in [13], the control inputs only work in some states. When all states are controlled, the event-triggered control degenerates into a conventional one. By utilizing the semi-tensor product of matrices, we combine reference k-valued logical systems with probabilistic k-valued logical systems, which is with time-delay. This gives rise to a brief algebraic form. Then, two effective methods are raised to conduct the controller of the output regulation problem underlying event-triggered control. Inspired by the overall discussions, the main contributions of this study are as follows.

- Based on the semi-tensor product method, we combine the probabilistic delayed system with the logical reference system and present the algebraic form of the output regulation problem.
- We offer three sufficient and necessary conditions for the output regulation problem of probabilistic k-valued logical systems with delays.
- Two efficient algorithms to solve the output regulation problem of probabilistic k-valued delayed logical systems are put forward.

Besides, the disturbance decoupling problem of mix-valued logical control networks via event-triggered control was also studied in [14]. A new method was proposed for the construction of output-friendly coordinate frame. The proposed method could directly obtain a logical coordinate transformation, and therefore was convenient to use. A generalized concept was introduced to the disturbance decoupling problem of mix-valued logical control networks. Based on the generalized concept, a new criterion was presented for the solvability of event-triggered disturbance decoupling problem. Compared with the triggering mechanism proposed in Sect. 6.3, the triggering times may be reduced by introducing the generalized concept. Moreover, in [15], the event-triggered control design for state/output synchronization of switched k-valued logical control networks was studied. Some necessary and sufficient conditions were presented for the event-triggered state/output synchronization of switched k-valued logical control networks. A constructive procedure was proposed to design state feedback event-triggered controllers for the synchronization of switched k-valued logical control networks.

References

1. Ljung, L., Söderström, T.: Theory and Practice of Recursive Identification. MIT Press, Cambridge (1983)
2. Li, R., Yang, M., Chu, T.: State feedback stabilization for Boolean control networks. IEEE Trans. Autom. Control **58**(7), 1853–1857 (2013)
3. Guo, P., Zhang, H., Alsaadi, F.E., et al.: Semi-tensor product method to a class of event-triggered control for finite evolutionary networked games. IET Control Theory Appl. **11**(13), 2140–2145 (2017)
4. Cheng, D., Qi, H., Li, Z.: Analysis and Control of Boolean Networks: A Semi-Tensor Product Approach. Springer, New York (2010)
5. Liang, J., Chen, H., Liu, Y.: On algorithms for state feedback stabilization of Boolean control networks. Automatica **84**, 10–16 (2017)
6. Edmonds, J.: Optimum branchings. J. Res. Natl. Bureau Standards B **71**(4), 233–240 (1967)
7. Yang, M., Li, R., Chu, T.: Controller design for disturbance decoupling of Boolean control networks. Automatica **49**(1), 273–277 (2013)
8. Li, H., Wang, Y., Xie, L., et al.: Disturbance decoupling control design for switched Boolean control networks. Syst. Control Lett. **72**, 1–6 (2014)
9. Obermaisser, R.: Event-Triggered and Time-Triggered Control Paradigms. Springer, New York (2004)
10. Chaves, M.: Methods for qualitative analysis of genetic networks. In: 2009 European Control Conference (ECC), pp. 671–676 (2009)
11. Heemels, W.P.M.H, Johansson, K.H., Tabuada, P.: An introduction to event-triggered and self-triggered control. In: Proceeding of 51st IEEE Conference on Decision and Control, pp. 3270–3285 (2012)
12. Pal, R., Datta, A., Bittner, M.L., et al.: Intervention in context-sensitive probabilistic Boolean networks. Bioinformatics **21**(7), 1211–1218 (2005)
13. Li, H., Wang, Y., Guo, P.: State feedback based output tracking control of probabilistic Boolean networks. Inf. Sci. **349**, 1–11 (2016)
14. Wang, S., Li, H., Li, Y., et al.: Event-triggered control for disturbance decoupling problem of mix-valued logical networks. J. Franklin Inst. **357**(2), 796–809 (2020)
15. Li, Y., Li, H., Duan, P.: Synchronization of switched logical control networks via event-triggered control. J. Franklin Inst. **355**(12), 5203–5216 (2018)

Correction to: Sampled-data Control of Logical Networks

Correction to:
Y. Liu et al., *Sampled-data Control of Logical Networks*,
https://doi.org/10.1007/978-981-19-8261-3

The book was inadvertently published with an incorrect copyright holder as "The Editor(s) (if applicable) and The Author(s), under exclusive license to Springer Nature Singapore Pte Ltd. 2023" whereas it should be "Higher Education Press, Beijing, China 2023". The copyright holder has now been updated in the book.

The updated original version for this book can be found at
https://doi.org/10.1007/978-981-19-8261-3

Y. Liu et al., *Sampled-data Control of Logical Networks*,
https://doi.org/10.1007/978-981-19-8261-3_7

Printed in the United States
by Baker & Taylor Publisher Services